Lecture Notes in Computer Sci

531

T0230026

Edited by G. Goos and J. Hartmanis

Advisory Board: W. Brauer   D. Gries   J.

E. M. Clarke   R. P. Kurshan (Eds.)

Computer-Aided Verification

2nd International Conference, CAV '90
New Brunswick, NJ, USA, June 18-21, 1990
Proceedings

Springer-Verlag
Berlin Heidelberg New York
London Paris Tokyo
Hong Kong Barcelona
Budapest

Series Editors

Gerhard Goos
GMD Forschungsstelle
Universität Karlsruhe
Vincenz-Priessnitz-Straße 1
W-7500 Karlsruhe, FRG

Juris Hartmanis
Department of Computer Science
Cornell University
Upson Hall
Ithaca, NY 14853, USA

Volume Editors

Edmund M. Clarke
School of Computer Science, Carnegie-Mellon University
Pittsburgh, PA 15213, USA

Robert P. Kurshan
AT&T Bell Laboratories, Mathematical Science Research Center
Murray Hill, NJ 07974, USA

CR Subject Classification (1991): C.2.2, C.3, D.2.4, F.3-4

ISBN 3-540-54477-1 Springer-Verlag Berlin Heidelberg New York
ISBN 0-387-54477-1 Springer-Verlag New York Berlin Heidelberg

Typesetting: Camera ready by author
Printing and binding: Druckhaus Beltz, Hemsbach/Bergstr.
45/3140-543210 - Printed on acid-free paper

Preface

This volume is devoted to the proceedings of the second workshop on **Computer-Aided Verification** (the first to use this specific title). The motivation for a workshop in computer aided verification is to bring together researchers working on effective algorithms or methodologies for formal verification (as distinguished, say, from attributes of logics or formal languages). The considerable interest generated by the first workshop (held in Grenoble, June 1989; see LNCS 407) motivated this meeting. As interest continued to grow, it was decided to hold CAV on an annual basis, and for that purpose the CAV Steering Committee was formed (see below). In view of this, we take the opportunity here to state the focus of CAV and briefly sketch the history of research in formal verification, as it relates to CAV.

It is the intention of the CAV Steering Committee that future CAV meetings will continue the current focus on the problem of making formal verification feasible for various models of computation. Present emphasis is on models associated with distributed programs, protocols, and digital circuits. A good test of algorithm feasibility is to embed it into a verification tool and exercise that tool on realistic examples. Thus, we have promoted special sessions for the demonstration of new verification tools. For the technical sessions, we seek theoretical results that lead to new or more powerful verification methods. We expect there will be less emphasis at CAV meetings on purely theoretical results in program logics – not because fundamental results of this type are unimportant but because this research is adequately covered in other conferences. Since we expect that a number of the results presented at CAV actually will be used by hardware and software designers, we seek to maintain a balance between presentations by researchers and practitioners.

Proofs of correctness of algorithms such as the Euclidean algorithm go far back into history. The importance of such proofs in computing apparently was realized by Turing. However, it was not until the 1960s and early 1970s that provably correct computation began to attract much attention as a self-contained area of research. Fundamental contributions in this period established the vehicles through which formal proofs of program correctness could be constructed from axioms and rules of inference in the same way that proofs in mathematics are constructed.

The first proofs of program correctness were hand constructed and, therefore, quite short and easy to follow. As a general approach, however, manually constructed proofs of correctness were beset by two fundamental problems. First was the problem of scalability: real programs tend to be long and intricate (compared with the statement of a mathematical theorem), so a proof of correctness could be expected to be correspondingly long. Under these circumstances, it was unclear to what extent a methodology based upon manually constructed proofs could be expected to be successful. The second problem was credibility: unlike published mathematics which may be expected to undergo extensive peer review, proofs of programs are more likely to be read only by the author. Much interesting work has continued in this direction, however, and through the mid-1980s most of the research on formal verification (as this area of research became known) remained focused upon manual proofs of correctness. Applications of the work did not overcome these two fundamental problems.

The purpose of CAV is to feature research specifically directed at overcoming these two problems of scalability and credibility. Presently, this thrust has become synonymous with computer-aided verification.

Initially, researchers thought that computer programs for theorem-proving could be used in automatic program verification. Logics emerged as a mechanism to formalize the manipulation of the properties to be proved. Out of fundamental work in logic and mathematics, automatic theorem-proving advanced rapidly. Automated theorem-proving had its own problems, however, stemming largely from its non-algorithmic nature, and basic problems of tractability and decidability. These difficulties seemed to provide obstacles which were much too difficult for early theorem-provers to overcome. Many of the pioneers in program verification became disillusioned and moved into other research areas where progress was more rapid.

Recent advances and the maturation of theorem-provers (and more specifically, proof-checkers) has renewed interest in their application to practical tasks such as hardware verification. Currently, this direction has demonstrated applicability for proving properties of data paths in hardware designs. Theorem-proving has been less successful in verifying properties related to *control*, particularly when concurrency and process synchronization are involved.

Together with the early disillusionment in theorem-provers emerged an intense interest in restricted logics for which formula satisfiability is decidable. Pre-eminent among these logics was propositional temporal logic.

However, this work had two significant deficiencies of its own. First, these logics invariably constituted a substantial abstraction of a restricted class of programs. In fact, abstraction was so great that formulas in the logic lost their connection to the programs they were meant to abstract. Second, as a purely computational matter, decision procedures still were largely intractable, being exponential in the size of the formulas.

This second problem was undercut in 1980 through the introduction of model-checking as an alternative to checking formula satisfiability. Not only was linear-time model-checking demonstrated (for branching-time temporal logic), but perhaps the first computer implementations of practical formal verification algorithms were produced, as well. Computational complexity nonetheless remained an issue: while model-checking could be done in time linear in the size of the model, the model itself grows exponentially in the number of model components; for 'real' models, complexity still was the gating issue.

This problem and the problem of bridging the gap from model to implementation were addressed soon after through the introduction of homomorphic reduction. This permitted checking complex models indirectly through checks on reductions which are relative to the respective properties under test. Homomorphism also served as a mechanism for stepwise refinement, relating implementations to design models. This led to compositional and hierarchical verification, as well as specialized reduction methods involving certain types of induction. Complexity still remained an issue, however, as homomorphic reductions may be difficult (or impossible) to produce, especially in the case of large data path models; even small data path models with many inputs are not made readily tractable by homomorphic reduction. The same difficulties applied in some degree to induction.

Significant inroads into these difficulties have been made through the work on symbolic model checking using binary decision diagrams (BDDs), as introduced at the first session of this workshop in Grenoble (1989). While not scalable (effective use of BDDs currently appears to be limited to around 150 binary variables, while applications often require thousands), use of BDDs in conjunction with homomorphic reduction and induction appears extremely promising, and has generated considerable excitement. Introduced at this same meeting was work on real-time verification, which continues to generate much interest, as well.

The current workshop (CAV'90) presents some very substantial progress using BDDs. In addition, this workshop introduces reductions based upon partial order representations; these reductions offer new potential for dealing with the tractability problem inherent in state-based models. Other papers presented here explore theorem-proving and proof-checking in controller verification, and the remaining papers present much vital work which continues and extends current verification methods.

We have witnessed a migration from theorem-proving to model-checking, and a recent renewed interest in proof-checking. This may represent the start of a swing back toward theorem-proving, especially through the combination of model-checking and symbolic techniques. If so, we may expect more general theorem-proving to become integrated into existing verification tools, providing a basis for static and dynamic reasoning from the same platform.

For example, verification of a (dynamic) property of a model through expansion of the model state space or BDD evaluation may be simplified by exploiting a symmetry or inductive property in the model; the symmetry or inductive property upon which the simplification is based may be verified through a (static) syntactic check on the model specification, using theorem-proving techniques.

Whatever the future may hold, our perception is that computer-aided verification has emerged from adolescence into a very exciting and promising adulthood.

Three more CAV workshops have been planned; the next will be held in Aalberg, Denmark in July 1991 under the direction of Kim Larsen.

These proceedings are derived from *Computer-Aided Verification '90*, E. M. Clarke, R. P. Kurshan (eds.), DIMACS Series in Discrete Mathematics and Theoretical Computer Science 3, American Mathematical Society/Association for Computing Machinery, 1991, which contains the full versions of the papers originally presented at CAV'90.

We would like to thank the other members of the Steering Committee and the members of the CAV'90 Program Committee for their invaluable help in making this workshop a success. We are appreciative to DIMACS for their sponsorship of the workshop and the contribution of their facilities at Rutgers University for the duration of the workshop (June 18-21). In particular, we thank Pat Toci of DIMACS for her extensive work organizing and managing the entire on-site logistics of the workshop, and we thank Thelma Pickell of AT&T Bell Laboratories for administrative management of the meeting.

Steering Committee: E. Clarke, R. Kurshan, A. Pnueli, and J. Sifakis

Program Committee: H. Barringer, G. Bochmann, R. Bryant, C. Courcoubetis, S. Dasgupta, D. Dill, A. Emerson, R. Gerth, B. Gopinath, Z. Har'El, G. Holtzmann, G. Milne, R. Platek, P. Sistla, M. Stickel, C. Stirling, P. Wolper, and M. Yoeli.

Rutgers University   E. M. Clarke
New Brunswick, NJ   R. P. Kurshan
May 1991   Program Co-Chair
                                                                    CAV'90

Contents

Temporal Logic Model Checking:
Two Techniques for Avoiding the State Explosion Problem
Edmund M. Clarke, Jr.
School of Computer Science
Carnegie Mellon

ABSTRACT

Verifying the correctness of finite-state concurrent programs has been an important problem for a long time. However, lack of any formal and efficient method of verification has prevented the creation of practical design aids for this purpose. Since all known techniques of simulation and prototype testing are time-consuming and not very reliable, there is an acute need for such tools. We describe an automatic verification system for such systems in which specifications are expressed in a propositional temporal logic. We use a procedure called temporal logic model checking to determine automatically if a specification is true of a large state-transition graph. In contrast to most other mechanical verification systems, our system does not require user assistance and will produce a counterexample trace to show how an error occurs if this is possible.

The main difficulty with our approach is that the state space we have to search may grow exponentially with the number of circuit components. Recently, however, we have made significant progress on this problem. In this talk we outline how our verification method works and describe two of the techniques that we have developed for dealing with the state explosion problem.

The first technique is an automatic procedure for reducing the state-space of a modular system by abstracting away details of a module's behaviour that are not relevant to the specification being checked. This method relies on a description of the system as a hierarchy of concurrently executing modules. To provide a vehicle for such descriptions, we have extended the facilities of the language we use for describing synchronous hardware controllers, to allow the design of modular systems. We illustrate the use of this approach to verify the controller of a CPU with decoupled access and execution units.

We have also modified the model checking algorithm to represent a state graph using binary decision diagrams (BDD's). Because this representation captures some of the regularity in the state space of sequential circuits with data path logic, we are able to verify circuits with an extremely large number of states. We demonstrate this new technique on a synchronous pipelined design with approximately 5 times 10^{20} reachable states. We give empirical results on the performance of the algorithm applied to other examples of both synchronous and asynchronous cirucits with data path logic.

Automatic Verification of Extensions of Hardware Descriptions

Hans Eveking
Institut für Datentechnik
Technische Hochschule Darmstadt
D-6100 Darmstadt, Fed. Rep. of Germany

The extension of a hardware description is a description where all properties of the original one are maintained. The concept applies to a variety of design and verification problems including logic-verification and the verification of behavioral vs. structural descriptions. For a systematic discussion, several classes of temporal behavior and HDL-constructs for their representation are introduced. The verification tool LOVERT is surveyed which allows for the automatic verification of several types of extensions.

1 Extensions of Descriptions

In the following, the correctness of finite state systems is discussed in terms of an HDL-based hardware specification technique [3]. One hardware description, the *specification*, defines the meaning of correctness for another one, the *implementation* (Fig. 1).

$$T(d1) \longrightarrow \boxed{\text{Description d1}} \longleftarrow \text{Specification}$$

$$T(d2) \longrightarrow \boxed{\text{Description d2}} \longleftarrow \text{Implementation}$$

Fig. 1: Basic situation of a hardware specification technique

Hardware specification techniques are based on the concept of the *axiomatiziation* of HDL descriptions [4]. The axioms associated with a hardware description d have two sources (i) the *model-specific axioms* which are due to the hardware-model involved, e.g., the axioms of boolean algebra, (ii) the *description-specific axioms* reflecting the properties of the specific description d. The model- and description-specific axioms associated with a hardware description characterize a *theory*, i.e., a formal system of the predicate calculus.

Definition 1: A formula A is a correct statement about a hardware description d iff it is a theorem of the associated theory $T(d)$, i.e.,

$$\vdash_{T(d)} A. \tag{1-1}$$

The relationship between the specification and the implementation can be discussed in terms of the relationship between the associated theories. We study classes of a particularly simple relationship between two descriptions d1 and d2:

Definition 2: A description d2 is an *extension* of d1 iff for all axioms of $T(d1)$, i.e., for $A(T(d1))$ holds

$$\vdash_{T(d2)} A(T(d1)). \tag{1-2}$$

Note that on the basis of 1-2 *all* correct statements about d1 are correct about d2, too. The limitations of the concept of extension are due to the fact that the underlying modelling concepts of d1 and d2 have to be the same since the model-specific axioms of $T(d1)$ have also to be theorems of $T(d2)$ as required by 1-2. Problems involving temporal abstraction or value homomorphisms for which an interpretation of the theory $T(d1)$ is necessary [3,5] are not covered by the concept of extension.

In Section 2, the semantics of finite state systems is defined in terms of some concepts of mathematical systems theory [9]. HDL-representations of several types of temporal behavior are proposed. The HDL-constructs are taken from the CONLAN family of HDL's [8]. An axiomatiziaton of all HDL-constructs will be given.

In Sections 3-5, several types of extensions will be introduced, and proof-procedures will be discussed.

2 Classes of Temporal Behavior

We study systems that can be characterized by means of *time-functions*. A time-function represents the values that can be observed at a *carrier*, i.e., a point of observation. We consider time-functions to be functions from the set of natural numbers representing the time into some range, e.g., the set of boolean values $\{L, H\}$. Let P be a set of n time-functions $f_1, \ldots, f_n$. $P(t)$ denotes the

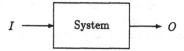

Fig. 2: A system with inputs and outputs

n-tupel $(f_1(t), \ldots, f_n(t))$, i.e., the tupel of all n values at point t of time. $P(< t_1, t_2 >)$ denotes the $(t_2 - t_1 + 1)$-tupel of all $P(t)$ in the interval $t_1 \leq t \leq t_2$.

A set P of time-functions associated with a system can be partitioned into a set I of input-functions and a set O of output-functions (Fig. 2). The classification of the temporal behavior of a system is based on the question:

Which information about $I(< 0, t >)$ and $O(< 0, t-1 >)$ determines $O(t)$ uniquely ?

In the rest of this Section, HDL-constructs for the representation of three classes of behavior will be presented.

2.1 Static behavior

A system has *static* behavior iff $O(t)$ is determined uniquely by $I(t)$ for all t. A typical example is the behavior of an AND-gate (Fig. 3).

L (low) and H (high) are the boolean constants. Static behavior is described by means of connections to carriers of type btml (boolean terminal)

g .= e

t	0	1	2	3	4	...
$a(t)$	L	L	H	L	L	...
$b(t)$	H	L	H	H	L	...
$g(t)$	L	L	H	L	L	...

Fig. 3: Static behavior of an AND-gate

with the meaning $\forall t : g(t) = e(t)$.
For the boolean functions & (and), | (or), || (exor) and ˜ (not), corresponding boolean functions are defined in the predicate calculus, e.g.

$$(and(a,b) = r) \iff (a = H) \land (r = b) \lor (a = L) \land (r = L).$$

The meaning of an HDL-expression is defined by means of a time-function, too. For instance, the time-function $\lambda(t)(and(a(t), b(t)))$ is associated with the boolean expression a & b of Fig. 3. As a result, the meaning of the statement g .= a & b of Fig. 3 is

$$\forall t : g(t) = (\lambda(t)(and(a(t), b(t)))) (t) = and(a(t), b(t)).$$

2.2 Transitional behavior

A system has *transitional* behavior iff $O(t)$ is uniquely determined by $I(t-1)$ and $O(t-1)$ for $0 < t$. The behavior of a Moore-machine [6] is an example of transitional behavior. This type of behavior is described by conditional transfers (Fig. 4) into carriers of type budv1 (boolean unit delay variable). If the transfer condition a is H at point $t-1$ then the value of x at point t becomes

```
IF a THEN x <- y ENDIF
```

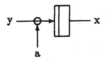

t	0	1	2	3	4	5	...
$a(t)$	L	L	H	L	L	H	...
$y(t)$	H	H	H	L	L	L	...
$x(t)$	L	L	L	H	H	H	...

Fig. 4: Transitional behavior of a transfer

the value of $y(t-1)$; otherwise, the old value of x is maintained. A default value L is assumed for point 0 of time.
The semantics of a conditional transfer is thus

$$x(0) = L,$$
$$\forall t : (0 < t) \land (a(t-1) = H) \Rightarrow (x(t) = y(t-1)),$$
$$\forall t : (0 < t) \land (a(t-1) = L) \Rightarrow (x(t) = x(t-1)). \tag{2-1}$$

There may be several conditional transfers into one carrier. Assume n conditional transfers into the carrier x:

```
IF a1 THEN x <- y1 ENDIF,

...,

IF an THEN x <- yn ENDIF
```

If transfer collisions are excluded then the meaning is:

$$x(0) = L,$$
$$\forall t: (0 < t) \wedge (a1(t-1) = H) \Rightarrow (x(t) = y1(t-1)),$$
$$\ldots$$
$$\forall t: (0 < t) \wedge (an(t-1) = H) \Rightarrow (x(t) = yn(t-1)),$$
$$\forall t: (0 < t) \wedge (a1(t-1) = L) \wedge \ldots \wedge (an(t-1) = L) \Rightarrow (x(t) = x(t-1)). \qquad (2\text{-}2)$$

If a transfer condition or source expression is a boolean expression, e.g.,

$$\text{IF a \& b THEN x <- y ENDIF}$$

then an anonymous time-function is associated with the boolean expression (Section 2.1). The time-functions of all carriers are thus bound to point $t-1$ of time. In the example, we obtain

$$x(0) = L,$$
$$\forall t: (0 < t) \wedge (and(a(t-1), b(t-1)) = H) \Rightarrow (x(t) = y(t-1)),$$
$$\forall t: (0 < t) \wedge (and(a(t-1), b(t-1)) = L) \Rightarrow (x(t) = x(t-1)).$$

2.3 Quasi-transitional behavior

A system has *quasi-transitional* behavior iff $O(t)$ is uniquely determined by $I(t)$ und $O(t-1)$ for $0 < t$. An example is the behavior of a latch (Fig. 5) described by a conditional assignment to a

$$\text{IF a THEN r := y ENDIF}$$

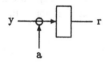

t	0	1	2	3	4	5	...
$a(t)$	L	L	H	L	L	H	...
$y(t)$	H	H	H	L	L	L	...
$r(t)$	L	L	H	H	H	L	...

Fig. 5: Quasi-transitional behavior of a latch

carrier of type bvar1 (boolean variable). Note that in the example of Fig. 5, the output r follows the input y directly at points 2 and 5 of time.

The meaning of the conditional assignment of Fig. 5 is defined by

$$x(0) = L,$$
$$\forall t: (0 < t) \wedge (a(t) = H) \Rightarrow (x(t) = y(t)),$$
$$\forall t: (0 < t) \wedge (a(t) = L) \Rightarrow (x(t) = x(t-1)). \qquad (2\text{-}3)$$

The input/output behavior of an automaton of Mealy-Type [6] is also quasi-transitional.

3 Static Descriptions vs. Static Descriptions

The first type of extensions applies to situations where the specification as well as the implementation are given by static descriptions.

Definition 3: A description is called *static* iff all carriers have static behavior.

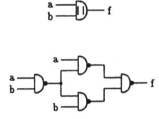

Fig. 6: An example of logic-verification

EXOR-function is implemented by means of a network of NAND-gates. To prove the correct implementation, the boolean terms of the output f are derived for the specification as well for the implementation; then the equivalence ≃ of both terms is shown:

$$a \mathbin{||} b \quad \simeq \quad \tilde{}(\tilde{}(a \mathbin{\&} \tilde{}(a\&b)) \mathbin{\&} \tilde{}(b \mathbin{\&} \tilde{}(a \mathbin{\&} b)))$$

Efficient procedures for the equivalence-proof of complex boolean expressions are available due to the work of Bryant [2] and Madre/Billon [7]. Since most HDL's provide bit-vectors as a basic type for the convenient description of complex circuits, the problem of logic-verification involves also the equivalence-proof of *vector-expressions*.

Definition 4: Two vector-expressions are equivalent, a ≅ b, iff for all elements i holds a[i] ≃ b[i].

The semantics of vectors and vector-operations can be defined by means of lists and list-operations [4]. Three examples of increasing complexity are shown in Fig. 7. The first problem is easily solved if the commutativity of the boolean and-function is extended for vector-functions. In the

$$a[1:8] \mathbin{\&} b[1:8] \quad \cong \quad b[1:8] \mathbin{\&} a[1:8]$$

$$adc(\ a[1:8],\ a[1:8],\ 0) \quad \cong \quad a[1:8]\ \#\ 0$$

$$gt(a,\ b) \quad \cong \quad \tilde{}(adc(\ 0\#a,\ 1\#\tilde{}b,\ 0)[2])$$

Fig. 7: Three examples of equivalent vector-expressions

second problem, the function adc is used which adds two n-bit vectors and a carry-input returning a normalized $n+1$-bit vector $1:n+1$. The left-most bit is the most-significant bit. For instance,

$$adc(\ a[1:8],\ b[1:8],\ c)$$

is a 9-bit vector; the carry-output is adc(a[1:8], b[1:8], c)[1]. The second equivalence of Fig. 7 is based on the fact that a plus a is equivalent to a multiplication of a by 2, i.e., a left-shift of a or the concatenation # of a and 0.[1] The third problem is even more difficult: in order to compare two vectors a and b, b is subtracted from a adding the complement of b to a; the most-significant bit of the sum has to be inverted.

To address such a variety of proof-complexity, the LOVERT approach [1] follows a two-step procedure:

- Step 1: two expressions are rewritten using a number of rewrite-rules. If the rewriting results in textually identical expressions then the equivalence is proven (Fig. 7). An example of a

Fig. 8: Transformation of an equivalence-proof into an identity-proof

rewrite-rule is the concatenation of the adc-function: two concatenated adders

```
adc( x, y, adc( v[1:n], w[1:n], cin) [1] ) #
adc( v[1:n], w[1:n], cin) [2:n]
```

are equivalent to one adder with catenated inputs:

```
adc( x # v[1:n], y # w[1:n], cin).
```

- Step 2: if the rewrite-technique fails, the vector-expressions are compiled into the basic boolean functions and, or and not. Vector expressions are sliced into single bits. The Madre/Billon tautology checking technique [7] is then applied (Fig. 9). Since the rewrite-rules used in Step 1 are not confluent, the second step ensures the completeness of the approach.

The following table shows the cpu-time of a SPARC-station needed to solve the third problem of Fig. 7 depending on wordlength:

Wordlength	8	16	32	64	128
CPU-time	0.8 sec.	0.9 sec.	1.0 sec.	1.8 sec.	3.4 sec.

The two-step procedure has a significant advantage in a situation where a design error is detected. The behavior of a verification tool in an error situation is an important aspect for its acceptance by a designer. In the example of Fig. 10, a 16-bit adder is implemented by means of four 4-bit adders; however, the carry chain is broken at the carry input of the instance a2 since the carry input is erroneously set to 0 rather than to the carry output a1.co of the first adder.

If the implementation of Fig. 10 is compared with the specification of an 16-bit adder

```
adc( a[1:16], b[1:16], 0)
```

---

[1]For convenience, the HDL constants 0 and 1 are overloaded and represent the boolean constants L and H as well as the integers 0 and 1

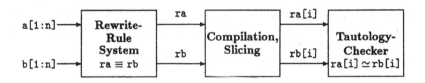

Fig. 9: Equivalence-proof of vector expressions

```
DESCRIPTION rt7483(IN  x, y: btml[1:4]; ci: btml;
                   OUT s: btml[1:4]; co: btml) BODY
          s  .=  adc( x, y, ci)[2:5],
          co .=  adc( x, y, ci)[1]
ENDrt7483
USE   a1( a[13:16], b[13:16], 0,      sum[13:16]),
      a2( a[9:12],  b[9:12],  0,      sum[9:12]),
      a3( a[5:8],   b[5:8],   a2.co,  sum[5:8]),
      a4( a[1:4],   b[1:4],   a3.co,  sum[1:4]): rt7483 ENDUSE
```

Fig. 10: Incorrect implementation of a 16-bit adder

then the response of LOVERT is the simplified expression of the implementation:

```
adc( a[1:12], b[1:12], 0) # adc( a[13:16], b[13:16], 0)[2:5]
```

The rewrite-rule system is able to simplify the expression for the three correctly chained 4-bit adders applying the simplification-rule for adders shown above; this results in an 12-bit adder which is concatenated with the last (and erroneously uncoupled) 4-bit adder. The expression gives thus a hint to the place where the problem is located.

This example shows also that LOVERT is able to cope with the problems involved in the aggregation of bit-sliced circuits.

4 Transitional vs. Structure-Oriented Static/ Transitional Descriptions

The second type of extensions applies to systems with sequential behavior.

Definition 5: A description is called *transitional* iff all carriers have transitional behavior.

Clearly, a description consisting of a collection of conditional transfers is transitional.

The main purpose of a transitional description is to display which transfers take place under which mutually exclusive conditions. The class of transitional descriptions comprises representations of simple state-diagrams as well as specifications of complex microprograms.

A further class of descriptions is the classical implementation of finite state machines as a composition of a transitional and a static subsystem (Fig. 11).

Definition 6: A *static/transitional description* is a combination of a static and of a transitional description.

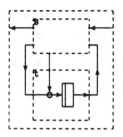

Fig. 11: Mixed transitional/static description

Definition 7: A *structure-oriented description* is a static/transitional description where each carrier of the transitional part occurs exactly once as destination.

Structure-oriented descriptions are amenable to an implementation by hardware since each transfer and each connection refers to one substructure.

We will now discuss a situation where the specification is a transitional description and where the implementation is a structure-oriented static/transitional description.

Assume several transfers into one carrier x in the transitional description d1:

```
DESCRIPTION d1 ...
    IF a1 THEN x <-  y1 ENDIF,
    ...,
    IF an THEN x <-  yn ENDIF
ENDd1
```

According to Definition 7, there is only one transfer into x in the description d2:

```
DESCRIPTION d2 ...
    IF ta THEN x <-  ty ENDIF,
    ta .=  ...,
    ty .=  ...,
    ...
    ASSERT c1, ..., cm END
ENDd2
```

The assertions c1, ... cm are used to specify don't care conditions, e.g., to exclude unreachable states or input combinations which must not occur.

To demonstrate 1-2, it must be proven for each transfer

```
        IF ai THEN x <-  yi ENDIF
```

of d1 that

1. the condition ai of d1 implies the condition ta of d2:

$$\vdash_{T(d2)} \forall t : (c1(t) = H) \wedge \ldots \wedge (cm(t) = H) \wedge (ai(t) = H) \Rightarrow (ta(t) = H) \qquad (4\text{-}1)$$

2. the condition ai implies the equality of the sources **yi** and **ty**, respectively:

$$\vdash_{T(d2)} \forall t: (c1(t) = H) \wedge \ldots \wedge (cm(t) = H) \wedge (ai(t) = H) \Rightarrow (ty(t) = yi(t)). \tag{4-2}$$

3. in the *storage situation* (see 2-2) the content of **x** remains unchanged if there is no ai active:

$$\vdash_{T(d2)} \forall t: (c1(t) = H) \wedge \ldots \wedge (cm(t) = H) \wedge (a1(t) = L) \wedge \ldots \wedge (an(t) = L) \Rightarrow$$
$$(ta(t) = L) \vee (ty(t) = x(t)). \tag{4-3}$$

Note that 4-1, 4-2 and 4-3 refer to the point t of time only and amounts thus to an equivalence proof of vector-expressions. The LOVERT system is also able to cope with this type of extensions. Typical examples of application are the proof of the correct implementation of microprograms by means of a register-bus structure and a microprogram-sequencer including a ROM. An average verification time of 1 - 2 sec. per transfer on a SPARC-station was observed. The correct implementation of a complete microprogram is thus proven in a few minutes.

5 Transitional vs. Quasi-Transitional Descriptions

In the third type of extensions, transitional ("buffered") behavior of transfers is implemented by means of quasi-transitional ("unbuffered") assignments.
An example-specification of transitional-behavior is given in Fig. 12.

```
DESCRIPTION d1 ...
    DECLARE r1, r2: budvl END
    IF p2 & b1(r2) THEN r1 <- v1(r2) ENDIF,
    IF p1 & b2(r1) THEN r2 <- v2(r1) ENDIF
ENDd1
```

Fig. 12: Specification of transitional behavior

The implementation by means of a quasi-transitional description using conditional assignments is shown in Fig. 13.

Definition 8: A description is called *quasi-transitional* iff all carriers have quasi-transitional behavior.

As an example of the proof of 1-2, we study one of the axioms associated with r1 of the specification (see 2-1):

$$\forall t: (0 < t) \wedge (p2(t-1) = H) \wedge (b1(r2(t-1)) = H) \Rightarrow (r1(t) = v1(r2(t-1))). \tag{5-1}$$

The corresponding axiom of the implementation of r1 is according to 2-3

$$\forall t: (0 < t) \wedge (p1(t) = H) \wedge (b1(r2(t)) = H) \Rightarrow (r1(t) = v1(r2(t))). \tag{5-2}$$

In order to accommodate 5-1 and 5-2, we have to require:

$$\forall t: (0 < t) \wedge (p1(t) = H) \Rightarrow (p2(t-1) = H) \tag{5-3}$$

and

$$\forall t: (0 < t) \wedge (p1(t) = H) \Rightarrow (r2(t) = r2(t-1)). \tag{5-4}$$

```
DESCRIPTION d2 ...
   DECLARE r1, r2: bvarl END
   IF p1 & b1(r2) THEN r1 := v1(r2) ENDIF,
   IF p2 & b2(r1) THEN r2 := v2(r1) ENDIF
ENDd2
```

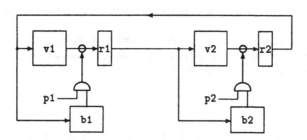

Fig. 13: Quasi-transitional implementation

The axioms associated with r2 are (see 2-3)

$$r2(0) = L,$$
$$\forall t: \ (0 < t) \wedge (p2(t) = H) \wedge (b2(r1(t)) = H) \Rightarrow (r2(t) = v2(r1(t))),$$
$$\forall t: \ (0 < t) \wedge ((p2(t) = L) \vee (b2(r1(t)) = L)) \Rightarrow (r2(t) = r2(t-1)).$$

Hence 5-4 is satisfied by

$$\forall t: \ (0 < t) \wedge (p1(t) = H) \Rightarrow (p2(t) = L). \qquad (5\text{-}5)$$

Considering 5-3 and 5-5 we see that the correct implementation is ensured by a two-phase clock with the two phases p1 and p2. An example behavior is given in Fig. 14.

t	0	1	2	3	4	...
$p1(t)$	H	L	H	L	H	...
$p2(t)$	L	H	L	H	L	...

Fig. 14: Two-phase clock

As a result, it was shown that the quasi-transitional description of Fig. 13 has also transitional behavior. Note that the specification d2 of Fig. 12 is an abstraction from real hardware: there is no circuit that corresponds, e.g., to the expression p2 & b1(r2).

6 Summary

The relatively simple concept of extension covers a number of relevant design and verification problems including the problem of logic-verification, the problem of the correct implementation of state-diagrams by means of structural resources, and the problem of the implementation of state-diagrams by means of two-phase clocked systems.

The introduction of several classes of temporal behavior represented by appropriate HDL-constructs allows for a systematic discussion of several types of extensions.

HDL-based hardware specification techniques offer a number of advantages: (i) hardware is represented in a convenient and user-friendly way, (ii) relevant classes of verification problems can be discussed in terms of the relationship between different forms of HDL-descriptions, (iii) the designer does not need proof-expertise since specialized verification tools support fully automatic verification.

The proof problems are complicated by the vector-functions provided by most HDL's for the compact representation of hardware. The LOVERT-approach proposes a combination of rewrite and tautology-checking techniques which makes the automatic verification of complex designs feasible.

References

[1] A. Bartsch, H. Eveking, H.-J. Faerber, M. Kelelatchew, J. Pinder, and U. Schellin. LOVERT - a logic verifier of register-transfer level descriptions. In L. Claesen, editor, *Proc. IMEC-IFIP Workshop on Applied Formal Methods for Correct VLSI Design*, pages 522–531, 1989.

[2] R.E. Bryant. Graph-based algorithms for boolean function manipulation. *IEEE C-35*, 677–691, 1986.

[3] H. Eveking. The application of CHDL's to the abstract specification of hardware. In Koomen/Moto-Oka, editor, *Proc. CHDL '85 (Tokio)*, pages 167–178, North-Holland, 1985.

[4] H. Eveking. Axiomatizing hardware description languages. *International Journal of VLSI Design*, 1990.

[5] H. Eveking. Formal verification of synchronous systems. In G. Milne and P.A. Subrahmanyam, editors, *Formal Aspects of VLSI Design*, pages 137–152, North-Holland, 1985.

[6] J. Hartmanis and R.E. Stearns. *Algebraic Structure Theory of Sequential Machines*. Prentice Hall, 1966.

[7] J.C. Madre and J.P. Billon. Proving circuit correctness by formally comparing their expected and extracted behavior. *Proc. 25th Design Autom. Conf.*, 205–210, 1988.

[8] R. Piloty, M. Barbacci, D. Borrione, D. Dietmeyer, F. Hill, and P. Skelly. *CONLAN Report*. Springer-Verlag, Berlin Heidelberg New-York Tokio, 1983.

[9] T. G. Windeknecht. *General Dynamical Processes*. Academic Press, 1971.

PAPETRI : Environment for the Analysis of PETRI nets

G. BERTHELOT (*), C. JOHNEN (**), L. PETRUCCI (***)

(*) CEDRIC-IIE, 18 Allée Jean Rostand, BP 77, 91002 EVRY CEDEX

(**) L.R.I., U.A. au CNRS 410, Bât. 490, Université Paris-Sud, 91405 ORSAY CEDEX

(***) MASI, tour 65, Université Pierre et Marie Curie, 4 place Jussieu, 75252 PARIS CEDEX 05

Work supported by ESPRIT project DEMON (BRA 3148).

Abstract

In this paper, we present PAPETRI, a general and integrated environment for editing and analysing Petri nets. PAPETRI allows us to work with different classes of nets. Several analysis tools are available for each of these classes.

The kernel of PAPETRI is the graphical and interactive editor, PETRIX. Five analysis tools are available : a graphical simulator of Petri nets, two constructors of covering graphs, a generator of semi-flows (COMBAG), a method using rewriting techniques (PETRIREVE), an analyser and simulator of algebraic nets (VAERA). Each of these tools, as well as PETRIX, is detailed hereafter.

1/ PAPETRI : overview

Petri nets[1] and their extensions are special classes of transitions systems where states are specified by tuples and transitions by a couple of tuples representing respectively precondition and modification of the state.

The goal of PAPETRI (in French, *Poste pour l'Analyse des réseaux de PETRI*) is to provide a friendly editing environment and to afford a great deal of analysis tools

---

[1] Basic notions and definitions on Petri nets can be found in [Rei85].

for several classes of Petri nets. These tools are not only based on nets classical methods but also on original approaches using rewriting techniques and abstract data types.

A particular care was accorded, in the conception of PAPETRI, on one hand to the flexibility of edition and on the other hand to the easy integration of new tools. The actual functionnalities are given hereafter, for each class of nets :

> • *Petri nets* : simulation, construction of covering graphs, computation of a generative family of semi-flows, analysis with rewriting techniques;

> • *Coloured nets* [Jen81] : construction of covering graphs, semi-flows calculus;

> • *Algebraic nets* [Vau85] : syntactic verification, simulation, skeleton analysis (Petri net), normed net analysis (algebraic net).

This list will be enlarged : the edition of nets may be extended to other classes, and other tools can be integrated in PAPETRI.

The main function of PAPETRI is PETRIX, a graphical and interactive environment for the edition of nets. The characteristics of PETRIX are detailed in section 2. The commands which may be used are called throughout a set of unfolding menus and dialog windows. They also allow us to call analysis tools for the class of the edited net.

Five analysis tools are presently integrated in PAPETRI :

• *Simulator* : graphical simulation of token game, described in section 2;

• *Covering graphs* : a first procedure builds the classical Karp and Miller covering graph and allows to check some net properties. A second one computes the minimal covering graph. These two tools are presented in section 3;

• *COMBAG* : semi-flows computation, introduced in section 4;

• *PETRIREVE* : analysis of Petri nets using rewriting techniques, described in section 5;

• *VAERA* : tool for syntactic verification, simulation and analysis of algebraic nets, detailed in section 6.

2/ PETRIX : the editor-simulator

The graphical and interactive editor of nets, PETRIX, uses the XWINDOWS environment on UNIX workstations.

PETRIX allows to draw a net, using the mouse and unfolding menus. It is possible to create, modify, drag and suppress the basic objects of a Petri net that are places, transitions, arcs and tokens. An arc joining a place and a transition together may be created as a broken arrow between the two objects. A token must be created inside a place but may be moved into another one. So, the graph obtained is a correct Petri net. Each of the objects may be named. It is possible to create texts which are either free (comments, title of the net, ...), or linked to another object. The shape and the size of graphical objects may be changed.

Several nets can be edited at the same time, each of them being displayed in separate windows. Operations such as cut (or copy)/paste allow the user to insert a part of a net into another net.

The size of edited nets is only limited by the available memory (several thousands of objects for 1Mb).

The simulation of an ordinary net allows to play the token game using the usual enabling rules. Evolutions can be viewed, as the marking of each place is modified in the window. The token game can be performed according to three modes : manual, semi-automatic or automatic. In manual mode, the simulator selects the enabled transitions and the user chooses the one he wants to occur. In semi-automatic mode, the evolution is done as long as only one transition can occur. Otherwise, all enabled transitions are highlighted and the user chooses the one to occur. In automatic mode, choices are randomly made. The simulation stops when no transition can occur. The user can stop the simulation whenever he wants, in any mode.

3/ Covering graphs

The more classical technique for the analysis of Petri nets consists in the construction of the *covering graph* : all the reachable markings are listed; when the set of reachable markings is infinite, a shortened list leads to build a bounded graph [KM69]. Important properties (boundedness of places, quasi-live transitions, regularity of the language) are decidable by procedures which analyze the graph.

A procedure to build this graph is available for places/transitions nets and coloured nets in PAPETRI. Once the graph is computed, an analyser of covering graphs provides diagnosis functions for the net as well as a visualisation of the graph. These functions are obtained through a menu.

The analysis functions concern the following properties : regularity of the language, existence of repetitive sequences of transitions, list of unbounded places and quasi-live transitions.

The analyser can compute the covering graph terminal connected components, which gives more information concerning liveness of transitions, home states, deadlocks of the net.

The mutual exclusion between places can be tested. The users specifies the places he wants to test the mutual exclusion of, then the analyser tests it on all nodes labels. Moreover, the analyser indicates whether these places can be empty all together.

The user can specify a sequence of transitions to occur, starting from the initial marking, and the analyser indicates the node obtained if the sequence labels a path in the graph. In the case of a bounded net, this function allows a behavioural study.

The graph may be scanned : the analyser lists all nodes labels.

The construction of a covering graph is expensive as well in time as in space. These difficulties may be reduced by the use of the minimal covering graph [Fin89] : are only kept the nodes with uncomparable labels. The number of nodes may thus be drastically decreased (several orders of magnitude). The construction of such a graph, both for places/transitions nets and coloured nets, was integrated in PETRIX [Mor89]. For the moment, the diagnosis functions are not available with this sort of graph.

For coloured nets, it could be possible to reduce furthermore the size of covering graphs, by the use of this technique combined with Jensen's, which are based on symmetries between colours ([HJJJ86]).

The construction of the covering graph allows to analyse the net for only one initial marking. However, a lot of nets model systems with various initial states. In our example, the protocol must be valid whichever the number of trains departing from West_Terminus or East_Terminus is. This goal can be reached by the study of semi-flows.

4/ COMBAG : semi-flows computation

In this section, we present COMBAG [Tre88] : a structural analysis method (i.e. which relies only on edges between places and transitions) which allows the validation of net properties independently of the initial marking. COMBAG is a tool for semi-flows computation in a Petri net. Two kinds of semi-flows are derived from the incidence matrix C :

- a *p-semi-flow* f weights every place such that $f^T.C = 0$. It leads to a net invariant : the weighted sum of tokens in places is constant for any reachable marking;

• a *t-semi-flow* f associates a number of occurrences with every transition such that C.f = 0. It denotes the repetitive and stationnary sequences of transitions.

In the following, we will use the term *semi-flow* to denote as well p-semi-flows as t-semi-flows.

The semi-flows over $\mathbb{N}$ (weights are positive integers) allow a finite generative family : every semi-flow over $\mathbb{N}$ can be expressed as a linear combination of the semi-flows in the generative family.

COMBAG calculates a generative family of semi-flows over $\mathbb{N}$, using Farkas' algorithm, the complexity of which is exponential w.r.t. time. To decrease the answering delays, COMBAG proposes, at each step, two optimization techniques. The first one consists in looking for the incidence matrix column that generates less operations, and the other one suppresses unusefull lines.

The use of coloured nets, which shorten Petri nets, leads to more concise nets. The arcs can be valued by variables (in this case, any token may be used, with no constraint as concerns its colour), or by colours (the token must have the same colour as the valuation).

COMBAG provides semi-flows for coloured nets having a finite number of colours. It calculates a generative family for three classes of semi-flows over $\mathbb{Z}$ (weights are signed integers) of a coloured net [VM84] :

• *type 1* : semi-flows f such that $f^T.|C| = 0$. They indicate invariant assertions on the number of tokens in places, without considering their colour.

• *type 2* : semi-flows f such that $f^T.C = 0$. They denote invariant assertions on the number of tokens of a colour, valid for every colour (except the distinguished neutral one).

• *type 3* : semi-flows f such that $f^T.\Pi_a(C) = 0$ (where $\Pi_a(C)$ is the projection of the incidence matrix on colour a). They represent invariant assertions on the number of tokens with colour a. This sort of semi-flows can be calculated for each colour (except the black one).

This technique enables us to find out structural properties of a net, but it does not give any piece of information concerning the net behaviour. In the next section, we introduce an original approach for the analysis of nets w.r.t. a class of initial markings.

5/ PETRIREVE : analysis with rewriting techniques

PETRIREVE [CJ85] analyses Petri nets with validation techniques based on rewriting systems. This technique studies the behaviour of a net independently of the initial martking, and thus validates structural properties of the net (termination).

PETRIREVE builts a set of oriented equations representing the behaviour of the net. Knuth-Bendix's completion transforms these equations into a rewriting system. To do so, PETRIREVE uses system REVE [FG83] - rewriting laboratory - developed by the CRIN (Nancy, France) and the MIT (Cambridge, USA). Studying the obtained rewriting system leads to straightly deduce some net properties (confluence, boundedness, ...). Other properties (invariants, quasi-liveness, reachability, termination, ...) may be validated by testing the completion of the rewriting system to which a specific equation is added.

We conceived a general technique to represent the behaviour of a net by a canonical rewriting system. A marking is represented by a term with operator *state* as header, this operator having as many arguments as there are places in the net. Each argument corresponds to a place : the value of the argument is the number of tokens in the place.

Each transition is represented by an equation. The left term denotes the precondition necessary for the transition to occur. Every term that unifies with the left term denotes a marking for which the transition can occur. The right term represents the state obtained after the occurrence of the transition.

PETRIREVE creates other equations in order to rewrite any term representing a deadlock as an identical term where the header is *deadlock*.

The rewriting system obtained after performing Knuth-Bendix's completion must converge to be used for proofs. Moreover, the equations have to be oriented as the occurrences of transitions. These two requirements leaded us to design an order on place-arguments of *state* to ensure termination, if possible, taking into account the orientation induced by transitions.

PETRIREVE automatically constructs the equations related to the net behaviour. These equations are transmitted to REVE which transforms them into a rewriting system with Knuth-Bendix's completion. The completion reduces the two members of an equation into their normal form and orientates the equation. Critical pairs eventually generate new equations. So, on one hand, rules can be redundant and thus suppressed of the system. On the other hand, these new equations may lead to reduction of already existing rules.

As the rewriting system is canonical, proofs which usually need a recurrent procedure can be performed. The equation to satisfy is added to the rewriting system. If the system (with this new equation) completion does not build any equation between generators (s, 0), then the equation is satisfied in the initial algebra [HH80]. To verify a p-semi-flow, the user adds a rule corresponding to the associated invariant to the system. It is as follows :

$$inv(state(before\ occuring)) == inv(state(after\ occuring)).$$

If these equations create rules between generators, the invariant is not coherent with the occurrence of the transition. As the completion does not create any equation between generators, the invariant is valid.

We deduce, with this technique, the termination of the net for a class of initial markings (and not for only one initial marking).

PETRIREVE is the first tool to verify the termination of a net for a class of initial states (a net terminates from an initial state if it always reaches a unique terminal state without successors). For the moment, PETRIREVE only analyses Petri nets. It may be extended to other classes of nets. However, the delays of completion will certainly be quite long.

6/ VAERA : Verification, Analysis and Evolution of algebRAic nets

The algebraic nets defined by J. Vautherin [Vau85] add another extent to Petri nets. The main interest of such a sort of high level nets is structuring data by the use of a specification of abstract data types. This is consistent with recent programming methods.

Let us shortly recall the way an algebraic net works. Such a net consists in the association of two components : a specification of abstract data types, which describes the data types used and the operations on them, and a Petri net. The places of the net are sorted, the tokens being constants of the specification, of the same sort as their containing place. The arcs are valued by terms of the specification.

The transitions are enabled according to the following criteria : first of all, the conditions on the amount of tokens in places must be valid, as in places/transitions nets, secondly, conditions between arcs valuations and tokens values must be satisfied. Each arc entering a transition is valued either by a term without variables, or by a single variable. In the first case, a token with this value must be used to fire the transition. In the second case, the variable takes the value of one of the tokens for the occurrence of the transition. Afterwards, if the transition occurs, the term associated with each arc exiting the transition is evaluated (variables occurring in the term have the value given when examining the entering arcs), and a token having this value is created in the corresponding place. Transitions may also have an auxiliary condition (sound equation) associated with them. This equation, the hand-sides of which are functions of the entering variables, must be valid for the transition to occur.

The analysis of algebraic nets is more complex than the analysis of places/transitions nets, due to the powerfulness of abstract data types. Tokens meaning is denoted as well by their presence in places as by their value. Two tokens within a same place may have different interpretations. We will first explain how PETRIX deals with the definition of an algebraic net, and we will detail afterwards the analysis tools we have designed.

When the user wants to design an algebraic net, he describes the Petri net part with PETRIX and uses VAERA for everything related to algebraic specifications. These must have been created by ASSPEGIQUE (in French, ASsistance à la SPEcification algébrIQUE, see [Cho88]), environment for algebraic specification developped at the LRI (Orsay, France), which can be called by VAERA.

Once the net is created, VAERA provides a tool for syntactic verification of the terms in the net and checks the coherence of the sorts used (tokens must be of sort of the place they are in, ...).

It is also possible to use the simulator included in VAERA to observe the behaviour of the net : the user selects, with the mouse, the transition to occur. The tool verifies that the conditions necessary for the transition to occur are satisfied, and if true, fires the transition as described before. As tokens may have different values, two modes of simulation are available : either the user chooses which tokens will be used for the transition to occur, or the tokens are randomly chosen by VAERA.

Let us now introduce our various analysis tools. It may be interesting, in order to analyse algebraic nets, to yield simpler models by ignoring some pieces of information. These models are the skeleton and the normed net. Indeed, the difficulty of analysis of algebraic nets countervails their expression power.

The skeleton of an algebraic net :

The first model that we will study is the underlying Petri net, named *skeleton* (see [Vau85]). VAERA can automatically generate the skeleton of an algebraic net : auxiliary conditions for transitions to occur, arcs valuations and tokens values are forgotten. Thus, we obtain a net which can be analysed by the places/transitions tools previously described.

Some properties such as the boundedness of a place, the non quasi-liveness of a transition and the termination of a net are true for the algebraic net if they are true for its skeleton.

<u>The normed net</u> :

Another model extracted from the algebraic net is the *normed net* (see [Pet88]). The normed net is a model between the algebraic net (the most structured model) and the skeleton (the less structured model). This net contains less information than the algebraic net, but more than the skeleton. The analysis of the normed net is indeed more complex than the analysis of the skeleton, but its importance leans upon the degree of information it contains. This is the reason why its analysis allows us to detect properties of the algebraic net which are not valid for the skeleton. The main analysis which can be performed concerns the termination of the net.

The normed net is derived from the algebraic net by forgetting the auxiliary conditions for transitions to occur and by changing the valuation of arcs as well as the values of tokens into their norm. The norm maps a term to its "size". For example, as concerns files, function *norm* counts the number of elements in the file.

To generate this net, the user will have to include in his specification the definition of function *norm*. The net obtained is then such that the values of the tokens are unsigned integers. A transition will not occur if this operation leads to create strictly negative tokens.

We have proved [Pet88] that the termination of the normed net implies the termination of the algebraic net. To validate this property, we observe the behaviour of the net transitions. If a transition cannot infinitely occur, we will say that it can be blocked. If all the transitions can be blocked, the net terminates. VAERA decides, in a lot of cases, of the blocking of transitions. It tells the user if the net terminates, and otherwise gives the list of blockable transitions.

For the moment, we only have semi-decision procedures. They cannot infirm termination. Hence, to improve the analysis of algebraic nets, we intend to add a tool for the construction of finite reachable markings graphs.

7/ Conclusion and future work

We presented, in this paper, PAPETRI. It offers, in a graphical environment, various analysis tools. Moreover, it deals with different classes of nets : ordinary nets, coloured nets, algebraic nets. The mixture of techniques from various theories allows to study all the aspects of a system and to fully validate it.

As concerns the examples we studied, the answering delays are reasonable w.r.t. the calculus complexity (a few minutes for the longer operations).

PAPETRI is an open system which may easily receive other tools and may be extended to other classes of nets. The next foreseen extensions deal with the construction of reachable markings graphs for algebraic nets and verification of home states.

Bibliography

[Cho88]   Choppy C. : *ASSPEGIQUE user's manual*, rapport de recherche L.R.I. n°452, Orsay, 1988.

[CJ85]   Choppy C., Johnen C. : *PETRIREVE : proving Petri net properties with rewriting systems*, LNCS vol.202, rewriting techniques and applications, Jouannaud J.P. Ed., Springer Verlag, 1985.

[FG83]   Forgaard R., Guttag J.V. : REVE : *a term rewriting system generator with failure resistant Knuth Bendix*, Proc. of an NSF Workshop on the rewrite rule laboratory, 1983.

[Fin89]   Finkel. A. : *A minimal coverability graph for Petri nets*, rapport de recherche L.R.I., Orsay, 1989.

[Jen81]   Jensen K. : *Coloured Petri nets and the invariants method*, T.C.S., 14-3, pp. 317-336, 1981.

[KM69]   Karp R.M., Miller R.E. : *Parrallel program schemata*, JCSS,4, pp. 147-195, 1969.

[HH80]   Huet G., Hullot J.J. : *Proofs by induction in equational theories with constructors*, JCSS, 25, pp. 239-266, 1982.

[HJJJ86] Huber P., Jensen M., Jensen K., Jepsen O. : *Reachability trees for High-Level Petri nets*, T.C.S., 45, 1986.

[Mor89]   Moreau J. M. : *Graphe de couverture minimal dans les réseaux de Petri*, rapport de stage de D.E.A., Université d'Orléans, 1989.

[Pet88]   Petrucci L. : *Etude d'un environnement d'exécution de réseaux de Petri algébriques*, rapport de stage de D.E.A., Université de Paris-Sud, Orsay, 1988.

[Rei85]   Reisig W. : *Petri Nets* , Springer Verlag, 1985.

[Tre88]   Treves N. : *A comparative study of different techniques of semi-flows computation in higher-level nets*, Proc. IX European Workshop on Applications and Theory of Petri nets, 1988.

[Vau85]   Vautherin J. : *Un modèle algébrique, basé sur les réseaux de Petri, pour l'étude des systèmes parallèles*, Thèse de doctorat d'ingénieur, Université Paris-Sud, Orsay, 1985.

[VM84]   Vautherin J., Memmi G. : *Computation of flows for unary-Predicate/Transition nets*, LNCS 188, G. Rozenberg ed., Springer Verlag, pp. 307-327, 1984.

Verifying Temporal Properties of Sequential Machines
Without Building their State Diagrams

Olivier Coudert   Jean Christophe Madre   Christian Berthet

Bull Research Center, PC 62 A13,
68, Route de Versailles
78430 Louveciennes FRANCE

Abstract

This paper presents the algorithm we have developed for proving that a finite state machine holds some properties expressed in temporal logic. This algorithm does not require the building of the state-transition graph nor the transition relation of the machine, so it overcomes the limits of the methods that have been proposed in the past. The verification algorithm presented here is based on Boolean function manipulations, which are represented by typed decision graphs. Thanks to this canonical representation, all the operations used in the algorithm have a polynomial complexity, expect for one called the computation of the "critical term". The paper proposes techniques that reduce the computational cost of this operation.

1 Introduction

Within the community of people working on the verification of sequential machines, the word "verification" has received two quite different meanings. For some people, to verify a machine is to prove that it holds some properties, such as "liveness" or "safety" properties [3] [4] [6] [7] [13]. For the others, to verify a machine is to prove that it is correct with respect to its behavioral specification, which is a program written in some high level hardware description language. This problem comes down to proving the behavioral equivalence between two machines [12]. Though both kinds of verification are needed to design complex circuits, these two verification problems have until recently been studied independently and the techniques that have been proposed to deal with them had few relations one with another.

The behavioral equivalence between two machines $\mathcal{M}_1$ and $\mathcal{M}_2$ can be proved by traversing the state-transition graph of the product machine $\mathcal{M}_1 \times \mathcal{M}_2$ [12]. The first techniques that have been proposed were based on a double enumeration of the states and the input patterns of the machines [11], so that in many cases the time needed to traverse the state-transition graph of the machine $\mathcal{M}_1 \times \mathcal{M}_2$ grows exponentially with the number of inputs of $\mathcal{M}_1$ and $\mathcal{M}_2$.

More recently we have presented [8] [9] a proof procedure that manipulates sets of states and inputs. The basic concepts that underlie this proof procedure are (1) to represent sets either by their *characteristic functions* or by *vectors of Boolean functions* [9] and (2) that the operations on these Boolean functions can be efficiently performed by denoting them with typed decision graphs [1]. This procedure can handle machines that could not be treated with the enumeration based methods referenced above.

It has been shown [3] [6] that these basic concepts can also be used to develop a procedure for proving that a sequential machine holds some temporal properties without building its state-transition graph. Such procedures could overcome the limits of the methods that have been developed in the past, which all required the partial or total building of this graph. However the procedure described in [3] [6] requires the construction of the *transition relation* of the sequential machine, which is not feasible for most of complex machines.

This paper presents the proof procedure that we have independently developed to check that a machine holds a temporal property. This technique, which does not use the transition relation of the machine, is based on specific operators called "restrict" and "expand". The paper is divided into 4 parts. Part 2 defines the model of machines and the temporal formulas that the proof system handles. Part 3 gives the proof algorithm used by the system and shows that it uses the standard Boolean operations in addition with the computation of a specific term. Part 4 is dedicated to the computation of this term that we call the "critical term" because its evaluation is the bottleneck of the algorithm. Part 5 gives some experimental results and discusses them.

2 Definitions

This section describes the inputs of the verification system presented in this paper. First it defines the model of sequential machines that the system handles. Then it gives the syntax and the semantics of the temporal formulas to be verified.

2.1 Model of Sequential Machines

The sequential machine $\mathcal{M}$ that must be verified is an uncompletely specified deterministic Moore Machine. This machine is defined by the 6-tuple $(n, m, r, \omega, \delta, \text{Init})$, where: the state space of the machine $\mathcal{M}$ is $\{0, 1\}^n$, its input space is $\{0, 1\}^m$, and its output space is $\{0, 1\}^r$; ω is the partial output function from $\{0, 1\}^n \times \{0, 1\}^m$ into $\{0, 1\}_\perp^r$; δ is the partial transition function from $\{0, 1\}^n \times \{0, 1\}^m$ into $\{0, 1\}^n_\perp$; Init is the characteristic function of the set of initial states of $\mathcal{M}$.

In this description, all the Boolean functions are denoted by typed decision graphs (TDG), which are a very compact graphical canonical form [1]. These Boolean functions are automatically computed from the behavioral description of the machine using symbolic execution [2]. The transition function δ is denoted by a couple (Cns, F). Cns denotes the domain where the transition function is defined, so it is a Boolean function from $\{0, 1\}^n \times \{0, 1\}^m$ into $\{0, 1\}$. It is built out of constraints given by the designer and additional constraints computed during the symbolic execution [2]. F is a vectorial function $[f_1 \dots f_n]$, where each f_j is a Boolean function from $\{0, 1\}^n \times \{0, 1\}^m$ into $\{0, 1\}$. The partial function δ is then defined by:

$$\delta = \lambda s.\lambda p.(\text{if Cns}(s, p) \text{ then } F(s, p) \text{ else } \perp)$$

For the sake of simplicity, we assume in the following that ω is completely specified. We also assume that $(\forall s \exists p \, \text{Cns}(s, p))$ is a tautology, which means that any state has at least one successor. This is not a restriction, because temporal formula have no meaning on states that have no successor. If $(\forall s \exists p \, \text{Cns}(s, p))$ is not a tautology, then $\lambda s.(\forall p \neg \, \text{Cns}(s, p))$ is the characteristic function of the set of states that have no successor.

2.2 State Formulas in Computation Tree Logic

The temporal formulas handled by the verification system are the *state formulas* of the computation tree logic CTL [7]. This logic is a formalism specially developed to express properties about the states and the computation paths of finite state systems. The different kinds of state formulas and their meanings with respect to a state of the machine $\mathcal{M} = (m, n, r, \omega, \delta, \text{Init})$ are the following:

1. $r_1, r_2, \dots, r_n, o_1, o_2, \dots, o_r$ are state formulas. For any state s of the machine, $s \models r_j$ if and only if the j-th component of s is 1. For any state s of the machine, $s \models o_j$ if and only if the j-th component of $\omega(s)$ is 1.

2. If f and g are state formulas then so are $(\neg f)$, $(f \wedge g)$, $(f \vee g)$, $(f \Leftrightarrow g)$, and $(f \Rightarrow g)$. The logical connectors have their usual meanings, for instance $s \models (f \wedge g)$ if and only if $s \models f$ and $s \models g$.

3. If f is a state formula, so are the formulas $EX(f)$ and $AX(f)$. For any state s of $\mathcal{M}$, $s \models EX(f)$ iff there exists at least one input pattern p such that $\delta(s, p) \models f$; by definition, $AX(f)$ is $\neg (EX(\neg f))$.

4. If f and g are state formulas, so are the formulas $E[f\,U\,g]$ and $A[f\,U\,g]$. For any state s of the machine $\mathcal{M}$, $s \models E[f\,U\,g]$ if and only if there exists at least one path $(s_0, s_1, ...)$ with $s_0 = s$, and: $\exists i\,((s_i \models g) \wedge (\forall j\,(0 \leq j < i \Rightarrow s_j \models f)))$.
For any state s of the machine $\mathcal{M}$, $s \models A[f\,U\,g]$ if and only if for all paths $(s_0, s_1, ...)$ such that $s_0 = s$, we have: $\exists i\,((s_i \models g) \wedge (\forall j\,(0 \leq j < i \Rightarrow s_j \models f)))$.

Abbreviations have been added to make the CTL formulas more legible: $EF(f) =_{def} E[True\,U\,f]$; $AF(f) =_{def} A[True\,U\,f]$; $EG(f) =_{def} \neg (AF(\neg f))$; $AG(f) =_{def} \neg (EF(\neg f))$, i. e. f holds on any state of every path. The reader can refer to [7] to find examples of properties of finite state systems expressed in CTL.

3 Verification Algorithm

It has been shown that "safety properties" of finite state systems [4] can be checked with the forward symbolic traversal procedure presented in [9]. The machine $\mathcal{M}$ holds a safety property f if and only if some sub-part $\mathcal{M}_f$ of the machine $\mathcal{M}$ has an observable behavior that is equivalent to the one of the most general model of the formula f, which is a finite state automaton $\mathcal{A}_f$. This automaton can be automatically built from the formula f. In this case the proof is made by traversing the state diagram of the machine $\mathcal{M}_f \times \mathcal{A}_f$ without building it [10]. For instance, to prove that the machine $\mathcal{M}$ satisfies the formula $f = AG\,(r_i \Rightarrow AX\,(AG\,(o_j)))$ comes down to proving that the observable behavior of $\mathcal{M}$ at its j-th output is equivalent to the behavior of the automaton $\mathcal{A}_f$ given in Figure 1.

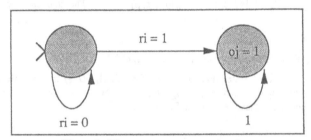

Figure 1. The automaton $\mathcal{A}_f$ associated to
the formula $f = AG\,(r_i \Rightarrow AX\,(AG\,(o_j)))$.

This part presents a more general proof algorithm that can handle all the state formulas described in the previous section. This algorithm does not require the building of the state-transition graph of the machine $\mathcal{M}$ under verification. The algorithm takes as inputs a machine $\mathcal{M}$ described by its 6-tuple $(m, n, r, \omega, (Cns, F), Init)$ and the formula f to be verified. It recursively computes the set of states $\mathcal{F}$ that satisfy the formula f from the sets of states that satisfy the sub-formulas of f. At each step there are only four basic cases to consider that correspond to the four kinds of formulas given in Section 2.2. Once the set of states $\mathcal{F}$ that satisfy the whole formula f is obtained, to check whether $s \models f$ for some state s of $\mathcal{M}$ comes down to checking whether s belongs to $\mathcal{F}$.

The verification algorithm manipulates sets represented by their characteristic functions, and functions are denoted by their TDG's. We will make no distinction between a function and its TDG, or between a set and the TDG of its characteristic function. The quantifiers that will be used in the sequel are handled using the following identities: if x is a propositional variable, $(\exists x\,f(x))$ is equivalent to $(f(0) \vee f(1))$, and $(\forall x\,f(x))$ is equivalent to $(f(0) \wedge f(1))$. For any n-tuple $x = [x_1 ... x_n]$ of propositional variables, $(Qx\,f(x)) =_{def} (Qx_1 ... Qx_n\,f(x_1, ..., x_n))$, where Q is either the existential or universal quantifier. The size of a TDG f is its number of vertices, and it will be noted $|f|$. The size of the TDG of a function is relative to a total ordering of its variables [1] [5].

The formulas of type (1) and (2) are trivial to treat. For instance, if $f = (f_1 \wedge f_2)$, and $\mathcal{F}_1$ and $\mathcal{F}_2$ are the characteristic functions of the sets of states that satisfy f_1 and f_2 respectively, then $\mathcal{F}$ is $\lambda s.(\mathcal{F}_1(s) \wedge \mathcal{F}_2(s))$.

The cost of computing the TDG of $\mathcal{F}$ is in $O(|\mathcal{F}_1| \times |\mathcal{F}_2|)$ [1] [5]. As soon as $\mathcal{F}$ is computed, it is trivial to determine whether $s \models f$, since this is equivalent to compute $\mathcal{F}(s)$, which is done in $O(n)$, and then to test whether $\mathcal{F}(s) = 1$. In the same way, $M \models f$ if and only if $\lambda s.(\text{Init}(s) \Rightarrow \mathcal{F}(s)) = 1$, which can be tested in $O(|\text{Init}| \times |\mathcal{F}|)$.

3.1 *EX* and *AX* Formulas

The sets of states that satisfy the formulas of types *EX* and *AX* are computed in a single step. Let f be a formula and $\mathcal{F}$ be the set of states that satisfy f. The set of states $\mathcal{EX}$ of the machine $\mathcal{M}$ that satisfy the temporal formula $EX(f)$ is $\{s \mid \exists p\ \delta(s, p) \in \mathcal{F}\}$. Its characteristic function is:

$$\mathcal{EX} = \lambda s.(\exists p\ \text{Cns}(s, p) \wedge \mathcal{F}(F(s, p))). \tag{1}$$

By definition, $AX(f) = \neg\ EX(\neg f)$, so the characteristic function $\mathcal{AX}$ of the set of states of the machine $\mathcal{M}$ that satisfy $AX(f)$ is:

$$\mathcal{AX} = \lambda s.(\neg\ (\exists p\ \text{Cns}(s, p) \wedge \neg\ \mathcal{F}(F(s, p)))). \tag{2}$$

3.2 *EU* and *AU* Formulas

The sets of states that satisfy the formulas of types *EU* and *AU* are computed using iterating algorithms [3]. Let f and g be two formulas and $\mathcal{F}$ and $\mathcal{G}$ be the sets of states that satisfy f and g respectively. The set $\mathcal{EU}$ of states that satisfy the formula $E[f\ U\ g]$ is the limit of the converging sequence $(\mathcal{E}_i)$ of sets defined by: $\mathcal{E}_0 = \mathcal{G}$, and $\mathcal{E}_{k+1} = \mathcal{E}_k \cup \{s \mid (s \in \mathcal{F}) \wedge (\exists p\ \delta(s, p) \in \mathcal{E}_k)\}$. The characteristic functions of these sets are the following:

$$\mathcal{E}_0 = \mathcal{G}, \text{ and}$$
$$\mathcal{E}_{k+1} = \lambda s.(\mathcal{E}_k(s) \vee (\mathcal{F}(s) \wedge (\exists p\ \text{Cns}(s, p) \wedge \mathcal{E}_k(F(s, p))))). \tag{3}$$

In the same way the set $\mathcal{AU}$ of states that satisfy the formula $A[f\ U\ g]$ is the limit of the converging sequence $(\mathcal{A}_i)$ defined by: $\mathcal{A}_0 = \mathcal{G}$, and $\mathcal{A}_{k+1} = \mathcal{A}_k \cup \{s \mid (s \in \mathcal{F}) \wedge (\forall p\ (\delta(s, p) \neq \perp) \Rightarrow (\delta(s, p) \in \mathcal{A}_k))\}$. The characteristic functions of these sets are the following:

$$\mathcal{A}_0 = \mathcal{G}, \text{ and}$$
$$\mathcal{A}_{k+1} = \lambda s.(\mathcal{A}_k(s) \vee (\mathcal{F}(s) \wedge (\forall p\ \text{Cns}(s, p) \Rightarrow \mathcal{A}_k(F(s, p))))). \tag{4}$$

3.3 The Critical Term

In the formulas (1) to (4) given in the preceding sections two terms appear that actually are the same one, and that has the form $(\exists p\ \text{Cns}(s, p) \wedge \chi(F(s, p)))$. Since the formula $(\forall x\ f)$ is equivalent to $(\neg\ (\exists x\ \neg f))$, the equation (4) can be rewritten into:

$$\mathcal{A}_{k+1} = \lambda s.(\mathcal{A}_k(s) \vee (\mathcal{F}(s) \wedge \neg\ (\exists p\ \text{Cns}(s, p) \wedge \neg\ \mathcal{A}_k(F(s, p))))). \tag{4'}$$

From the computational point of view this means that the four basic cases of CTL formulas can be treated with the standard Boolean operations (negation, conjunction, and disjunction), the test of equivalence (to test whether the sequences $\mathcal{E}_k$ and $\mathcal{A}_k$ have converged), in addition with the evaluation of the *critical term* $(\exists p\ \text{Cns}(s, p) \wedge \chi(F(s, p)))$, where χ is a TDG denoting a Boolean function from $\{0, 1\}^n$ into $\{0, 1\}$. Next section explains why we call this term the "critical term", and discusses how it can be computed.

4 Computing the Critical Term

This section shows that computing the critical term is the bottleneck in the verification algorithm. Then it explains why the critical term can be easily computed when the TDG of the transition relation of the machine can be build [3] [6]. At last, it presents the techniques we have developed to perform this computation when it is not possible to build this TDG, which happens for most of complex circuits.

4.1 Is the Computation of the Critical Term a Difficult Problem?

Since typed decision graphs are canonical, testing the equivalence of two TDG's is in $O(1)$. All the standard Boolean operations on TDG's are performed in polynomial time [1] [5]. This means that the computational cost of the formal verification of CTL formulas depends on the evaluation of the critical term.

The complexity of the critical term computation can be studied using the composition problem. This problem is to compute the TDG of $\lambda x.g(f_1(x), ..., f_n(x))$, from the TDG's of the Boolean functions $f_1, ..., f_n$, and g. The three following theorems show that computing the critical term is the most difficult operation that appears in the algorithm presented in Section 3.

Theorem 1. *Composition is $O(1)$-reducible to the computation of the critical term.*

Proof. It is immediate that composition is reducible in $O(1)$ to the evaluation of the critical term $(\exists p \; Cns(s, p) \wedge \chi(F(s, p)))$, where $Cns = 1$, $\chi = g$, and $\lambda s.\lambda p.F = [\lambda s.\lambda p.f_1(s) \; ... \; \lambda s.\lambda p.f_n(s)]$. ❑

Theorem 2. *Composition is NP-hard.*

Proof. Let $C = \wedge_j c_j$ be a 3-conjunctive normal form made of n clauses c_j. Each clause c_j is a disjunction $(q_{1j} \vee q_{2j} \vee q_{3j})$ of three literals, so the TDG's of these clauses are built in $O(n)$. Let g be the function $\lambda[y_1 ... y_n].(\wedge_j y_j)$, whose TDG is a n-vine built in $O(n)$. Since C is satisfiable if and only if the TDG of $g(c_1, ..., c_n)$ is not equal to 0, which can be tested in $O(1)$, composition is NP-hard. ❑

The two precedings theorems show that the computation of the critical term is NP-hard. By using the hypothesis that the order of the variables cannot be modified, we obtain the more precise following theorem.

Theorem 3. *Composition is a no polynomial problem if the ordering of the variables is fixed. More precisely, there exist some TDG's $f_1, ..., f_n$ and g, using $O(n)$ variables, with $\Sigma_j |f_j| + |g| = O(n)$, and such that $|g(f_1, ..., f_n)| = O(2^n)$.*

Proof. We consider 2n variables $y_1 < ... < y_{2n}$, and the function $g = \lambda[x_1 ... x_n].(\wedge_j x_j)$. Its TDG is a n-vine of size n. We define n functions $f_j = \lambda[y_1 ... y_{2n}].(y_j \Leftrightarrow y_{2n-j+1})$. The TDG's of the functions f_j have all two vertices, so $\Sigma_j |f_j| = 2n$. Thus $|g| + \Sigma_j |f_j| = O(n)$. The composition $g(f_1, ..., f_n)$ is the function $\lambda y.(\wedge_j f_j(y))$, that is $\lambda[y_1 ... y_{2n}].(\wedge_j (y_j \Leftrightarrow y_{2n-j+1}))$, but the TDG of the term $((y_1 \Leftrightarrow y_{2n}) \wedge (y_2 \Leftrightarrow y_{2n-1}) \wedge ... \wedge (y_n \Leftrightarrow y_{n+1}))$ with the order $y_1 < ... < y_{2n}$ has a size in $O(2^n)$. ❑

This last theorem shows that computing the critical term on TDG's is at least exponential in the worst case. However, one can object that the TDG of the term $((y_1 \Leftrightarrow y_{2n}) \wedge (y_2 \Leftrightarrow y_{2n-1}) \wedge ... \wedge (y_n \Leftrightarrow y_{n+1}))$ is in $O(n)$ with the variable ordering $y_1 < y_{2n} < y_2 < y_{2n-1} < ... < y_n < y_{n+1}$. We may think that composition can be computed more efficiently if we are able to find a variable ordering that minimizes the sizes of the manipulated TDG's. The following remarks show that the problem is not such simple. First, finding such a variable ordering is itself a NP-hard problem. Second, there exist some functions of n variables whose TDG has a no polynomial size with respect to n, whatever the variable ordering. The last remark shows that the composition, and so the computation of the critical term, is *certainly* no polynomial even if an oracle would provide dynamically a "good order".

4.2 Using the Transition Relation to Compute the Critical Term

The transition relation Δ of the machine $\mathcal{M}$ is a subset of $\{0, 1\}^n \times \{0, 1\}^n$. For any states s and s', (s, s') belongs to Δ if and only if $(\exists p \; (s' = \delta(s, p)))$. This means that $\Delta =_{def} \lambda s.\lambda s'(\exists p \; Cns(s, p) \wedge (s' \equiv F(s, p)))$. Using the transition relation Δ, the critical term $(\exists p \; Cns(s, p) \wedge \chi(F(s, p)))$ can be rewritten into:

$(\exists s' \; \Delta(s, s') \wedge \chi(s'))$.

Given the TDG's of Δ and χ, the TDG of $(\Delta(s, s') \wedge \chi(s'))$ can be built in $O(|\Delta| \times |\chi|)$, and the critical term is obtained by eliminating from this TDG the n atoms associated to s'. Note that this elimination is not polynomial with respect to n in the worst case, but experience shows it remains polynomial on practical examples.

Assuming that the TDG of Δ can be built, this technique is very efficient [6], because it uses only the standard logical operators, which have relatively low computational costs. But building Δ requires the computation of the TDG of $(Cns(s, p) \wedge (\wedge_j (s_j' \equiv f_j(s, p))))$, which is done in $O(|Cns| \times 2^n \times \Pi_j |f_j|)$. Experience shows that for complex machines, it is not possible to build the TDG of the transition relation Δ, so others techniques are needed.

4.3 Using the "Restrict" and the "Expand" Operators to Compute the Critical term

The term $(\exists p\ Cns(s, p) \wedge \chi(F(s, p)))$ can be computed in two steps. The first step consists in building the TDG of the formula $(Cns(s, p) \wedge \chi(F(s, p)))$, and the second step in eliminating from this TDG the atoms associated to the input pattern p. The term $\chi(F(s, p))$ is the composition of χ with the vectorial function $F = [f_1 ... f_n]$, whose computation is exponential in the worst case, as shown in Section 4.1. More precisely, it can be easily shown, using a proof similar to the one of Theorem 3, that the computation of the TDG $\chi(f_1, ..., f_n)$ from the TDG's of $f_1, ..., f_n$ and χ is at least in $O(|\chi| \times \Pi_j |f_j|)$.

Section 4.3.1 shows that it is not necessary to build explicitly the TDG of $\chi \circ F$ to compute the term $(\exists p\ Cns(s, p) \wedge \chi(F(s, p)))$. It presents the "expand" operation that avoids this construction. Section 4.3.2 presents the "restrict" operator that further reduces the computational cost of the composition by reducing the sizes of the TDG's f_j used in the term $\chi(F(s, p))$.

4.3.1 The "Expand" Operation

The idea that underlies the "expand" operation is to express the term $\chi(F(s, p))$ as a sum of functions whose TDG's have less vertices than the TDG of $\chi(F(s, p))$. Using these functions noted $h_1, ..., h_k$, the critical term can be rewritten into $(\exists p\ Cns(s, p) \wedge (\vee_j h_j(s, p)))$, that can be further transformed into $(\exists p \vee_j (Cns(s, p) \wedge h_j(s, p)))$ by distributing the conjunction over the disjunction. The existential quantifier commutes with the disjunction, so that the critical term finally becomes:

$(\vee_j (\exists p\ Cns(s, p) \wedge h_j(s, p)))$.

The final form of this term expresses that its computation can be decomposed into a sequence of computation of simpler terms.

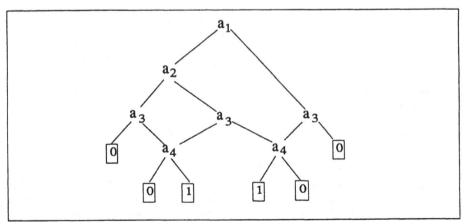

Figure 2. Application of the Expand Function on a BDD.

Consider the functions $f_1, ..., f_n$, and g. The computation of a list of Boolean functions $h_1, h_2, ..., h_k$ whose sum is equal to the function $g(f_1, ..., f_n)$ is made by the "expand" operation. For clarity's sake, we present here the "expand" operation on the binary decision diagrams (BDD) [5].

Each path in the BDD of the function g starting from the root and leading to the leaf 1 defines a *cube* c_j of the function g, and the function g is equal to the sum of its cubes. This means that we could take as functions h_j the functions $c_j(f_1, ..., f_n)$. However this would lead to perform many redundant computations. Consider for instance the BDD drawn in Figure 2. In this BDD there are four paths from the root to 1 that define the four cubes $c_1 = (\neg a_1 \wedge \neg a_2 \wedge a_3 \wedge a_4)$, $c_2 = (\neg a_1 \wedge a_2 \wedge \neg a_3 \wedge a_4)$, $c_3 = (\neg a_1 \wedge a_2 \wedge a_3 \wedge \neg a_4)$, and $c_4 = (a_1 \wedge \neg a_3 \wedge a_4)$. The four functions that would be computed are $h_1 = (\neg f_1 \wedge \neg f_2 \wedge f_3 \wedge f_4)$, $h_2 = (\neg f_1 \wedge f_2 \wedge \neg f_3 \wedge f_4)$, $h_3 = (\neg f_1 \wedge f_2 \wedge f_3 \wedge \neg f_4)$, $h_4 = (f_1 \wedge \neg f_3 \wedge f_4)$.

It is obvious that during the computation of the functions h_1, h_2, h_3, and h_4, the product $(\neg f_1 \wedge f_2)$ is computed twice, as well as the product $(\neg f_3 \wedge f_4)$. Some of these redundant computations can be eliminated by storing partial results in the vertices of the graph that are shared. This is done by the function "expand" that performs a top-down traversal of the BDD of g, and stores in each vertex v of g, the BDD of the function $C_v(f_1, ..., f_n)$, where C_v is the sum of all the cubes represented by the paths starting from the root of the BDD of g and leading to v. Each time the top-down traversal reaches a leaf equal to 1, the function "expand" produces one of the functions h_j.

For the BDD drawn in Figure 2, the partial functions stored in the vertices are: (1) for the vertex (1), $(\neg f_1)$ for (2), $(\neg f_1 \wedge \neg f_2)$ for (3), $(\neg f_1 \wedge f_2)$ for (4), $((\neg f_1 \wedge \neg f_2 \wedge f_3) \vee (\neg f_1 \wedge f_2 \wedge \neg f_3))$ for (5), (f_1) for (6), and $((\neg f_1 \wedge f_2 \wedge f_3) \vee (f_1 \wedge \neg f_3))$ for (7), and the "expand" operation produces two functions:

$$h_1 = (f_4 \wedge ((\neg f_1 \wedge \neg f_2 \wedge f_3) \vee (\neg f_1 \wedge f_2 \wedge \neg f_3))), \text{ and}$$

$$h_2 = (\neg f_4 \wedge ((\neg f_1 \wedge f_2 \wedge f_3) \vee (f_1 \wedge \neg f_3))).$$

Experience shows that the TDG's of the functions $h_1, ..., h_k$ generated by the "expand" operation are smaller than the TDG of the function $\chi(F(s, p))$. The time needed to compute each of these functions directly depends on the sizes of the TDG's of the functions $\delta_j(s, p)$. Next section presents a new Boolean operator that can be used to reduce the sizes of the TDG's used in the term $\chi(F(s, p))$.

4.3.2 The "Restrict" Operator

This section presents the Boolean operator "restrict" [8], noted "$\Downarrow$", that gives a means to reduce the sizes of the TDG's of the functional vector $F(s, p)$ used in the formula $\chi(F(s, p))$. This directly reduces the computational cost of the critical term.

The idea that led to the "$\Downarrow$" operator is made clear by the following remark. In the equation (3) given in Section 3.2:

$$\mathcal{E}_{k+1} = \lambda s.(\mathcal{E}_k(s) \vee (\mathcal{F}(s) \wedge (\exists p \; Cns(s, p) \wedge \mathcal{E}_k(F(s, p))))), \tag{3}$$

as soon as the term $\mathcal{E}_k(s)$ is equal to 1 or the term $\mathcal{F}(s)$ is equal to 0, the value of $\mathcal{E}_{k+1}(s)$ does not depend on the value of $(\exists p \; Cns(s, p) \wedge \mathcal{E}_k(F(s, p)))$. In the same way, when $Cns(s, p)$ is equal to 0, then so is $(Cns(s, p) \wedge \mathcal{E}_k(F(s, p)))$. This means that in the term $\mathcal{E}_k(F(s, p))$, the vectorial function F can be replaced with its restriction $F|_D$ to the domain D, where D is the set whose characteristic function is $\lambda s.\lambda p.(\neg \mathcal{E}_k(s) \wedge \mathcal{F}(s) \wedge Cns(s, p))$.

The same remark holds for the equations (1), (2), and (4'). For these equations the characteristic functions of the domains D to which the vectorial function F can be restricted are the following:

$D = Cns$ for (1) and (2),

$D = \lambda s.\lambda p.(\neg \mathcal{E}_k(s) \wedge \mathcal{F}(s) \wedge Cns(s, p))$ for (3),

$D = \lambda s.\lambda p.(\neg \mathcal{A}_k(s) \wedge \mathcal{F}(s) \wedge Cns(s, p))$ for (4').

The "restrict" operator has two arguments f and D that are Boolean functions. Its definition [8] on Shannon's canonical form is given in Figure 3. In this figure, we note f.root the atom that occurs at the top of Shannon's canonical form of f, and (f/¬a, f/a) the Shannon's decomposition of f with respect to the atom a. We have the identity: $f = (\neg a \wedge f/\neg a) \vee (a \wedge f/a)$.

```
function restrict(f, D);
if D = 0 then error;
if D = 1 or f = 0 or f = 1 then return f;
let a = D.root in {
    if D/¬a = 0   then return restrict(f/a, D/a);
    if D/a = 0    then return restrict(f/¬a, D/¬a);
    if f/¬a = f/a then return restrict(f, D/¬a ∨ D/a);
    return ((¬a ∧ restrict(f/¬a, D/¬a)) ∨ (a ∧ restrict(f/a, D/a)));
}
```

Figure 3. Semantics of the Restrict Operator.

The main properties of "⇓" are expressed by the following theorems. Their proofs are made by induction on Shannon's canonical form of f and D, and by case analysis.

Theorem 4. *Let f and D ≠ 0 be two Boolean functions. Then D(x) = 1 implies that (f ⇓ D)(x) = f(x).*

Theorem 5. *Let f and D be two Boolean functions, with D ≠ 0. If (D ⇒ f), then (f ⇓ D) = 1. If (D ⇒ ¬f), then (f ⇓ D) = 0.*

Theorem 6. *Let f and D be two Boolean functions, with D ≠ 0. Amongst the functions g such that (D ⇒ (g ≡ f)) is a tautology, there is a unique function whose Shannon's canonical form has a minimal number of vertices, and this function is (f ⇓ D).*

Corollary. *Let f and D be two Boolean functions, with D ≠ 0. Shannon's canonical form of (f ⇓ D) has less or the same number of vertices than Shannon's canonical form of f.*

Remark. Each time the case (D/¬a = 0) or (D/a = 0) is used to recursively compute Shannon's canonical form of (f ⇓ D), one branch of Shannon's canonical form of f is deleted. This means that the sooner a "0" occurs on a branch of Shannon's canonical form of D, the larger the number of vertices of Shannon's canonical form of f that are deleted is. This expresses that the smaller the set D is, the greater the reduction is [8].

The operator "⇓" defined above on Shannon's canonical form can also be defined on TDG's. Passing from the canonical tree representation to the canonical graph representation gives rises to the following problem. It can happen that during the application of the operator "⇓" to the TDG's of the function f and of D, some vertex that is shared in the TDG of the function f gets restricted several times, with different vertices from D. This can lead to create a TDG that has more vertices than the one of f. This means that Theorems 4 and 5 hold for this operator on TDG's, but not Theorem 6. To assure that the application of the operator "⇓" on the TGD's of f and D does not return a larger TDG than the one of f, it is sufficient to compare the sizes of the TDG's of f and of (f ⇓ D), and to return the smallest TDG, which is done in $O(\max(|f|, |D|))$. The function that returns the TDG of (f ⇓ D) uses Theorem 5 to speed up the computation, and uses a cache to avoid redundant computations. The complexity of the operator "⇓" is $O(|f| \times |D|)$.

4.4 Discussion of the Method

The computation of the critical term depends mainly on the cost of the composition. The "expand" operation can be basically seen as a method to compute $\chi(f_1, ..., f_n)$, whose computational cost is at least

$O(|\chi| \times \Pi_j |f_j|)$. Theorem 4 assures that, for the domains D associated to the equations given in Section 4.3.2, the application of the operator "$\Downarrow$" on each component of the vectorial function $[f_1 \dots f_n]$ is valid. Applying the "$\Downarrow$" operator on F with the domain D is made in $O(|F| \times |D|)$. The TGD's of the functions $(f_1 \Downarrow D), \dots, (f_n \Downarrow D)$ have less vertices than the ones of the functions $f_1, \dots, f_n$. Assuming that each TGD f_j has a size s, and that each TGD $(f_j \Downarrow D)$ has a size bound by $(s \times R)$, where $0 \le R \le 1$ is the reduction ratio, the composition $\chi \circ (F \Downarrow D)$ is made with a gain of complexity of at least $(1/R)^n$. This means that the quadratic operation "$\Downarrow$" can reduce exponentially the computational cost of the critical term.

5 Experimental Results - Discussion

The verification algorithms given above have been written in LISP, and the CPU times given here have been obtained on a BULL DPX5000 mini computer with 32 megabytes of main memory.

CLM is a part of an interface board designed at BULL. It is made of 33 state registers, and has 14 inputs. The TDG's that represent its vector of transition functions has more than 10000 vertices and the largest of them has more than 2500 vertices. This machine has about 377000 valid states out of the 2^{33} possible states. The CPU time needed to symbolically traverse the state diagram of this machine using the procedure given in [9] is 1227 seconds. We did not succeed in computing the TDG of the transition relation of this machine, so the method of [3] and [6] cannot be applied. The property to be proved valid required the computation of only one fixed point that was obtained in 38 steps that took 4000 seconds of CPU time. The composition $\chi(F(s, p))$ of the critical term stopped to be computable using the standard composition algorithm after a very small number of steps. The "restrict" operation was very useful since it reduced the TDG's of $f_1, \dots, f_{33}$ used in the "expand" operation in such a way that during the iteration, none of them had more than 186 vertices. The largest TDG representing one of the sets in the sequence had 352 vertices.

The sequential circuit *Sync* is a synchronizer that has 21 state registers and 4 inputs. The property to be verified was that the value of the output *OK* is equal to 1 in any valid state of the machine. This property is expressed by the formula Init $\models AG(OK)$. The CPU time needed to prove this formula valid was 4160 seconds and the fixed point was found in 9 steps. The "restrict" operator was not very useful since it did not reduce the sizes of the TDG's of the transition function. The largest TDG found in the sequence of sets had more than 2000 vertices.

The property *AG(OK)* is a safety property. It can be checked by traversing the state diagram of the machine *Sync* using the symbolic traversal procedure given in [9]. At each step during the traversal that starts from the initial state Init of the machine, it is sufficient to check that the output *OK* is equal to 1. Since the traversal stops when all the valid states of the machine have been reached, the proof is complete. Verifying the property in this manner takes 140 seconds, and the traversal is done in 20 steps. This example seems to show that whenever the formula to be proved is a safety property, it is better to use this procedure rather than the general procedure described in this paper.

In fact, the technique proposed here cannot give to the symbolic backward traversal procedure the performance of the forward symbolic traversal procedure [9]. The symbolic backward traversal procedure is essentially based on the computation of the reverse image of the vectorial function δ on a set χ. The symbolic forward traversal procedure is based on the computation of the image of the vectorial function δ on a set χ. Given a machine specified by the 6-tuple $(n, m, r, \omega, \delta, \text{Init})$, we can show that the reverse image symbolic computation is *intrinsically* more difficult than the image symbolic computation. It is easy to show that image symbolic computation is NP-hard (for instance, composition is linearly reducible to image symbolic computation), and that it is $O(n)$-reducible to reverse image symbolic computation. The problem is that reverse image symbolic computation is not polynomially reducible to image symbolic computation, because reducing the former to the latter requires to inverse the vectorial function δ, which is a NP-hard problem (image symbolic computation itself is $O(1)$-reducible to vectorial Boolean function inversion).

Conclusion

This paper has presented a proof algorithm that automatically checks whether some uncompletely defined deterministic Moore machine $\mathcal{M}$ holds some property expressed in the CTL formalism. This algorithm does not require the building of the state-transition graph of the machine $\mathcal{M}$, so it overcomes the limits of the previous methods based on this construction. This means that sequential machine with a very large number of states and transitions can be handle with this method.

Moreover, this proof algorithm does not require the building of the TDG of the transition relation of the machine, which is too large for many practical examples. The proof algorithm is based on the algorithms that were initially developed for proving the equivalence between two machines, in addition with a procedure that essentially computes the reverse image of a vectorial Boolean function. This procedure is the bottleneck of the proof algorithm.

References

[1]   J. P. Billon, "Perfect Normal Forms for Discrete Functions", *BULL Research Report N°87019*, 1987.

[2]   J. P. Billon, J. C. Madre, "Original Concepts of PRIAM, an Industrial Tool for Efficient Formal Verification of Combinational Circuits", in *The Fusion of Hardware Design and Verification*, G. J. Milne Ed., North Holland, 1988.

[3]   S. Bose, A. Fisher, "Automatic Verification of Synchronous Circuits Using Symbolic Logic Simulation and Temporal Logic", in *Proc. of the IFIP Int. Workshop, Applied Formal Methods for Correct VLSI Design*, Leuven, Nov. 1989.

[4]   A. Bouajjani, J. C. Fernandez, N. Halbwachs, "An Executable Temporal Logic for Expressing Safety Properties", July 1990.

[5]   R.E. Bryant, "Graph-based Algorithms for Boolean Functions Manipulation", *IEEE Transaction on Computers*, Vol C35 N°8, 1986.

[6]   S. Burch, E. M. Clarke, K. L. McMillan, "Sequential Circuit Verification Using Symbolic Model Checking", in *Proc. of Design Automation Conference (DAC)*, Orlando FL, USA, July 1990.

[7]   E. M. Clarke, O. Grumbreg, "Research on Automatic Verification of Finite-State Concurrent Systems", *Annual Revue Computing Science*, vol. 2, pp 269-290, 1987.

[8]   O. Coudert, C. Berthet, J. C. Madre, "Verification of Synchronous Sequential Machines Based on Symbolic Execution", in *Lecture Notes in Computer Science: Automatic Verification Methods for Finite State Systems*, Volume 407, J. Sifakis Editor, Springer-Verlag, pp 365-373, June 1989.

[9]   O. Coudert, C. Berthet, J. C. Madre, "Verification of Sequential Machines Using Boolean Functional Vectors", in *Proc. of the IFIP Int. Workshop, Applied Formal Methods for Correct VLSI Design*, Leuven, November 1989.

[10] O. Coudert, C. Berthet, J. C. Madre, "Formal Boolean Manipulations for the Verification of Sequential Machines", in *Proc. of the First European Design Automation Conference (EDAC)*, Glasgow, March 1990.

[11] G. J. Holtzman, "Algorithms for Automated Protocol Validation", in *Lecture Notes in Computer Science: Automatic Verification Methods for Finite State Systems*, Volume 407, J. Sifakis Editor, Springer-Verlag, June 1989.

[12] Z. Kovahi, *Switching and Finite Automata Theory*, McGraw- Hill Book Edition, 1978.

[13] J. P. Queille, J. Sifakis, "Fairness and Related Properties in Transition Systems", *Acta Informatica*, pp 195-220, 1983.

Formal Verification of Digital Circuits Using Symbolic Ternary System Models*

Randal E. Bryant
School of Computer Science
Carnegie Mellon University
Pittsburgh, PA 15213 USA

Carl-Johan H. Seger
Department of Computer Science
University of British Columbia
Vancouver, B.C. V6T 1Z2 Canada

Abstract

Ternary system modeling involves extending the traditional set of binary values $\{0, 1\}$ with a third value X indicating an unknown or indeterminate condition. By making this extension, we can model a wider range of circuit phenomena. We can also efficiently verify sequential circuits in which the effect of a given operation depends on only a subset of the total system state.

This paper presents a formal methodology for verifying synchronous digital circuits using a ternary system model. The desired behavior of the circuit is expressed as assertions in a notation using a combination of Boolean expressions and temporal logic operators. An assertion is verified by translating it into a sequence of patterns and checks for a ternary symbolic simulator. The methodology has been used to verify a number of full scale designs.

1 Introduction

Most formal models for hardware verification assume that every signal always has a well-defined, discrete value. For example, a *binary* model assumes that each signal must be either 0 or 1. In this paper we present a methodology for formal verification in which a third value X is added to the set of possible signal values, indicating an unknown or indeterminate logic value. By shifting to a ternary system model, we gain several advantages.

As a first advantage, this extension makes it possible to model an increased range of circuit phenomena. For example, we can deal with circuits in which nondigital voltages are generated in the course of normal circuit operation. This occurs frequently when modeling circuits at the switch-level [4], due to (generally transient) short circuits or charge sharing. We can also deal with circuits in which indeterminate behavior occurs due either to timing hazards or to circuit oscillation.

---

*This research was supported by the Defense Advanced Research Projects Agency, ARPA Order Number 4976, and by the National Science Foundation, under grant number MIP-8913667.

In all of these cases, the modeling algorithm expresses this uncertainty by assigning value X to the offending circuit nodes, indicating that the actual digital value cannot be determined [6, 7].

As a second advantage, we can efficiently verify many aspects of digital circuit behavior by representing the circuit with a ternary system model. We do this by *ternary symbolic simulation*, in which a simulation algorithm designed to operate on scalar values 0, 1, and X, is extended to operate on a set of symbolic values. Each symbolic value indicates the value of a signal for many different operating conditions, parameterized in terms of a set of symbolic Boolean variables. Since the value X indicates that a signal could be either 0 or 1 (or a non-digital voltage), we can often represent many different operating conditions by the constant value X, rather than with a more complex symbolic value.

Simulators that support ternary modeling intentionally err on the side of pessimism for the sake of efficiency. That is, they will sometimes produce a value X even where exhaustive case analysis would indicate that the value should be binary (i.e., 0 or 1). On the other hand, symbolic simulation avoids this pessimism, because it can resolve the interdependencies among signal values. By combining the expressive power of symbolic values with the computational efficiency of ternary values, we can trade off precision for ease of computation.

In earlier work, we demonstrated the utility of ternary modeling for verifying a variety of circuits [1, 5]. This earlier work demonstrated the viability of circuit verification by symbolic simulation, but it fell short in terms of generality, ease of use, and degree of automation. In this paper, we correct this shortcoming by presenting a formal state transition model for a ternary system, a formal syntax for expressing desired properties of the system, and an algorithm to decide whether or not the system obeys the specified property. Our state transition system is quite general, and is compatible with a number of circuit modeling techniques. The specifications take the form of *symbolic trajectory formulas* mixing Boolean expressions and the temporal *next-time* operator. Finally, our decision algorithm is based on ternary symbolic simulation. It tests the validity of an assertion of the form $[A \implies C]$, where both A and C are trajectory formulas. That is, it determines whether or not every state sequence satisfying A (the "antecedent") must also satisfy C (the "consequent"). It does this by generating a symbolic simulation sequence corresponding to the antecedent, and testing whether the resulting symbolic state sequence satisfies the consequent.

An important property of our algorithm is that it requires a comparatively small amount of simulation and symbolic manipulation to verify an assertion. The restrictions we impose on the formula syntax guarantee that there is a unique weakest symbolic sequence satisfying the antecedent. Furthermore, the symbolic manipulations involve only variables explicitly mentioned in the assertion. Unlike other symbolic circuit verifiers [2], we do not need to introduce extra variables denoting the initial circuit state or possible primary inputs. Finally, the length of the simulation sequence depends only on the depth of nesting of temporal next-time operators in the assertion.

By modifying the COSMOS symbolic switch-level simulator[3], we have been able to implement the algorithm described in this paper and to verify several full scale circuit designs. The following table indicates the performance of our prototype verifier on several different circuits. All CPU times were measured on a DEC 3100 (a 10–20 MIPS machine). We also list the maximum memory requirement of the process, as this is more often the limiting factor in symbolic manipulation than is CPU time.

Circuit	Transistors	CPU Time	Memory
64 × 32 bit moving data stack	16,470	1.25 min.	3.1 MByte
64 × 32 bit stationary data stack	15,873	7.5 min.	5.7 MByte
1K static RAM	6,875	3.7 min.	9.5 MByte

2 Ternary System

Let $B = \{0, 1\}$ be the set of the binary values and let $T = \{0, 1, X\}$. The value X is introduced to denote an "unknown", or "don't care" value.

Define the partial order $\sqsubseteq$ on T as follows: $a \sqsubseteq a$ for all $a \in T$, $X \sqsubseteq 0$, and $X \sqsubseteq 1$. The partial ordering orders values by their "information content." That is, X indicates an absence of information while 0 and 1 represent specific, fully-defined values.

We say that ternary values a and b are *compatible*, denoted $a \sim b$, when there is some value $c \in T$ such that $a \sqsubseteq c$ and $b \sqsubseteq c$. Also, given two compatible ternary values a and b, the *join* between them, denoted $a \sqcup b$, is defined to be the smallest element $c \in T$ in the partial order such that $a \sqsubseteq c$ and $b \sqsubseteq c$.

It is convenient to define an algebra over T with operators $\sqcup$, $\cdot_t$, $+_t$, and $-^t$, where the latter are the obvious extensions of the corresponding Boolean operations $\cdot$ (product), $+$ (sum), and $-$ (complement).

Let T^n, $n \geq 1$, denote the set of all possible vectors of ternary values of length n, i.e., $\{\langle a_1, \ldots, a_n \rangle | a_i \in T, 1 \leq i \leq n\}$. The partial order $\sqsubseteq$, the binary relation $\sim$, and the operation $\sqcup$ are all extended to T^n pointwise.

A ternary function, $f : T^n \to T$, is said to be *monotone* when for any $\vec{a} \in T^n$

and $\vec{b} \in T^n$ we have

$$\vec{a} \sqsubseteq \vec{b} \;\Rightarrow\; f(\vec{a}) \sqsubseteq f(\vec{b})$$

This definition is extended pointwise to vector functions, $\vec{f}: T^n \to T^m$.

The above monotonicity definition is consistent with our use of information content. If a function is monotone, we cannot "gain" any information by reducing the information content of the arguments to the function. In other words, changing some signals from binary values to X will either have no effect on the output values, or it will change some binary values to X.

To express the behavior of a circuit operating over time, we must reason about *sequences* of states. Conceptually, we will consider the state sequences to be infinite, although the properties we will express can always be determined from some bounded length prefix of the sequence. Define a the set $\mathcal{S}^n$ to consist of all sequences $[\vec{a}_0, \vec{a}_1, \ldots]$ where each $a_i \in T^n$. The relations $\sqsubseteq$ and $\sim$ are extended from vectors to sequences pointwise. That is, two sequences $[\vec{a}_0, \vec{a}_1, \ldots]$ and $[\vec{b}_0, \vec{b}_1, \ldots]$ are ordered (compatible) if and only if each pair $\vec{a}_i$ and $\vec{b}_i$ is ordered (compatible), for all $i \geq 0$.

For vector $\vec{a}$ and sequence S, the expression $\vec{a}S$ denotes the sequence consisting the vector $\vec{a}$ followed by the vectors in S.

3   Circuit Model

The underlying model of a circuit we use is quite simple, as well as general. A circuit C is a triple $(\mathcal{N}, \vec{Y}, \mathcal{V})$, where $\mathcal{N}$ is a set of nodes (let $s = |\mathcal{N}|$), $\vec{Y}$ is a vector of excitation functions, and $\mathcal{V}$ is a set of symbolic Boolean variables with which parameterized properties of the circuit are to be expressed.

The excitation functions are defined in a non-traditional way. We view them as expressing "constraints" on the values the nodes can take on one time unit later given the current values on the nodes. By constraint we mean specific binary values, whereas the value X indicates that no constraint is imposed. Since the value of an input is controlled by the external environment, the circuit itself does not impose any constraint on the value; hence the excitation of an "input node" is X. More formally, if node n_i corresponds to an input to the circuit then $Y_{n_i}(\vec{a}) = X$ for every $\vec{a} \in T^s$. Nodes that do not correspond to inputs are called *function nodes*. For a function node n_i the excitation function is a monotone ternary function $Y_{n_i}: T^s \to T$ determined by the circuit topology and functionality.

State sequences are useful when reasoning about circuit behaviors. However, not all state sequences represent possible behaviors of a circuit. The excitation functions generally restrict the possible state sequences significantly. We formalize this property by introducing the concept of a circuit trajectory. Given a circuit C and an arbitrary sequence $[\vec{a}_0, \vec{a}_1, \ldots] \in \mathcal{S}^s$ we say that the sequence is a *circuit trajectory* if and only if

$$\vec{Y}(\vec{a}_i) \sqsubseteq \vec{a}_{i+1} \text{ for } i \geq 0.$$

The set of all trajectories of circuit C is denoted $S(C)$. The above rule for trajectories is consistent with our definition of an excitation function, i.e., a function computing a constraint on the possible value of a node one time unit later. Thus if the current excitation of a node is binary, say a, then the node must take on the value a in the next state in a valid trajectory. On the other hand, if the excitation is X, then the node value is not constrained.

4 Specification Language

Our specification language describes a property of the circuit as an *assertion* of the form $[A \implies C]$, where both A and C are *symbolic trajectory formulas* expressing constraints on the circuit trajectory.

Before we can define our language, we need to introduce some notation and definitions. If $\mathcal{V}$ is a set of symbolic Boolean variables then an *interpretation*, ϕ, is a function $\phi: \mathcal{V} \to \mathcal{B}$ assigning a binary value to each variable. Let Φ be the set of all possible interpretations, i.e., $\Phi = \{\phi: \mathcal{V} \to \mathcal{B}\}$. A *domain constraint*, $\mathcal{D} \subseteq \Phi$, defines a restriction on the values assigned to the variables. We will denote such domain constraints by Boolean expressions. That is, let E be a Boolean expression over elements of $\mathcal{V}$.[†] This expression defines a Boolean function $e: \Phi \to \mathcal{B}$ and thus denotes the domain constraint $\mathcal{D} = \{\phi | e(\phi) = 1\}$. The set of all interpretations Φ is denoted by the Boolean function $\mathbf{1}$, defined as yielding 1 for all interpretations. Expressing domain constraints by Boolean expressions allows us to compactly specify many different circuit operating conditions with a single formula.

4.1 Symbolic Trajectory Formulas

A trajectory formula expresses a set of constraints on a circuit trajectory. When the formula contains Boolean expressions, each interpretation of the variables yields a different set of constraints. A *step-level* symbolic trajectory formula is defined recursively as:

1. **Constants:** TRUE is a trajectory formula.

[†]For the sake of brevity, we omit a formal syntax of Boolean expressions. Any standard expression syntax suffices.

2. **Atomic propositions:** for $n_i \in \mathcal{N}$ both $(n_i = 1)$ and $(n_i = 0)$ are trajectory formulas.

3. **Conjunction:** $(F_1 \wedge F_2)$ is a trajectory formula if F_1 and F_2 are trajectory formulas.

4. **Domain restriction:** $(E \rightarrow F)$ is a trajectory formula if E is a Boolean expression over $\mathcal{V}$ and F is a trajectory formula.

5. **Next time:** $(\mathbf{X}_s F)$ is a trajectory formula if F is a trajectory formula.

We say that a formula is *instantaneous* when it does not contain any next time operator $\mathbf{X}_s$. For convenience, we often drop parentheses when the intended precedence is clear.

The truth of a formula F is defined relative to a circuit, an interpretation ϕ of the variables in $\mathcal{V}$, and a circuit trajectory. The truth of F, written $\mathcal{C}, \phi, S \models F$, is defined recursively. In the following, assume that both S and $\vec{a}S$ are trajectories of $\mathcal{C}$.

1. $\mathcal{C}, \phi, S \models$ TRUE holds trivially.

2. (a) $\mathcal{C}, \phi, \vec{a}S \models (n_i = 1)$ iff $a_i = 1$.
   (b) $\mathcal{C}, \phi, \vec{a}S \models (n_i = 0)$ iff $a_i = 0$.

3. $\mathcal{C}, \phi, S \models (F_1 \wedge F_2)$ iff $\mathcal{C}, \phi, S \models F_1$ and $\mathcal{C}, \phi, S \models F_2$

4. $\mathcal{C}, \phi, S \models (E \rightarrow F)$ iff $e(\phi) = 0$ or $\mathcal{C}, \phi, S \models F$, where e is the Boolean function denoted by the Boolean expression E.

5. $\mathcal{C}, \phi, \vec{a}S \models \mathbf{X}_s F$ iff $\mathcal{C}, \phi, S \models F$.

For an instantaneous formula, its truth can be defined relative to a single state. For instantaneous formula F, the notation $\mathcal{C}, \phi, \vec{a} \models F$ indicates that F holds under interpretation ϕ for state $\vec{a}$. A formal definition of this notation can be derived by a straightforward adaptation of rules 1–4 above.

4.2 Assertions

Our verification methodology entails proving *assertions* about the model structure. These assertions are of the form $[A \implies C]$, where the *antecedent* A and the *consequent* C are trajectory formulas. The truth of an assertion is defined relative to a circuit $\mathcal{C}$ and an interpretation ϕ. Unlike a formula, however, an assertion is considered true only if it holds for all trajectories. That is, $\mathcal{C}, \phi \models [A \implies C]$, when for every $S \in \mathcal{S}(C)$ we have that $C, \phi, S \models A$ implies that $C, \phi, S \models C$. Given a circuit and an assertion, the task of our checking algorithm is to compute the Boolean function expressing the set of interpretations under which the assertion is true. For most verification problems, this should simply be the constant function 1, i.e., the assertion should hold under all variable interpretations.

We have intentionally chosen to introduce only a heavily restricted trajectory formula syntax for our base logic. By imposing these restrictions, we can guarantee the following key property:

Proposition 1 *For any trajectory formula F, and any interpretation ϕ, one of the following cases must hold:*

1. *There is no trajectory $S \in \mathcal{S}(C)$ for which $C, \phi, S \models F$, or*

2. *There exists a unique trajectory $S_{F,\phi} \in \mathcal{S}(C)$ such that for every $S \in \mathcal{S}(C)$ we have $C, \phi, S \models F$ if and only if $S_{F,\phi} \sqsubseteq S$.*

In the first case above, we say that the formula F is not satisfiable under interpretation ϕ. In the second case, we refer to the sequence $S_{F,\phi}$ as the *weakest trajectory* satisfying formula F under interpretation ϕ.

Note that this proposition expresses a very strong property of our logic. It demonstrates the reason why we can verify an assertion by simulating a single symbolic sequence, namely the one encoding the weakest trajectories allowed by the antecedent for every interpretation. It is stronger than the simple monotonicity condition that if $C, \phi, S \models F$ and $S \sqsubseteq S'$, then $C, \phi, S' \models F$.

The logic, as described above, is convenient for deriving the underlying theory. Unfortunately, expressing "interesting" assertions about real circuits using only the constructs above is very tedious. Two shortcomings make using the logic cumbersome: the fine granularity of the timing, and the lack of more powerful logical constructs. It is convenient to add extensions that do not add any expressive power, but make it easier to write assertions.

This basic structure of starting with a minimal basic logic and then adding more elaborate structures as extensions also mirrors our current implementation. The implementation consists of two parts. The underlying logic, with some few extensions, is taken care of by our modified version of the COSMOS symbolic switch-level simulator. The syntactic extensions are supported by a front-end written in SCHEME. The user writes SCHEME code that, when evaluated, generates a file of low-level simulation commands which are then evaluated by the simulator.

5 Symbolic Simulation

In creating a symbolic model, we extend the scalar model defined in terms of the binary and ternary domains $\mathcal{B}$ and $\mathcal{T}$, to one defined in terms of binary- and ternary-valued functions over the variables $\mathcal{V}$. Define the symbolic domain $\mathcal{B}(\mathcal{V})$ (respectively, $\mathcal{T}(\mathcal{V})$) as denoting the set of functions mapping an interpretations in Φ to $\mathcal{B}$ (resp., $\mathcal{T}$). More formally $\mathcal{B}(\mathcal{V}) = \{f: \Phi \to \mathcal{B}\}$ and $\mathcal{T}(\mathcal{V}) = \{f: \Phi \to \mathcal{T}\}$. We then extend the operations defined over scalar values to create a symbolic algebra.

We can also extend the vector and sequence algebra defined over scalar values to their counterparts defined over symbolic values. That is, define the vector domain $\mathcal{T}(\mathcal{V})^n$ as

$$\mathcal{T}(\mathcal{V})^n = \{\langle a_1, \ldots, a_n \rangle | a_i \in \mathcal{T}(\mathcal{V})\}.$$

In implementing a symbolic simulator, we in effect extend the excitation function $\vec{Y}$ to the symbolic domain as $\vec{Y}: \mathcal{T}(\mathcal{V})^s \to \mathcal{T}(\mathcal{V})^s$. For $\vec{a} \in \mathcal{T}(\mathcal{V})^n$, let $\vec{a}(\phi) \in \mathcal{T}^n$ denote the vector with each element i equal to $a_i(\phi)$. In this way, we can view the symbolic vector $\vec{a} \in \mathcal{T}(\mathcal{V})^n$ either as a vector of symbolic elements, or as a symbolic value which for a given interpretation yields a scalar vector.

We extend most operations from scalar to symbolic domains in a uniform way. Consider an operation $op: \mathcal{D}_1 \times \mathcal{D}_2 \to \mathcal{D}_3$, defined over vectors, single elements, or a combination of the two. Its symbolic counterpart $op: \mathcal{D}_1(\mathcal{V}) \times \mathcal{D}_2(\mathcal{V}) \to \mathcal{D}_3(\mathcal{V})$ is defined such that for all $a \in \mathcal{D}_1(\mathcal{V})$ and $b \in \mathcal{D}_2(\mathcal{V})$, we have $(a \ op \ b)(\phi) = a(\phi) \ op \ b(\phi)$. We use this method to extend the ternary algebraic operations $\cdot_t$, $+_t$, and $-^t$, as well as the operation $\sqcup$.

When extending a relation R symbolically, we define the result to be a function specifying the interpretations under which its arguments are related. That is, given a binary relation $R \subseteq \mathcal{D}_1 \times \mathcal{D}_2$, define $R: \mathcal{D}_1(\mathcal{V}) \times \mathcal{D}_2(\mathcal{V}) \to \mathcal{B}(\mathcal{V})$ as $(a \ R \ b)(\phi) = 1$ if and only $a(\phi) \ R \ b(\phi)$. We use this method to define operations $\sim$ and $\sqsubseteq$ over both single elements and vectors.

We require one operation that is extended to vectors in a nonstandard way. Define the infix operator $?: \mathcal{B} \times \mathcal{T} \to \mathcal{T}$ as $a \ ? \ b$ equals b if a is 1, and equals X otherwise. When extending this operation to vectors, only the second argument is vector-valued. That is the operation $?: \mathcal{B} \times \mathcal{T}^n \to \mathcal{T}^n$ is defined as $(a \ ? \ \vec{b})_i = a \ ? \ b_i$. This operation is then extended symbolically in the manner described above.

As a final operation, we define a variant of the join operation that is defined even when for some $\phi \in \Phi$, we have $\vec{a}(\phi) \not\sim \vec{b}(\phi)$. When using this operation, we will separately keep track of the conditions under which the arguments are compatible. Define the operation $\overset{*}{\sqcup}: T(\mathcal{V})^n \times T(\mathcal{V})^n \to T(\mathcal{V})^n$ as

$$(\vec{a} \overset{*}{\sqcup} \vec{b})(\phi) = \begin{cases} \vec{a}(\phi) \sqcup \vec{b}(\phi), & \vec{a}(\phi) \sim \vec{b}(\phi) \\ \vec{X}, & \text{otherwise} \end{cases}$$

where $\vec{X}$ denotes a vector with all elements equal to X.

5.1 Translating Instantaneous Formulas to Symbolic Vectors

Given the above definitions and an instantaneous formula F, we derive a "domain" function OK_F, and a "weakest" symbolic vector $\vec{a}_F$ as follows:

1. If F is TRUE then $OK_F = 1$, and $\vec{a}_F = \langle X, \ldots, X \rangle$.

2. (a) If F is $(n_i = 1)$ then $OK_F = 1$, and $\vec{a}_F = \langle X, \ldots, X, 1, X, \ldots, X \rangle$, where the 1 is in position i.

   (b) If F is $(n_i = 0)$ then $OK_F = 1$, and $\vec{a}_F = \langle X, \ldots, X, 0, X, \ldots, X \rangle$, where the 0 is in position i.

3. If F is $(F_1 \wedge F_2)$ then $OK_F = OK_{F_1} \cdot OK_{F_2} \cdot (\vec{a}_{F_1} \sim \vec{a}_{F_2})$, and $\vec{a}_F = \vec{a}_{F_1} \overset{*}{\sqcup} \vec{a}_{F_2}$.

4. If F is $(E \to F_1)$ then $OK_F = \overline{e} + OK_{F_1}$, and $\vec{a}_F = e ? \vec{a}_{F_1}$, where e is the Boolean function denoted by the expression E.

The following proposition summarizes the main properties of OK_F and $\vec{a}_F$.

Proposition 2 *Given a circuit C, let F be an instantaneous formula and OK_F and $\vec{a}_F$ be derived as above. Then $OK_F(\phi) = 1$ iff there exists some state $\vec{b} \in T^s$ such that $C, \phi, \vec{b} \models F$. Furthermore, if $OK_F(\phi) = 1$, then $C, \phi, \vec{b} \models F$ iff $\vec{a}_F(\phi) \sqsubseteq \vec{b}$.*

5.2 Checking Assertions

Our first step in verifying an assertion is to rewrite the antecedent and consequent into a normal form where all next-time operators are collected together. It is easy to show that a trajectory formula F can be rewritten into $F_0 \wedge \mathbf{X}_s F_1 \wedge \mathbf{X}_s^2 F_2 \wedge \ldots \wedge \mathbf{X}_s^{k-1} F_{k-1}$, for some $k \geq 1$, where each F_i is instantaneous. Note that some of the F_i's might be the trivial formula TRUE. Note also that such a sequence can be extended by appending $\mathbf{X}_s^i \text{TRUE}$ for $i \geq k$. Hence, without any loss of generality, we will henceforth assume that the antecedent and the consequent in an assertion are trajectory formulas in normal form containing the same number of terms.

Given an assertion $[A \implies C]$ of the form

$$[A_0 \wedge \mathbf{X}_s A_1 \wedge \ldots \wedge \mathbf{X}_s^{k-1} A_{k-1} \implies C_0 \wedge \mathbf{X}_s C_1 \wedge \ldots \wedge \mathbf{X}_s^{k-1} C_{k-1}]$$

define a sequence of symbolic ternary vectors $\vec{x}_0, \ldots, \vec{x}_{k-1}$ as follows:

$$\vec{x}_i = \begin{cases} \vec{a}_{A_0}, & i = 0 \\ \vec{Y}(\vec{x}_{i-1}) \overset{\bullet}{\sqcup} \vec{a}_{A_i}, & i > 0. \end{cases}$$

Define the Boolean function $OK_A = \Pi_{0 \leq i < k} OK_{A_i}$, where Π denotes Boolean product. This function yields 0 for those interpretations for which the antecedent contains some internal inconsistency. For example, the formula $A = (n_i = a) \wedge (n_i = b)$ would have $OK_A = \overline{a \oplus b}$, because this formula cannot be satisfied when $\phi(a) \neq \phi(b)$. Define the Boolean function $Traj = \Pi_{1 \leq i < k}[\vec{Y}(\vec{x}_{i-1}) \sim \vec{a}_{A_i}]$. This function yields 0 for those interpretations where an incompatibility arises in the trajectory.

We can show that A is satisfiable under some interpretation ϕ if and only if $OK_A(\phi) \cdot Traj(\phi) = 1$. Furthermore, we can extend the sequence $\vec{x}_0, \ldots, \vec{x}_{k-1}$ to be an infinite sequence by defining $\vec{x}_i = \vec{Y}(\vec{x}_{i-1})$ for all $i \geq k$. It can then be shown that for interpretation ϕ the sequence $\vec{x}_0(\phi), \vec{x}_1(\phi), \ldots$ is the weakest trajectory satisfying A under interpretation ϕ. This construction then provides a proof of Proposition 1. This demonstrates how our symbolic simulator can set up the weakest allowable conditions allowed by the antecedent under all possible interpretations.

To check the consequent, define the Boolean function $OK_C = \Pi_{0 \leq i < k} OK_{C_i}$. This function yields 0 for those interpretations for which the consequent contains some internal inconsistency. Finally, define the Boolean function $Check = \Pi_{0 \leq i < k}[\vec{a}_{C_i} \sqsubseteq \vec{x}_i]$. This function yields 0 for those interpretations where some trajectory satisfying the antecedent may violate the consequent.

Now define $OK_{[A \implies C]}$ as: $\overline{OK_A} + \overline{Traj} + (OK_C \cdot Check)$. Informally, this equation states that the assertion is true under those interpretations for which the antecedent is unsatisfiable (due either to internal inconsistencies or to an incompatibility in the trajectory), as well as those for which the consequent holds (i.e, it is both internally consistent and is satisfied.)

The main result of this paper is captured in the following theorem:

Theorem 1 *Given a circuit C and an assertion $[A \implies C]$ let $OK_{[A \implies C]} \in \mathcal{B}(\mathcal{V})$ be derived as above. Then*

$$C, \phi \models [A \implies C] \quad \text{if and only if} \quad OK_{[A \implies C]}(\phi) = 1.$$

Hence, determining whether a circuit satisfies $[A \implies C]$ is reduced to determining whether $OK_{[A \implies C]} = 1$.

6 Conclusions

In terms of mathematical sophistication, the problem solved by our algorithm is far less ambitious than what is attempted by full-fledged temporal logic model checkers. However, we believe that our language is rich enough to be able to describe many important properties of a circuit and to provide a direct path by which such properties may be automatically verified. By keeping the goals of our verifier simple, we obtain an algorithm that is capable of dealing with much larger circuits. We are currently applying these ideas to larger and more complex circuits.

References

[1] D. L. Beatty, R. E. Bryant, and C.-J. H. Seger, "Synchronous Circuit Verification by Symbolic Simulation: An Illustration," *Sixth MIT Conference on Advanced Research in VLSI*, 1990.

[2] S. Bose, and A. L. Fisher, "Automatic Verification of Synchronous Circuits using Symbolic Logic Simulation and Temporal Logic," *IMEC-IFIP International Workshop on Applied Formal Methods for Correct VLSI Design*, 1989, pp. 759–764.

[3] R. E. Bryant, D. Beatty, K. Brace, K. Cho, and T. Sheffler, "COSMOS: a Compiled Simulator for MOS Circuits," *24th Design Automation Conference*, 1987, 9–16.

[4] R. E. Bryant, "Boolean Analysis of MOS Circuits," *IEEE Transactions on Computer-Aided Design of Integrated Circuits and Systems*, Vol. CAD-6, No. 4 (July, 1987), 634–649.

[5] R. E. Bryant, "Formal Verification of Memory Circuits by Switch-Level Simulation," To appear in *IEEE Transactions on Computer-Aided Design of Integrated Circuits and Systems*, 1990.

[6] J. A. Brzozowski, and M. Yoeli. "On a Ternary Model of Gate Networks." *IEEE Transactions on Computers C-28*, 3 (March 1979), 178–183.

[7] C-J. Seger, and R. E. Bryant, "Modeling of Circuit Delays in Symbolic Simulation", *IMEC-IFIP International Workshop on Applied Formal Methods for Correct VLSI Design*, 1989, pp. 625–639.

Vectorized Model Checking
for Computation Tree Logic

Hiromi Hiraishi, Shintaro Meki and Kiyoharu Hamaguchi
Department of Information Science, Kyoto University
Sakyo-ku, Kyoto, 606, Japan.

Abstract

The aim of this paper is to show how big model checking problems for Computation Tree Logic (CTL) can be handled by using current powerful vector processors. Although efficient recursive model checking algorithms for CTL, which run in time proportional to both the size of Kripke structures and the length of formulas, have been already proposed [7, 2], their algorithms cannot be vectorized due to recursive procedure calls. In this paper we propose a new model checking algorithm, called a vectorized model checking algorithm, for CTL which is suitable for the execution on vector processors. It can handle more than 1 million state Kripke structure derived from a deterministic sequential machine.

1 Introduction

Recently various kinds of formal methods for automatic verification have been widely studied. Among them, the model checking approach based on a branching time temporal logic called CTL (Computation Tree Logic) [2, 3, 5, 6, 7] is one of the most efficient approaches. In verification of a system which consists of several machines communicating with each other, however, there is a problem so-called a state explosion problem. Although there are several works trying to avoid this problem [8, 10], it seems to be difficult to avoid the problem in general, and there are strong requirements for verification of large systems[11].

We mainly aim at clarifying how big machines can be verified based on the model checking algorithm for CTL by using current powerful vector processors. As the first step to this purpose, we challenged to vectorize model checking for CTL on Kripke structures this time. Although the model checking algorithms in [7] are efficient and runs in time proportional to both the size of Kripke structures and the length of CTL formulas, they are not suitable for vector processors because it is difficult to vectorize them due to their recursive procedure calls. In order to extract high performance of vector processors, we need an algorithm using repeated uniform operations on array type data. It is easy to develop such a model checking algorithms based on fixpoint calculations of CTL semantics, but the direct implementation of the fixpoint calculations would easily lead to an algorithm whose time complexity is proportional to $|S|^3$ or $|S|^4$, where $|S|$ is the number of states of Kripke structure.

The new model checking algorithm called vectorized model checking algorithm proposed here can be vectorized for executions on vector processors. It runs in time linear to both the size of Kripke structures (i.e. $|S| + |R|$, where $|R|$ is the number of edges of Kripke structure) and the length of CTL formulas. We also implemented the algorithm on a vector processor FACOM VP400E. The analysis of storage requirement of the implementation shows that it can manipulate more than 1 million state Kripke structure derived from a deterministic sequential machine. We also present the result of an experiment which shows the efficiency of our implementation.

This paper is organized as follows: Section 2 summarizes the definition of CTL. Section 3 describes the vectorized model checking algorithm for CTL in conjunction with explanations about vector processors. In Section 4 we explain the implementation of the algorithm on a vector processor FACOM VP400E and show its experimental result. Section 5 concludes this paper with summarizing remaining future problems.

2 Computation Tree Logic

Computation Tree Logic (CTL)[6] is a branching time temporal logic. Let AP be a set of atomic propositions. CTL formulas are inductively defined as follows:

- If $p \in AP$, p is a CTL formula.
- If η is a CTL formula, then so are $\neg\eta$, $EX\eta$ and $EG\eta$.
- If η and ξ are CTL formulas, then so are $\eta \vee \xi$ and $E[\eta\,\mathcal{U}\xi]$.

The semantics of CTL is defined over a Kripke structure $K = (S, R, I)$, where

- S is a non-empty finite set of states.
- $R \subseteq S \times S$ is a total binary relation on S (i.e. for $\forall s \in S$, there exists $s' \in S$ such that $(s, s') \in R$).
- $I : S \rightarrow 2^{AP}$ is an interpretation function which labels each state with a set of atomic propositions true at that state.

An infinite sequence of states $\pi = s_0 s_1 s_2 \ldots$ is called a *path* from s_0 if $(s_i, s_{i+1}) \in R$ for $\forall i \geq 0$. $\pi(i)$ denotes the i−th state of the sequence π (i.e. $\pi(i) = s_i$).

The truth-value of a CTL formula is defined at a state of a Kripke structure and $K, s \models \eta$ denotes that a CTL formula η hold at a state s of a Kripke structure K. If there is no ambiguity, we will omit K and just write as $s \models \eta$. The relation $\models$ is recursively defined as follows:

- $s \models p\ (\in AP)$ iff $p \in I(s)$.
- $s \models \neg\eta$ iff $s \not\models \eta$.
- $s \models \eta \vee \xi$ iff $s \models \eta$ or $s \models \xi$.
- $s \models EX\eta$ iff there exists some next state s' of s (i.e. $(s, s') \in R$) such that $s' \models \eta$.
- $s \models EG\eta$ iff there exists some path π on K starting from the state s such that $\pi(i) \models \eta$ for $\forall i \geq 0$.
- $s \models E[\eta\,\mathcal{U}\xi]$ iff there exists some path π on K starting from the state s such that $\exists i \geq 0$, $\pi(i) \models \xi$ and $\pi(j) \models \eta$ for $0 \leq \forall j < i$.

DO 10 $I = 1, N$
10 $A(I) = B(I) + C(I)$

(a) contiguous access

DO 10 $I = 1, N, K$
10 $A(I) = B(I) + C(I)$

(b) constant strided access

DO 10 $I = 1, N$
10 $A(I) = B(L(I))$

(c) indirectly addressed access

Figure 1: Three types of vector accesses.

DO 10 $I = 1, N$
10 **IF** $(A(I)$ **.GT.** $0.0)$ $B(I) = B(I) + C(I)$

(a) DO loop with a conditional statement

$K = 0$
DO 10 $I = 1, N$
    IF $(A(I)$ **.GT.** $0.0)$ **THEN**
       $K = K + 1$
       $C(K) = B(I)$
    ENDIF
10 **CONTINUE**

(b) vector compress function

Figure 2: DO loops containing conditional statements

3 Vectorized Model Checking Algorithm

3.1 Vector processors

Vector processors are supercomputers for large-scale computations. They achieve more than several hundred MFLOPS (Million FLoating-point Operations Per Seconds) by vector instructions which execute uniform operations on array-structured data using pipelined functional units, and they usually have large main memory of several hundred mega bytes. In conjunction with floating-point operations, they also support integer and bit-wise logical operations.

Although the maximum speed of vector processors are very high, the following two points are very important in programming for vector processors to achieve their maximum performance:

Vectorization ratio: Vectorization ratio is the rate of the operations executed by vector instructions to the whole operations in a program. This ratio should be more than 90% to obtain high performance of vector processors.

Vector length: Since there are some overheads for setting up vector instructions, the length of operands (vector length) of vector instructions should be large enough; it should be larger than several hundreds to get maximum performance of vector processors.

As for data transmission between the main memory and vector registers, there are load/store pipelines which support basically the following three types of vector accesses: *contiguous vector access, constant strided vector access* and *indirectly addressed vector access* (see Figure 1).

Furthermore, vector processors support **DO** loop with conditional statements and *vector compress function* shown in Figure 2. These vectorized functions are very powerful in implementing vectorized model checker for CTL.

```
1. procedure  Verify_Not(¬η)          1. procedure  Verify_Or(η ∨ ξ)
2. for all  s ∈ S do                  2. for all  s ∈ S do
3.   Label(¬η, s) := Not(Label(η, s));  3.   Label(η ∨ ξ, s) := Or(Label(η, s), Label(ξ, s));
4. return;                             4. return;
5. end of procedure                    5. end of procedure
```

Figure 3: Vecorized model checking algorithms for Boolean operators

3.2 Vectorized model checking algorithm

The vectorized model checking algorithm for CTL runs in bottom-up way by labeling each state with the truth value of each sub-formula contained in a given CTL formula. $Label(\eta, s)$ denotes the truth value of a CTL formula η at a state s in the following.

Case Boolean operations ($\neg\eta$ or $\eta \vee \xi$): We need only to calculate logical negation or disjunction of $Label(\eta, s)$ and $Label(\xi, s)$ for each state s as $Verify\_Not(\neg\eta)$ and $Verify\_Or(\eta \vee \xi)$ in Figure 3.

Case $EX\eta$: First, it assumes that $EX\eta$ is *false* at all the states. For all states where η holds, it labels $EX\eta$ to be *true* at their all predecessor states (see $Verify\_EX(EX\eta)$ in Figure 4.).

Case $E[\eta\,U\,\xi]$: From the definition of $E[\eta\,U\,\xi]$, it holds at the states where ξ is *true* and it also holds at the states which are reachable to such states only through the states where η holds. Therefore, this is a kind of reachability problem.

In the procedure $Verify\_EU(E[\eta\,U\,\xi])$, the sets of states N_1 and N_2 are used to keep track of the states where the truth value of $E[\eta\,U\,\xi]$ is newly determined to be *true*. It first initializes the set N_1 and then it calculates the initial values of $Label(E[\eta\,U\,\xi], s)$ at each state (lines $3 \sim 11$) as follows: if ξ holds at the state, it is 1 (i.e. *true*) because $E[\eta\,U\,\xi]$ holds at the state apparently and the state is inserted to N_1; if neither ξ nor η hold at the state, it is 0 (i.e. *false*) because $E[\eta\,U\,\xi]$ does not hold at the state apparently; if η holds but ξ does not, it is temporarily assigned to 2 indicating that it will be determined later by checking the reachability.

Next, for all the states labeled 2, the procedure checks the reachability to the state labeled 1 through the states labeled 2, and the reachable states are labeled 1 and the unreachable states are labeled 0. This step is done as follows (lines $12 \sim 27$): For each state where $E[\eta\,U\,\xi]$ is newly determined to be *true*, the labels of its predecessor states are checked, and if they are 2, then they are relabeled as 1 because they are the reachable states and they are added to the set which keeps track of the states newly labeled as 1. This step is repeated until no more states are newly labeled as 1 (lines $12 \sim 23$). Finally, the states whose labels are still 2 are labeled as 0 because they are the unreachable states (lines $24 \sim 27$).

Case $EG\eta$: From the definition of $EG\eta$, it holds at the states on the loops which are constructed only by the states where η holds. It also holds at the states reachable to such loops through the states where η holds. In order to find out such states, the procedure $Verify\_EG((\eta)$ (Figure 4) first constructs a sub-graph logically by extracting all the states where η holds. Then it keeps removing the states from the sub-graph whose out degrees are 0 while such states exist. It is almost clear that $EG\eta$ holds at the states contained in

```
1.  procedure  Verify_EX(EXη)
2.  for all  s ∈ S do
3.   Label(EXη, s) := 0;
4.  for all  s' ∈ S do
5.   if  Label(η, s') = 1 then
6.    for all  s such that (s, s') ∈ R do
7.     Label(EXη, s) := 1;
8.  return;
9.  end of procedure


1.  procedure  Verify_EU(E[η U ξ])
2.  N₁ := ∅
3.  for all  s ∈ S do
4.   if  Label(ξ, s) = 1 then
5.   begin
6.    Label(E[η U ξ], s) := 1;
7.    N₁ := N₁ ∪ {s};
8.   end
9.   else if  Label(η, s) = 0 then
10.   Label(E[η U ξ], s) := 0;
11.  else  Label(E[η U ξ], s) := 2;
12. while  N₁ ≠ ∅ do
13. begin
14.  N₂ := ∅;
15.  for all  s' ∈ N₁ do
16.   for all  s such that (s, s') ∈ R do
17.    if  Label(E[η U ξ], s) = 2 then
18.    begin
19.     Label(E[η U ξ], s) := 1;
20.     N₂ := N₂ ∪ {s};
21.    end
22.  N₁ := N₂;
23. end
24. for all  s ∈ S
25.  if  Label(E[η U ξ], s) = 2 then
26.   Label(E[η U ξ], s) := 0;
27. return;
28. end of procedure
```

```
1.  procedure  Verify_EG(EGη)
2.  N₁ := ∅;
3.  for all  s ∈ S do
4.   if  Label(η, s) = 1 then
5.   begin
6.    Label(EGη, s) := 1;
7.    N₁ := N₁ ∪ {s};
8.   end
9.   else
10.   Label(EGη, s) := 0;
11. if  N₁ ≠ ∅ then
12. begin
13.  for all  s' ∈ N₁ do
14.   for all  s such that (s, s') ∈ R do
15.    if  Label(EGη, s) ≥ 1 then
16.     Label(EGη, s) := Label(EGη, s) + 1;
17.  N₂ := ∅;
18.  for all  s ∈ N₁ do
19.   if  Label(EGη, s) = 1 then
20.   begin
21.    Lable(EGη, s) := 0;
22.    N₂ := N₂ ∪ {s}
23.   end
24.  N₁ := N₂;
25. end
26. while  N₁ ≠ ∅ do
27. begin
28.  N₂ := ∅;
29.  for all  s' ∈ N₁ do
30.   for all  s such that (s, s') ∈ R do
31.   begin
32.    if  Label(EGη, s) = 2 then
33.     N₁ := N₁ ∪ {s};
34.    Label(EGη, s) := Label(EGη, s) − 1;
35.   end
36.  N₁ := N₂;
37. end
38. for all  s ∈ S do
39.  if  Label(EGη, s) ≥ 2 then
40.   Label(EGη, s) := 1;
41.  else
42.   Label(EGη, s) := 0;
43. return;
44. end of procedure
```

Figure 4: Vectorized model checking algorithms for temporal operators

the resulting sub-graph and it does not hold at all other states.

More precisely, after initializing the set of states N_1 to be empty, the procedure labels the states as 1 where η holds; it labels the states as 0 where η does not hold (lines 3 ∼ 10). For each state labeled 1, if the labels of its predecessor states are greater than 0, then they are incremented (lines 13 ∼ 16). At this point, the label 0 means that $EG\eta$ does not hold at the state; the label greater than 0 means that η holds at the state and it has its label - 1 successor states where η holds. Next, for each state labeled 1 (i.e. the state which has no successor states where η holds), the label is relabeled to 0 and the state is inserted to the set N_2 which keeps track of the states newly labeled as 0 (lines 18 ∼ 23). In lines 27 ∼ 37, for each state newly determined that $EG\eta$ does not hold on it, the labels of its predecessor states are decremented. This step is repeated for those states that become to have no successor states until no more such states exist. Finally, the states whose labels are greater than 1 are relabeled to 1 because such states have at least one infinite path on which η always holds; the other states are labeled as 0 (lines 38 ∼ 43).

3.3 Time complexity

It is clear that *Verify_Not*($\neg\eta$) and *Verify_Or*($\eta \vee \xi$) runs in time proportional to $|S|$.

In the case of *Verify_EX*($EX\eta$), the lines 6 ∼ 7 are executed only $|R|$ times in total by adopting a data structure which assigns a list of its predecessor states to each state. Therefore, the time complexity is $O(|S| + |R|)$.

In the case of *Verify_EU*($E[\eta\,\mathcal{U}\xi]$), the *for loop* from the line 3 to the line 11 is executed $|S|$ times. The lines 16 ∼ 21 are executed only $|R|$ times in total because it never checks the predecessor states of the same state twice. The last *for loop* (lines 24 ∼ 26) is repeated $|S|$ times. Therefore, its time complexity is $O(|S| + |R|)$.

In the case of *Verify_EG*($EG\eta$), it is easy to see that the lines 3 ∼ 10, 18 ∼ 23, and 38 ∼ 43 are executed $|S|$ times in total, and the lines 14 ∼ 16 and 30 ∼ 35 are repeated $|R|$ times in total in the same way as in the case of *Verify_EU*($E[\eta\,\mathcal{U}\xi]$). Therefore, its time complexity is $O(|S| + |R|)$.

Since one of these procedures is executed for each sub-formula of a given CTL formula η, the time complexity of the vectorized model checking algorithm is $O((|S| + |R|)|\eta|)$, where $|\eta|$ denotes the length of η.

3.4 Fairness constraints

Fairness constraints can be handled efficiently by labeling *fair states* which have at least one *fair path* [7, 9]. This can be done by first obtaining *fair strongly connected components* and then getting reachable states to the fair strongly connected components.

There is a well known linear time algorithm to get strongly connected components of a directed graph based on the depth first search [1]. It seems to be difficult to vectorize this algorithm. We leave the vectorization of this part as a future problem and decided to use the non-vectorized well known algorithm.

Once the strongly connected components have been obtained, it is easy to vectorize the decision procedure if they are fair or not. The reachability problem can be also vectorized in the same way as model checking of $E[\eta\,\mathcal{U}\xi]$.

The labeling for fair states should be done once before starting model checking and it should be also done when evaluating EG operator.

4   Vectorized Model Checker

4.1   Implementation

We have implemented the vectorized model checking algorithm on a vector processor FACOM VP400E as a vectorized model checker. In the VP400E, three pipelined vector functional units, each of which consists of 4 pipelined units, can operate in parallel with 7 nano second cycle time. Its peak performance is about 1714 MFLOPS. It has a 256 M byte main memory, in which we can use 200 M bytes as a user area.

The input of the vectorized model checker is a Moore type deterministic sequential machine and CTL formulas to be verified. It creates the corresponding Kripke structure internally from a given Moore type deterministic sequential machine.

Let $M = (X, Z, \Sigma, \delta, \lambda, s_0)$ be a Moore type deterministic sequential machine, where

- X is a finite and nonempty set of binary input signals (atomic propositions);

- Z is a finite and nonempty set of binary output signals (atomic propositions);

- Σ is a finite and nonempty set of states;

- $\delta : 2^X \times \Sigma \to \Sigma$ is the state transition function;

- $\lambda : \Sigma \to 2^Z$ is the output function;

- s_0 is the initial state.

Then, the corresponding Kripke structure $K = (S, R, I)$ becomes as follows:

$$
\begin{aligned}
S &= \{s'_{i,j} | s_i \in \Sigma, j \in 2^X \text{ and } \delta(j, s_i) \text{ is defined.}\} \\
R &= \{(s_{i,j}, s_{i',j'}) | s_{i,j}, s_{i',j'} \in S, \delta(j, s_i) = s_{i'}\} \\
I(s'_{i,j}) &= \{x | x \in j\} \cup \{z | z \in \lambda(s_i)\}
\end{aligned}
$$

Intuitively, there is a one to one correspondence between the transition edges of the Moore type deterministic sequential machine M and the states of the corresponding Kripke structure K. The size of the Kripke structure becomes as follows:

$$
\begin{aligned}
O(|S|) &= O(|\Sigma| \times 2^{|X|}) \\
O(|R|) &= O(|E| \times 2^{|X|} \\
O(|S| + |R|) &= O((|\Sigma| + |E|) \times 2^{|X|}),
\end{aligned}
$$

where $|E|$ represents the number of transition edges of M and $O(|E|) = O(|\Sigma| \times 2^{|X|})$.

Almost all parts of the model checker are vectorized except the fairness constraints handling. The most time consuming parts of the vectorized model checker are the calculations of labels of the predecessor states for each state (lines $16 \sim 21$ of $Verify\_EU(E[\eta U \xi])$ and lines $30 \sim 35$ of $Verify\_EG(EG\eta)$ in Figure 4). The average vector length for these parts becomes the average number of predecessor states of each state, and it is $|R|/|S|$ or $2^{|X|}$. That is, if the machine has 10 input signals, the average vector length becomes 1024 and it is enough large to extract high performance of a vector processor.

In order to represent a transition relation R of a Kripke structure, we use 2 integer arrays Q and R. Since there is a one to one correspondence between the edges of a sequential machine and the states of the corresponding Kripke structure, it is possible to number the states of the Kripke structure so that each state s_i has its predecessor states $s_{Q(i)} \sim s_{R(i)}$. By using this numbering method, we can use the efficient *contiguous*

Table 1: Kripke structures for benchmark tests

Name	|Spec|		#States	#Edges
SR8	Total:	306	131,072	67,108,864
	Boolean:	212		
	Temporal:	94		
SR9	Total:	368	524,288	536,870,912
	Boolean:	254		
	Temporal:	114		

access (see Figure 1) to calculate the labels of predecessor states which is the most time consuming parts of the model checker. The sizes of the arrays Q and R are both $|S|$. These 2 integer arrays Q and R can be easily constructed directly from a given Moore type deterministic sequential machine in proportional time to the size of the corresponding Kripke structures by using one additional integer array of size $|S|$.

As for N_1 and N_2 used in *Verify_EU*($E[\eta U\xi]$) and *Verify_EG*($EG\eta$), we also use integer arrays with the corresponding index variables. Initialization of N_i (i.e. $N_i := \emptyset, i = 1, 2$) can be done by just substituting 0 to the corresponding index variable. Insertion of a state s to N_i (i.e. $N_i := N_i \cup \{s\}$) can be done by just storing the state s at the place in N_i pointed by its corresponding index variable and updating the index variable. As for copying data from N_2 to N_1 (i.e. $N_1 := N_2$), we just exchange the role of N_1 and N_2 instead of copying data actually. The maximum required sizes of the arrays N_1 and N_2 are $|S|$.

In order to store the truth value of each sub-formula at each state, we use 1 bit each. Therefore, the required memory in total for this purpose is $|S| \times |\eta|/8$ bytes. In addition, we use a working integer array of size $|S|$ to store a label for each state in checking temporal operators.

We also use $|S| \times 8$ words to handle fairness constraints.

Note that 1 integer word consists of 4 bytes. Therefore, the total amount of required memory is

$$|S| \times 56 + |S| \times |\eta|/8 \text{ bytes.}$$

For CTL formulas which contains 256 and 1024 operators, the vectorized model checker can manipulate Kripke structures of 2.3 million and 1.1 million states respectively with main memory of 200 M bytes.

4.2 Example

In order to measure the efficiency of the vectorized model checker, we applied it to the verification of two large Kripke structures (SR8 and SR9) corresponding to synchronous shift registers with parallel load and serial output. SR8 consists of 131,072 states and 67,108,864 edges (each state has 512 edges), and SR9 consists of 524,288 states and 536,870,912 edges (each state has 1024 edges) as shown in Table 1. The CTL formulas which give their full specifications contain more than 300 different sub-formulas with no fairness constraints.

The benchmark results without fairness handling are shown in Table 2. Both SR8 and SR9 are verified to be *true* on the vector machine of VP-400E in about 5 seconds

Table 2: Results of benchmark tests

Name	Scalar Execution (m sec.)	Vecotr Execution (m sec.)	Acceleration Ratio	Average Vector Length	Used Memory
SR8	132,126	4,998	26.4	512	13 MB
SR9	1,131,039	28,985	39.0	1,024	52 MB

and 29 seconds by using 13 MB and 52 MB of memory respectively. This means that our vectorized model checker evaluates about 7 ~ 8 states in a second, which implies that it will be able to verify 1 million state Kripke structure in a couple of minutes. Furthermore, the acceleration ratio obtained by our algorithm is around 26 ~ 39, which is extremely high ratio in non numeric application programs.

We also verified SR8 by using the CTL model checker B1.0 developed by Clarke et al. [5] installed on Sun-3/80. It took about 5,068 seconds to verify SR8. By considering that some benchmark tests show that the scalar unit of FACOM VP-400E is about 6.7 times faster than Sun-3/80, it will still take about 750 seconds to verify SR8 even if we install the CTL model checker B1.0 on VP-400E because their algorithm cannot be vectorized due to recursive calls.

5 Conclusion

We proposed a vectorized model checking algorithm for CTL and implemented the vectorized model checker on a vector processor FACOM VP400E. Almost all parts of the algorithms are vectorized except the parts to obtain strongly connected components for fairness constraints handling.

It can handle about 2.3 million or 1.1 million state Kripke structures derived from a Moore type deterministic sequential machine when the length of a give CTL formula is less than 256 or 1024 respectively.

We also presented examples which show that a CTL formula of 368 different subformulas is verified to be true in 29 seconds on a Kripke structure with 524,288 states and 536,870,912 edges.

There are several remaining future problems:

The first is to devise a vectorized algorithm for obtaining strongly connected components of a directed graph, and we need more consideration.

The second is a vectorization of model checking which can handle sequential machines directly [2]. We think this is not so difficult and would like to implement it in the near future.

The third is a vectorization of model checking based on Binary Decision Diagrams (BDD) because the BDD is a powerful technique to reduce necessary amount of memory for model checking dramatically in some cases [4].

Although our current version of the vectorized model checker takes a single Moore type sequential machine as input, it would be interesting to challenge the vectorization of the procedure to create direct product of several concurrent/parallel sequential machines. It would be also interesting to devise vectorized algorithm for model checking without creating direct product explicitly.

Acknowledgments The authors would like to express their appreciations to Prof. E. M. Clarke of CMU for his valuable suggestions. They also would like to express their appreciations to Prof. S. Yajima, Dr. N. Takagi, Mr. N. Ishiura and Mr. H. Ochi of Kyoto University for their precious discussions and comments.

References

[1] A. V. Aho, J. E. Hopcroft, and J. D. Ullman. *The Design and Analysis of Computer Algorithms*. Addison Wesley, 1974.

[2] M. C. Browne. An improved algorithm for the automatic verification of finite state systems using temporal logic. Technical Report CMU-CS-86-156, Carnegie Mellon University, 1986.

[3] M. C. Browne, E. M. Clarke, D. L. Dill, and B. Mishra. Automatic verification of sequential circuits using temporal logic. *IEEE Transactions on Computers*, C-35(12):1035–1044, December 1986.

[4] J. R. Burch, E. M. Clarke, K. L. McMillan, D. L. Dill, and J. Hwang. Symbolic model checking: 10^{20} states and beyond. In *Proc. Logic in Computer Science*, 1990.

[5] E. M. Clarke, S. Bose, M. C. Browne, and O. Grumberg. The design and verification of finite state hardware controllers. Technical Report CMU-CS-87-145, Carnegie Mellon University, July 1987.

[6] E. M. Clarke and E. A. Emerson. Synthesis of synchronization skeletons for branching time temporal logic. In *Proc. Workshop on Logic of Programs*, pages 52–71. Springer-Verlag, 1981.

[7] E. M. Clarke, E. A. Emerson, and A. P. Sistla. Automatic verification of finite state concurrent systems using temporal logic specifications: A practical approach. Technical Report CMU-CS-83-152, Carnegie Mellon University, 1983.

[8] E. M. Clarke and O. Grumberg. Avoiding the state explosion problem in temporal logic model checking algorithms. In *Proc. 6th Annual ACM Symposium of Principles of Distributed Computing*, pages 294–303, August 1987.

[9] E. M. Clarke and O. Grumberg. Research on automatic verification of finite-state concurrent systems. Technical Report CMU-CS-87-105, Carnegie Mellon University, January 1987.

[10] E. M. Clarke, D. E. Long, and K. L. McMillan. A language for compositional specification and verification of finite state hardware controllers. In *Proc. 9th IFIP Symposium on Computer Hardware Description Languages and their Applications*, pages 281–295, June 1989.

[11] S. Graf, J. L. Richier, C. Rodriguez, and J. Voiron. What are the limits of model checking methods for the verification of real life protocols ? In *Proc. Workshop on Automatic Verification Methods for Finite State Systems*, June 1989.

Introduction to a Computational Theory and Implementation of Sequential Hardware Equivalence*

Carl Pixley

Microelectronics and Computer Technology Corporation (MCC)

Computer Aided Design Program

Abstract

A theory of sequential hardware equivalence [1] is presented, including the notions of gate-level model (GLM), hardware finite state machine (HFSM), state equivalence ($\sim$), alignability, resetablility, and sequential hardware equivalence ($\approx$). This theory is motivated by (1) the observation that it is impossible to control the initial state of a machine when it is powered on, and (2) the desire to decide equivalence of two designs based solely on their netlists and logic device models, without knowledge of intended initial states or intended environments.

Binary decision diagrams are used to represent predicates about pairs of harware designs. Algorithms are given for computing pairs of equivalent states and sequential hardware equivalence as implemented in the MCC CAD Sequential Equivalence Tool (SET).

1 Introduction

A problem often encountered in commercial hardware design is to map an existing design from one technology to another (in some way, superior) technology. Differences in physical characteristics of the new and old technologies (e.g., different relative speeds or area characteristics) often cause designers to reimplement parts of the design to exploit the characteristics of the new technology. When reimplementation involves only purely combinational parts of the machine, tautology-checking algorithms can be used to decide equivalence.

However, sometimes sequential parts are reimplemented as well (e.g., combinational logic might be moved across storage elements). Unfortunately, the designer may not have an accurate specification (other than the existing design) of the individual part that is being replaced. For example, he may not know a reset state or reset sequence for the part. Furthermore, the designer may not have a specification of the part's intended environment to know what signals the environment will emit and how it should interact with the part. Therefore, the theory of sequential hardware equivalence presented in [1] (and elaborated here) is motivated by the desire to decide whether two gate-level designs are equivalent without reference to the intended environment, knowledge of initial (reset) states, and knowledge of reset (homing) sequences.

---

*This report is a revision of MCC Technical Report: CAD 448-89 (Q).

1.1 Related Research

In the classical theory of finite state machines [5], two FSMs are defined to be equivalent provided they "accept" the same input sequences. A FSM "accepts" a valid input sequence if, starting from a preferred initial state, the sequence leads to one of a set of designated final output states. Because, the theory presented here deals with comparison of two gate-level models (as defined by interconnections of logic gates and primitive storage elements), there is no notion of initial state, final state, or valid input to a state (though this notion of equivalence can be modified to reflect input constraints).

In the present theory, "state" is any one of the 2^n assignments of boolean values to the n many primitive storage elements of a design, and hardware is modeled in terms of its sequential input/output behavior from any initial state. This viewpoint is required because the initial state of a machine, when it is powered up, cannot be reliably predicted.

Also, the present theory is primarily concerned with equivalence of two hardware designs, rather than comparison of a design to a specification, though the latter can be accomplished with similar computational techniques. This viewpoint is dictated by the desire to understand when one design can be safely replaced by another, "equivalent", design.

Coudert, Berthet, and Madre [2] present a verification method based on characteristic functions of sets of tuples of boolean values. Starting from a pair of initial states, they compute the set of reachable states until either a discrepancy between output functions is recognized or all reachable state pairs have been examined. By contrast, the method presented here extracts the set of all pairs (possibly empty) for which the two machines have the same input-output behaviors. Surprisingly, our experiments indicate that our algorithm generally converges faster. However, their more incremental approach should be able to handle larger designs.

Devadas et al. [8] describe a method for comparing the state transition graphs of two finite automata (derived either from gate, register transfer, or ISP-like specifications) with boolean simulation. In contrast to the theory presented here, their comparison algorithm presupposes that the machines start in a valid initial state. The efficiency of their algorithm results from their ability to extract "don't care" information from the RTL or logic-level description, which allows them to greatly reduce the number of cases to simulate. In contrast, it is a remarkable property of the BDD representation of the state transition relation employed in the current work, that only dependencies among logic variables which affect the characteristic function are explicitly represented.

Symbolic simulation has been shown by Randal E. Bryant and his students [4] to be an effective method for checking that a hardware implementation responds correctly to a sequence of symbolic inputs. This approach can avoid the combinational explosion that would normally occur in boolean simulation when evaluating circuit operation over many combinations of input and initial state. Symbolic simulation can effectively compare the output behaviors of two designs with the same (finite) sequence of symbolic inputs but cannot, by itself, establish the equivalence of output behaviors for all possible sequences of inputs of arbitrary length as in the present theory. On the other hand, symbolic simulation can complement the present approach. When two designs are found to be inequivalent, symbolic simulation can find a sequence of inputs that cause the outputs to diverge starting from a pair of initial states that are thought to be equivalent.

Bose and Fisher [10] use BDDs to characterize states of a machine and the machine's next-state function. They then check temporal logic properties of the design using algorithms similar to those in [1].

It is well known that the size of an ordered binary decision diagram is sensitive to the ordering of boolean variables employed. For combinational circuits, orderings of circuit inputs derived from the circuit interconnections are described in [6] and [7]. In the present work, there is the added complication of ordering both circuit inputs and storage element outputs (representing both state and next-state variables). In the present work this is accomplished by a rather simple depth-first heuristic. However, better ordering of variables is a continuing subject of investigation.

None of the previous works presents a theory of equivalent sequential hardware design based solely on netlists and logic models of devices.

2 Theory

2.1 Gate-Level Models and HFSMs

Synchronous designs are modeled at the gate level in terms of combinational elements and primitive storage elements. A *primitive storage element (PSE)* is a device that transports its input to its output on a clock event and holds the output value until the next clock event. An example of a primitive storage element is a simple D flip-flop (without enable or reset). Fortunately, most real storage devices, such as D flip-flops with enable, reset and both Q and Qn outputs, can be modeled as a network of these primitive storage elements and combinational logic.

A *gate-level model (GLM or design)* is defined to be an interconnection of purely combinational elements and primitive storage devices. Each interconnection (i.e., net) is required to have exactly one driver (design input or device output). A design may have no loops of purely combinational elements.

A *state* of a GLM is an assignment of boolean values (0 or 1) to the output of each primitive storage element of the design. Suppose a design (GLM), D, has i many inputs, n many PSEs, and o many outputs. Design D has 2^n many states. For each state, each output is a boolean function of the inputs of D (if, for each state, each output function is a constant function, the design is called a Moore machine; otherwise, it is called a Mealy machine). To account for quotient designs, which are defined later, we present the following notion.

Definition 1 *A Hardware Finite State Machine (HFSM) is a quadruple, (Ins, States, Transition, Outputs) where Ins is a non-empty set of symbols, States is any non-empty set, Transition is a total function from $\{0,1\}^I \times S$ into S, and Outputs is an n-tuple of functions ($n \geq 0$), each of which has domain $\{0,1\}^I \times S$ and range $\{0,1\}$.*

We will think of the transition function as a relation, $Transition(\vec{q}, \vec{in}, next\text{-}q)$, that holds if and only if the circuit has state $next\text{-}q$ when it receives inputs, $\vec{ins}$, while in state, $\vec{q}$. Note that this definition does not mention initial states or accepting states as in classical finite state machine theory. We define two designs to be compatible if they have the same set of inputs and outputs. This notion of hardware equivalence is defined only for compatible designs. We now define several notions that are useful in stating the theory.

Definition 2 *Let s0 be any state of design D, and let SEQ be any finite sequence of boolean input vectors.* **SEQ(s0)** *is defined to be the state of design D following n machine cycles with inputs SEQ starting at s0.*

Definition 3 *A set of states, S, is* closed under (all) inputs *means that if s is an element of S and $\vec{in}$ is an input vector, then $\vec{in}(s)$ is an element of S.*

2.2 Equivalent States and the Quotient Design

Definition 4 *Suppose s0 and s1 are states of compatible designs D0 and D1, respectively. We define s0 to be* **equivalent** *to s1 (i.e., s0 $\sim$ s1) to mean that for the state pair (s0,s1) and for all state pairs reachable from (s0,s1) by sequences (of arbitrary length) of identical inputs, all corresponding output functions for the two designs are the same.*

Clearly $\sim$ is an equivalence relation among states of compatible designs. An algorithm for computing the set of all equivalent state pairs of two compatible design is given later.

Definition 5 *The* **quotient machine** $(D/\sim)$ *of design D, is the HFSM with the same inputs as D, with states being the set of equivalence classes induced by $\sim$, and with pthe induced transition relation and output functions.*

The proof that this definition makes sense is based upon the observation that if two states are equivalent, they have the same output functions, and for any input vector, their successor states are equivalent. The equivalence class of state $\vec{q}$ (i.e., set of states equivalent to $\vec{q}$) is denoted $[\vec{q}]$.

2.3 Alignability and Design Equivalence

Given that one cannot predict what state a design will be in when it is powered on, knowing that two designs have some pair of equivalent states is not enough to infer design equivalence. It must be possible to force the designs to behave the same no matter what their initial states are.

Definition 6 *A pair of states (s0,s1) of a design pair (D0,D1) is* **alignable**, *if there is a sequence, SEQ, of inputs (called an* **aligning sequence***) such that $SEQ(s0) \sim SEQ(s1)$.*

Definition 7 *Given a compatible design pair (D0,D1),* **Alignable-State-Pairs(D0,D1)** *is the set of all pairs of alignable states.*

An algorithm for computing Alignable-State-Pairs(D0,D1) is given later. We now define a pair of designs to be equivalent if they have some equivalent states, and for every pair of initial states, there is some sequence of inputs (called an *aligning sequence*) that will force them to behave the same.

Definition 8 *Two designs, D0 and D1, are* equivalent *($D0 \approx D1$) if and only if all state pairs are alignable (i.e., Alignable-State-Pairs(D0,D1) is the set of all pairs of states).*

The following fundamental theorem shows that if every state pair is alignable with some aligning sequence (that may depend upon the pair), then there is a *single* aligning sequence that will align all state pairs. More detailed proofs of the following theorems are given in [1].

Theorem 1 (fundamental alignment theorem) *If two design are equivalent, then there is a single aligning sequence (called a* universal aligning sequence*) for all state pairs.*

The proof of the fundamental theorem is based upon the observation that the set of equivalent state pairs is closed under all inputs. So if p is a pair of non-equivalent states, let SEQ0 be a sequence that forces p to an equivalent pair (SEQ0 exists by hypothesis). Note that other pairs that were not equivalent may have been forced into equivalent states, but each pair that was already equivalent moved to equivalent states. Therefore, the number of non-equivalent pairs in the image of SEQ0 is fewer by at least one than the original number of non-equivalent pairs. This process is repeated until there are no more non-equivalent state-pairs. The concatenation of these sequences is the desired universal aligning sequence.

We intend to define hardware equivalence ($\approx$)in such a way that it is an equivalence relation (i.e., reflexive, symmetric, and transitive) on the set of reasonable designs.

Theorem 2 *The relation $\approx$ is symmetric and transitive.*

An application of this theorem is the observation that if a design is equivalent to any other design, then it must be equivalent to itself.

Corollary 1 *If $A \approx B$, then $A \approx A$.*

The following theorem shows the necessity of the hypothesis that all state pairs are alignable in Theorem 1.

Theorem 3 *There is a pair of machines having equivalent states such that no single aligning sequence will drive all alignable state pairs into the set of equivalent states.*

In fact, the example in [1] is of a single design such that not all state pairs are alignable. This example shows that, in general, hardware equivalence ($\approx$) is not reflexive. The next section characterizes reflexivity.

2.4 Resetability

The notion of resetability turns out to be fundamental to the present theory of sequential hardware equivalence.

Definition 9 *A state, s0, of a machine is a* **reset state** *if and only if there exists a sequence of inputs, SEQ, (called a* **reset sequence***) such that if s is any state then SEQ(s) = s0. The set of (all) reset states of machine D is denoted* **Reset-States(D)**. *A design is* **resetable** *if it has a reset sequence.*

The following theorem characterizes the set of reset states as the set of states reachable from any reset state.

Theorem 4 *Let D be any HFSM and let s0 be any reset state of design D. Then Reset-States(D) is the set of states reachable from s0.*

Note that this theorem may be applied when D is the quotient of an HFSM. The following theorem characterizes self-equivalence.

Theorem 5 (Reset Theorem) *Any design, D, is equivalent to itself (i.e. $D \approx D$) if and only if the quotient $(D/ \sim)$ is resetable. Furthermore, an input sequence, SEQ, aligns all pairs of $D \times D$ if and only if SEQ is a reset sequence for $(D/ \sim)$*

Definition 10 *A design is* **essentially resetable** *means that $D/ \sim$ is resetable.*

Corollary 2 *(to Corollary 1) If $A \approx B$, then A and B are essentially resetable.*

Theorem 6 (Equivalence Theorem) *The relation $\approx$ is an equivalence relation on the set of essentially resetable designs.*

2.5 The Isopmorphism Theorem

We now present without proof an alternate characterization of hardware equivalence. This characterization requires the notion of isomorphism for HFSMs. As usual, isomorphism just means that the two structures are identical up to renaming. Two HFSMs $H0$ and $H1$ are **isomorphic** if and only if they are compatible and there is a one-to-one function F from the states of $H0$ onto the states of $H1$ satisfying the following two properties. For each state $S0$ of $H0$, corresponding output functions of the two designs are the same for states $S0$ and $F(S0)$. For any input vector $\vec{in}$ and state $S0$ of $H0$,

$$transition_{H1}(\vec{in}, F(S0)) = F(transition_{H0}(\vec{in}, S0))$$

The following theorem characterizes hardware equivalence ($\approx$) for the set of essentially resetable designs.

Theorem 7 (Isomorphism Theorem) *Suppose that D0 and D1 are essentially resetable designs, then the following are equivalent:*

1. *$D0 \approx D1$.*

2. *There exists some equivalent state pair for D0 and D1.*

3. *State equivalence ($\sim$) is an isomorphism from Reset-States($D0/ \sim$) onto Reset-States($D1/ \sim$).*

From the point of view of the theory of sequential hardware equivalence presented here, the essence of a design is captured by the reset states of its quotient, i.e., Reset-States($D/ \sim$). For any essentially resetable design, D0, any other design, D1, is equivalent to D0 if and only if D1 is essentially resetable and the set of reset states of the quotient of D1 modulo $\sim$ is isomorphic to Reset-States($D0/ \sim$). In fact, Reset-States($D/ \sim$) is minimal in the following sense.

Theorem 8 *For any essentially resetable design, D, the Reset-States(D/ $\sim$) is a design with the smallest number of states that is equivalent to D.*

3   Algorithms

Binary decision diagrams (BDDs) [3] are used to represent characteristic functions of sets of n-tuples of boolean values. For example, suppose boolean values are assigned to the variables, $\vec{q}$, $\vec{in}$, and $\vec{nxt\text{-}q}$. Then *transition*($\vec{q}, \vec{in}, \vec{nxt\text{-}q}$) has value TRUE if and only if the values assigned to $\vec{nxt\text{-}q}$ are the next values of the primitive storage elements of the circuit when the current values are $\vec{q}$ and the inputs are $\vec{in}$. The predicate *qs-are-next-qs* is true if and only if corresponding variables *qs* and *nxt-qs* have the same value. If *foo* is a predicate involving variables *qs*, then *exist qs:foo&qs-are-next-qs* is the same predicate in terms of variables *nxt-qs*. The programs below illustrate how to calculate the BDDs for the predicates *Transition*, *Equivalent-Outputs*, *Equivalent-State-Pairs*, and Equivalent-Designs.

3.1   Calculating *Transition* and *Equivalent-Outputs*

The transition predicate is derived as follows from a netlist for a design. For the input to each primitive storage element, q, the input to q is expressed as a boolean function, *in-q* of variables, $\vec{q}$ and $\vec{in}$. For each variable *nxt-q*, let *nxt-predicate*($\vec{q}, \vec{in}, \vec{nxt\text{-}q}$) be the predicate, *nxt-q* $\equiv$ *in-q*. The transition predicate, *transition*($\vec{q}, \vec{in}, \vec{nxt\text{-}q}$), is then the conjunction of all the *nxt-predicate* predicates.

For corresponding outputs, Out_0 and Out_1, of the two designs, the output is expressed as a function, $Out\text{-}fun_0(\vec{in}, \vec{q_0})$ and $Out\text{-}fun_1(\vec{in}, \vec{q_1})$ of the inputs and q-values. Let $Out(\vec{q_0}, \vec{q_1})$ be the predicate $\forall \vec{in}, (Out_0(\vec{q_0}, \vec{in}) \equiv Out_1(\vec{q_1}, \vec{in}))$. *Equivalent-Outputs*($\vec{q_0}, \vec{q_1}$) is the conjunction of all *Outs*.

3.2 Calculating *Equivalent-State-Pairs*

Let A_0 be the set of state pairs, $(\vec{q_0}, \vec{q_1})$, for which *Equivalent-Outputs* $\equiv 1$, i.e., corresponding outputs of the two machines agree. In general, let state pair $(\vec{q_0}, \vec{q_1})$ belong to A_{i+1} if and only if *Equivalent-Outputs*$(\vec{q_0}, \vec{q_1}) \equiv 1$ and the set of all states immediately reachable from $(\vec{q_0}, \vec{q_1})$ is in A_i. A simple induction argument shows that a state-pair, $(\vec{q_0}, \vec{q_1})$, belongs to A_i if and only if *Equivalent-Outputs*$(\vec{q_0}, \vec{q_1}) \equiv 1$ and for any sequence of inputs, $in_i, in_{i-1}, \ldots, in_1$, having length i, all of the states-pairs reached by that set of inputs satisfies Equivalent-Outputs. Furthermore, each A_{i+1} is a subset of A_i. Hence, the intersection of all A_is is *Equivalent-State-Pairs*.

```
char := 1
next_char := Equivalent-Outputs
loop until (char = next_char)
  char := next_char
  correct-next-qs := (exist qs: char&qs-are-next-qs)
  ins-qs-point-to-correct-next-qs
      := (exist nxt-qs:transition&correct-next-qs)
  qs-always-point-to-correct-next-qs
      := (all ins:ins-qs-point-to-correct-next)
  next_char := Equivalent-Outputs&qs-always-point-to-correct-next-qs
end loop
equivalent-state-pairs := char
equivalent-state-pairs-non-empty := exist qs:equivalent-state-pairs
```

3.3 Calculation of *Alignable-Pairs*

Suppose that the set of equivalent state pairs is non-empty (i.e., the characteristic function, *equivalent-state-pairs*, is not FALSE). Let B_0 be the set of equivalent state pairs. For all $i > 0$, let B_{i+1} be the union of B_i and the set of state-pairs, p, such that for some vector of inputs, $\vec{in_p}$, if $\vec{in_p}$ is applied to p, then the resulting state is in B_i. The set B_i is the set of state-pairs that can be transformed into equivalent-state-pairs in i or fewer cycles. Since B_i is a subset of B_{i+1}, let the union of all the B_is be called the **alignable-pairs**. The following algorithm computes the predicate *alignable-pairs*$(\vec{q})$ from *equivalent-state-pairs*.

```
char := 0
next_char := equivalent-state-pairs
loop until (char = next_char)
  char := next_char
  new-align := (exist in's nxt-qs:transition &
                    (exist qs: char & qs-are-next-qs))
  next_char := char or new-align
```

```
end loop
alignable-pairs := char
Equivalent-Designs := (all qs:alignable-pairs)
```

Two designs are equivalent if and only if *Equivalent-Designs* is True.

3.4 Experimental Results

The above algorithms together with an algorithm that computes all minimal reset sequences [1] are implemented in the MCC CAD Sequential Equivalence Tool (SET). The results using the 1988 version of the BDD routines in COSMOS system by R. Bryant of Carnegie Mellon University are reported in [1]. The longest comparison was of two four-bit serial multipliers (multipliers are known to generate large BDDs) with 18 and 15 storage elements, respectively, which took less than four cpu minutes on a Sun 4. For all 44 of the Berkeley/MCNC examples (from bbara through train4) two designs were synthesized from KISS2 descriptions by a commercial synthesis tool using different state encodings. Equivalence for each pair was decided in less than 23 cpu seconds and most were decided in less than two cpu seconds.

SET is being modified to use the more efficient Brace-Rudell-Bryant [11] program from Carnegie-Mellon University and Synopsys Inc.

4 Conclusion

An intuitively appealing definition of equivalence of gate level models (i.e., sequential hardware designs) was formulated which, for essentially resetable designs, is an equivalence relation. Several theorems were presented. A design is equivalent to itself if and only if its quotient is resetable. For two essentially resetable designs the following are equivalent. (1) The designs are equivalent. (2) The set of reset states of their quotients are isomorphic. (3) Their quotients have at least one pair of equivalent states. Algorithms were presented for computing the transition relation, the set of state pairs with equivalent outputs, the set of pairs of equivalent states, the set of alignable state pairs, and design equivalence. Experimental results show that deciding design equivalence for relatively small sequential designs is tractable.

References

[1] C. Pixley, A Computational Theory and Implementation of Sequential Hardware Equivalence, *DIMACS Technical Report 90-31, volume 2, Workshop on Computer-Aided Verification June 18-21*, Robert Kurshan and E.M. Clarke, eds.

[2] O. Coudert, C. Berthet, Jean Christolphe Madre, Verification of Sequential Machines Using Boolean Functional Vectors, *Proceedings of the IMEC-IFIP International Workshop on Applied Formal Methods For Correct VLSI Design*, November 13-16, 1989.

[3] R.E. Bryant, Graph-Based Algorithms for Boolean Function Manipulation, *IEEE Transactions on Computers*, Vol. C35 No. 8, August 1986.

[4] R.E. Bryant, Can a Simulator Verify a Circuit?, *Formal Aspects of VLSI Design*, eds. G.J.Milne et al., North-Holland, 1966.

[5] J.E.Hopcroft, J.D.Ullman, *Formal Languages and Their Relation to Automata*, Addison-Wesley, 1969.

[6] M. Fujita, H. Fujisawa, N. Kawato, Evaluation and Improvements of Boolean Comparison Method Based on Binary Decision Diagrams, *ICCAD '88*, pp. 2-5.

[7] S. Malic, A.R.Wang, R.K.Braton, A. Sangiovanni-Vincentelli, Logic Verification using Binary Decision Diagrams in a Logic Synthesis Environment, *ICCAD '88*, pp. 6-9.

[8] S. Devadas, H-K. T. Ma, A.R. Newton, On The Verification of Sequential Machines At Differing Levels of Abstraction, *24th ACM/IEEE Design Automation Conference*, Paper 16.2, pp. 271-276.

[9] E.M. Clarke, E.A. Emerson, A.P. Sistla, Automatic Verification of Finite State Concurrent Systems Using Temporal Logic Specifications: A Practical Approach, *Proceedings of the 10th Symposium on Principles of Programming Languages*, Austin, Texas, Jan. 1983, pp. 117-126.

[10] S.Bose, A.L.Fisher, Automatic Verification of Synchronous Circuits Using Symbolic Logic Simulation and Temporal Logic, *Proceedings of the IMEC-IFIP International Workshop on Applied Formal Methods For Correct VLSI Design*, November 13-16, 1989.

[11] K. Brace, R. Bryant, and R. Rudell, Efficient Implementation of a BDD Package, *Proceedings of the 27th ACM/IEEE Design Automation Conference*, June 1990, pp. 40-45.

64

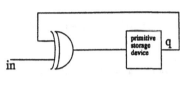

transition(q, in, next-q) = in&(q xor next-q) or ~in&(q xnor next-q)
                          = ~(in xor q xor next-q)

extension = {   q in next-q
                (1, 1, 0),
                (1, 0, 1),
                (0, 1, 1),
                (0, 0, 0) }

Figure 1

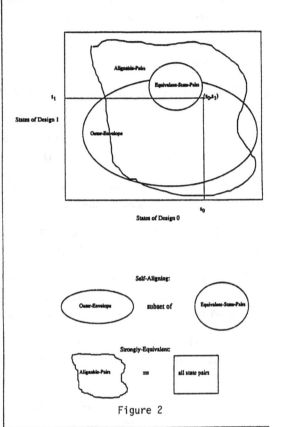

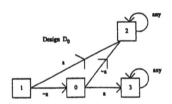

Figure 2

Figure 3

Auto/Autograph

Valérie Roy *
Robert de Simone †

Abstract

We describe the Auto and Autograph tools for verification and analysis of concurrent systems in their more recent developments. Auto is dedicated to a philosophy of "verification by reduction", based on automata morphisms and quotients. Autograph provides a graphical lay-out on which to display both terms and informations on terms, back and forth to Auto. We stress the openness aspects of both systems and their interface formats to the outside world. We see this as a contribution to the evergrowing collaborative trends in between similar tools, mostly under the pressure of national or european Esprit projects.

1 Introduction

Collaborative work in the field of verification tools for process calculi is steadily increasing, mostly under the pressure of european Esprit Projects. Interconnections in between such tools, which appear to be highly complementory, is now becoming a reality which should altogether avoid anarchical outburst and strict bureaucratic overplanning. While describing advances in the design of Auto and Autograph, we insist on their particular forms of interface formats, some of them already common with other tools[GLZ 89, Fer 87, BG 87, CPS 89, Arn 89, BFHP 89].

But the major part of this paper is aimed at describing new features of the Auto and Autograph systems, mostly concerned with the way they deal with recollection of results from analysis after reductions. These informations should be reported on the algebraic process terms themselves (or even better: on their graphical display), while experiments are conducted on their underlying transition systems.

2 Auto

2.1 Presentation

Auto [SV 89] is a support system for verification, and more generally analysis, of process calculi terms. It deals only with terms with finite model representations in the semantics

---

*ENSMP-CMA Sophia-Antipolis
†INRIA Sophia-Antipolis

of transition systems, and so bases itself on automata transformation algorithms, hereafter called **reductions**. The main activities in Auto may be split in four: Structural Automata Construction, Automata Reduction (with Abstraction), Formal Analysis: Equivalence and other Comparisons and Retrieval of Results and Diagnostics.

Of course those activities are to a large extend entwined, and separated only for the sake of description. Reductions may take place at intermediate levels of construction, due to congruence properties of the equivalence directing the reduction. Partial results lifted back from certain directions of analysis may be reinjected along other lines. By *retrieval of results* we mean the uplifting of informations from the reduced structures back to the early processes. We shall stress this aspect further in a following section.

The general syntaxic means of reduction in Auto is embodied in *abstraction*, using so-called *abstract actions*. Abstract actions represent an higher level class of behaviours. The idea is to consider certain (terminated) sequences of concrete behaviours (signal emissions/receptions) as altogether equivalent and atomic, and to call such a set simply an abstract action. A process which may succeed in completing any of these "runs" at the more concrete level will perform the abstract action then. This framework generalises *observational behaviours* defining weak bisimulation, where any sequence from $\tau^*.a.\tau^*$, with τ a private actions, are to be seen as equal to a simple a.

Abstract actions gather in a new alphabet, called an *Abstraction Criterion*. Abstracting a process results in a system which is conceptually far simpler, and where meaningful activities have been isolated. Pragmatical use of abstraction in "everyday" verification often follows these lines: abstract actions are split in between "positive" ones, leading to configurations of the system to be inquired; then some "negative" ones do represent faulty behaviours which should certainly not take place there. The abstraction is then a refutation attempt. But if the dreaded behaviours do nevertheless occur, they identify a number of (abstract) questionable states, and also questionable behaviours leading to them.

2.2 Data Objects and Types

Auto uses but a restricted number of objects classes, specially meaningful in the approach of *verification by reduction*. They are: Actions and Processes, Abstract Actions and Criteria, Automata, Partitions, Pathes.

Processes and *Abstract Actions* are user-provided and enjoy an external syntax for input to the system. Current algebraic *processes* syntax is adapted from Meije[Bou 85], a language that extends CCS[Mil 80] to cope with the possibility of unrelated signals taking place simultaneously. This leads to a full commutative group of actions in a single step duration. On the other hand efforts are made to allow any process calculus as legible syntax to Auto [MV 90]. Current *abstract actions* syntax is based on regular expressions notations, with as generators predicates on signal names. Words from the entailed rational languages should be seen as sequences of behaviours. Full details may be found in [BRSV 89]. These syntaxes may be output by the Autograph interface (described later), from simple graphical representations.

Pathes are just sequences of transitions, alternating states with performed actions, while starting and ending with a state.

Partitions are (most often disjoint) sets of sets of states, representing for instance equivalence classes amongst all states of a given automaton. A partition is thus almost always linked to a related automaton. In case the partition does not recover all states, missing ones are supposed to share in the same additional class. A one-class partition thus represents a predicate (or property) selecting a subset of an automaton state.

The relation from data structures to conceptual Auto activities is as follows: *Processes* are expanded into *automata* by **structural automata construction**, as defined in the SOS rules formalism; *Abstract actions* help define the **reductions** to be applied on these automata. Fortunately many short-hand notations for specific useful abstractions are also implemented in Auto, so one rarely needs defining them from scratch. Still, dedicated abstractions add power to the expressiveness of reductions; One subsequently extract either *partitions* or *pathes* through **Analysis** and **Equivalence** checking, as informative structures. *Partitions* collect classes of equivalent states, or lists of states enjoying a given behavioural property. *Pathes* are generated in particular as indications of faulty behaviours by application of "abstraction as refutation"; *Partitions* and *Pathes* are reinterpreted and carried back to ther original process algebraic description by **Retrieval of Results**. In this sense both should have in the future a syntactic description to be handed back to graphical display. Another structure which is important to picture is the reduced automaton itself, so that one checks whether the abstract system conforms the expectation, or not.

2.3 Auto as a system

Auto is a small command language consisting of a main toplevel loop, storing results of parsings and computations in global identifiers. *Parsing* commands, are used for entering processes and abstract actions into Auto, possibly from Autograph; *set* commands are used for computing (composition) of functions on objects already stored in the environment, binding the result to new identifier names. All this was described more thoroughly in [BRSV 89].

2.4 Input/output

We stress here to a certain level of detail the formats Auto may use to exchange files with various companion systems. Some were devised in the framework of the french C^3 project, while others bridge and adapt to various european verification tools in the framework of Esprit projects.

Auto supports textual "easily" readable syntaxic formats for reading and writing terms and automata in Meije (or CCS) syntax. While pleasant to deal with, this format suffers two drawbacks: at reading, the parsing is time-consuming –specially on large automata–; at writing, pretty-printing procedures make it space-consuming instead. This format is the one currently used from Autograph to Auto.

Concerning automata, there exists a specific "non-human-readable" format containing exactly the lisp object, which is used for saving and restoring automata, or from the reactive real-time language Esterel to Auto [Res 90]. But future directions lay in the definition of a common format (called fc in french) with the tools Mec[Arn 89] and Aldebaran[Fer 87]. The format needs hardly parsing and proves fairly complete to carry additional optional informations (and in the case of Auto to embed pathes and partitions in their related automaton structure).

Now the situation is far less clear in the case of networks of processes, specially since there is a broad variety of operators used there. First considerations on basic needs shall be expressed while describing Autograph.

2.5 Reductions

All reductions are performed by merging states (of the same automaton), constructing quotients from behavioural equivalence relations. These relations are possibly computed on transformed automata, after abstraction of behaviours has taken place.

2.5.1 Abstractions

Behaviours abstraction functions in Auto are:

1. abstract, which uses an abstract criterion as parameter.

2. tau-sature, which completes transitions according to prefix or postfix *tau* transitions (building τ^* and $\tau^*.a.\tau^*$).

Abstract results in an automaton on a new abstract alphabet, constructed while searching iteratively the concrete one for sequences matching the abstract actions. tau-sature is only a useful short-hand for computing weak bisimulation.

2.5.2 Quotients

Semantical equivalences are defined on these refined automata and their behaviours. They use fixpoints of relations, like bisimulations and other branching time semantics, or treat full sequences of behaviours as experiments, like trace languages and other linear time semantics [Gla 88]. All such semantics shall be congruences for all operators of the form of [GV 89]. Such functions in Auto are:

1. mini for strong bisimulation congruence.

2. refined-mini, starting from an initial partition.

3. obs for weak observational bisimulation congruence.

4. tau-simpl, which collapses τ-cycles behaviours.

5. trace for trace language congruence.

6. dterm for trace determinisation (without minimisation). congruence.

All these reductions take optional signal lists parameters, indicating which subset stays visible. Other signals are renamed down to τ. We could add a refined-trace functions, using as parameter a one-class partition indicating terminal states. Obs is not primitive and could be obtained as mini(tau-sature(tau-simpl(....

2.6 Comparisons and Equivalences

We already described some equivalences used for minimizing automata. The same functions may theoretically be used to compare two transition systems, depending whether both initial states lie equivalent.

But here we want to collect informations in case of non equivalence. Supposing both processes were first reduced to normal form, then the final partition retains only singleton or couples as equivalent classes: couples represent equivalent states (one from each automaton), and singletons represent states without match (from either automaton). This structure should be further queried, possibly interactively, or the subset of equivalent states could be displayed graphically, or longest strings of unmatched behaviours running only through singletons could be searched. More actual experiments are still due here.

The functions providing equivalence informations are few in Auto. One should remember that they may be combined with previous reduction functions to compose richer equivalence checking.

1. **eq** provides a "yes/no" answer on bisimulation equivalence.

2. **strong-partition** provides the list of equivalence classes.

Equivalence is only an instance of comparaison, observation and testing on an automata under analysis. Promising other directions are *graphical specifications* [BL 89].

2.7 Retrieval of Results and Diagnostics

Results put to light by reductions and further analysis should now be lifted back to the initial process terms. Pathes and state partitions will thus be provided "inverse images", so to speak. As a way to attain this, we build a side structure from the shape of the results and the actual combinations of reductions performed. We shall call *guideline* such a structure.

A guideline is a generalized automaton-like observer, safe that it includes additional info in its states, possibly drawn from the process itself. In contrast, an observer is usually required to be defined independantly from its observed system. Typical such info consists in states that are allowed as configurations of the observed process while the guideline reaches this state. Guidelines may also contain another sort of state information, namely places it allows as terminated endpoints of abstract behaviours.

$$
\begin{array}{lcl}
\text{Term or Automaton} & \xrightarrow{\;reduction\ function\;} & \text{Reduced Automaton} \\
\Downarrow & & \Downarrow\ analysis \\
\text{Source info} & & Pathes\ or\ Partition \\
\Uparrow & & \Downarrow\ synthesis \\
\text{Uplifted Guideline} & \xleftarrow{\;\;retrieval\;\;} & \text{Guideline}
\end{array}
$$

As pictured above, the guideline shall be built going backward stucturally along the chain of reductions performed. Some reductions will change its state space, others shall add or transform transitions. In the end there shall always be the possibility of **applying** it on the corresponding observed system of the same stage. Applying a guideline on a process (or an automata) will consist in "simulating" the observed process in a way that will satisfy the guideline, or equivalently in a way constrained by the guideline. This will identify which states, or behaviours, of the term under analysis, shall indeed project themselves by reduction down to the chosen pathes or partition classes.

With this reconstruction philosophy the only information which needs be kept through reductions is the relation from a state to the class of states that were mapped to it taking the quotient. These states in Auto are formally encoded in a vector description of states in the sequential automata components occurring in the early algebraic process description.

The operation induced by reduction functions on guidelines are actually quite simple: **quotients** replace in each state the set of allowed configurations of the analysed system. It is replaced by its inverse image for the morphism associated with the quotient.

Abstraction replaces actual actions by their concrete actions regular sets counterparts. This induces new states in the guideline, and a need to indicate which sets may be terminal as representative behaviours of the analysed process.

3 Autograph

3.1 Presentation

Autograph [RS 89] is a plain, non syntaxic graphical editor for pictures that are later interpreted into process calculi terms. By this we mean that the drawings produced are not structural –safe in particular spots–, and that coherency and structure are checked only on demand, for translation into textual algebraic form. This allows the graphical description to depart from strict ties to the textual one, including elliptic representation conventions. This allow also semantics of drawings generated from few simple graphical object types to be endowed with scarcely variant semantics, depending on the target formalism and language. Autograph was started with the Meije algebra in mind, but could give graphical versions to (process algebraic simple kernels of) Esterel[BG 87] and Lotos[BB 89].

3.2 Graphical Data Objects and Meanings

3.2.1 Graphics

Autograph works from menus and mouse selection of functions to be applied on graphical objects. As in Auto, the object classes were kept to a bare minimum, highly common, and endowed with graphical links properties which are fairly *ad-hoc*. They are: Boxes, Ports, Vertices, Edges and Labels (not a graphical object)

Roughly, these few graphical elements represent all that was informally used in graphical description of parallel systems (communicating automata) even before efforts were made to embed them into an algebraic syntax formally, which was to lead to process calculi theory.

Boxes are rectangles and represent subprocesses. Boxes next to one another are supposedly set in parallel. Ports are circles, which may only be located on the boxes frames.

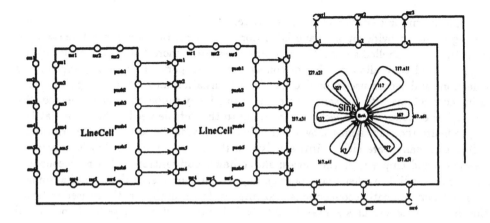

Figure 1: a loosing and duplicating line of the Stenning protocol

They figure the connexion points for communication. Possible communications are represented by edges drawned in between ports. So boxes, ports and edges build up *networks of processes*, and recollect the *structural operators* features from process calculi. Figure 1 shows a network and automaton example.

Vertices are round-shaped and figure automata states. Transitions are figured as edges (just like connections in between ports). So edges are broken lines, whose ends imperatively fall both into states or ports. Vertices and edges thus create *regular individual processes*, and recollect the *dynamic behavioural operators* features from process calculi.

3.2.2 Namings

All these elements may be named (or tagged) using labels. Labels appear on the screen but are not conceived as graphical objects, that is they do not share the same uniform internal treatment as other objects (in particular they may not be themselves named!). Naming is not always imperative. Names serve different purposes depending on the type of object they are applied on:

• on boxes, they provide a global or local naming facility so that a description may be shared in between several windows, and subterms instanciated in several locations. Recursive terms (using themselves as subterms) may also be generated this way;

• on ports, they indicate the local name of the signal to be communicated upon. In particular, all ports of named boxes referring to subsystems drawned elsewhere should be named, and no other need be. This implies that Autograph takes care of all renamings to avoid names clash and generate proper communications at translation into textual form. Naming other ports still may help control this translation;

• on vertices, they help control the translation by providing user-defined identifiers for the state names, which then become the variables occurring in the (guarded right linear) recursive definition of automata term; they may also occur several time in the same automaton (if only once with outgoing edges) and save drawing long backheaded arrows through a multirepresentation of the same state.

• on edges, they provide on one hand labels for transitions, that are imperative (an edge may actually be labelled several times, thus representing more than one transition); on the other hand they allow naming of connections inter ports, to save drawings of line connections and secure user-provided names at translation altogether. All ports locally sharing connections with the same names are transitivally supposed to be connected. See [Roy 89] for details. All connections pervading to the outside need to be named at this level, to figure the way signals show externally and provide the sort of the description.

To recall, names are either: **imperative**, in the case of transition labels, innermost boxes ports and empty subterms boxes; **short-hand conventions** allowing recourse to textual pointers links rather than common physical line joining, in the case of automata states and communication connections; **naming controls** on the identifiers used at translation into process calculus syntax in the case of states, intermediate levels ports and connections.

3.2.3 Additional semantics

There is a default semantics (expressed as Meije or CCS terms) to the drawings, at least the correct ones. Notions of correct drawings, together with all abuse of representation allowed by Autograph, are out the scope of this paper and described in [Roy 89]. Briefly: Boxes next to one another represent parallel subterms, containment of boxes represent the subterm relation, edges without outside ports generate signal restrictions and edges in between ports with different names generate relabellings. Automata are even more simply interpreted.

But this semantics may be altered, using a system of semantical annotations by functions from a special menu, each time dedicated to a specific new calculus to represent. This holds mostly of networks.

Justification for these annotations is drawned from the Ecrins [MdS 87] system, where a syntaxic formalism is defined for creating new calculi through new operators. A new operator is introduced by syntaxic considerations (name, binding power, type) and SOS rules for semantics. The type is always of the form $(Signals^m \otimes Process^n) \longrightarrow Process$. So we may superpose a semantics (and even several if they may be disambiguated by binding powers) with semantical markings: on a box (the target process). its ports (the *Signals* parameters), its son boxes (the process arguments), their ports (these arguments behaviours).

This part of Autograph is in development and shall be developped hopefully in the final paper.

3.3 Autograph's Interface

3.3.1 Menus and Functions

Autograph contains a number of menus (in its current version). Half are classical screen and window management menus, including cut-and-paste of subdrawings and Postscript dump of improved representations of both networks and automata (with edges splined up for instance). The other half is more specific to the edition of objects. There are four such menus: Nets for boxes and ports, Automata for vertices, Edges for transitions and connections, and Labels for namings. They all contain functions like select, move, create, kill and so on...

There is a completion function in the Windows menu of Autograph, which checks for coherency of drawings and construct the following records at nodes, whose fields may be variant according to the Type field:

1. **Type**, the name of the corresponding operator (*e.g.* PAR, SEQ, AUTOMATON, VAR, LOCALDECL)

2. **Name** (mostly in case of variables and local process declarations).

3. **Fathernode** for navigation.

4. **SigParameters** for the relevant arguments to the operator.

5. **SigRename** because signals renamings, which introduce process instantiation, have a specific representation in Autograph (by edges drawing).

6. **Internals** for signals local scope binding, as in CCS restriction.

7. **Sort** giving optionally the set of all visible signals at this subterm level

8. **SubProcessesGraph** identifying the subprocesses and optionally a relation in between them

9. **LocalDeclaredProcesses** for local process definitions that are used in the subprocesses graph (mostly in case of LOCALDECL operators).

10. **Automata** for sequential automata components. These automata will correspond after actual translation to entries in a "automata fc" table.

11. **Pragmas** for uninterpreted info (like graphical positions remainders).

3.3.2 Input/output

There is an additional variable menu in Autograph, or more exactly there are distinct versions of Autograph, depending on a menu providing the semantic annotating and translation functions to a given calculus format, or a companion verification systems (*e.g.* Auto). In the future translation functions should all produce instances of a "common format", prealgebraic. This is largely started for automata with the format described in annex, far less for networks. But globally the format should closely follow the records structure produced by the previous completion function.

There is also a specific explore function, whose mode is in particular started while inputting a description file which lacks geometrical features, for instance as provided by Auto. Exploration of automata is an established functionality of Autograph. Exploration of networks is being completed, and should be described in the full paper, as well as exploration of *pathes* and *partitions* on already displayed process descriptions. These functions are immediate, provided an "fc" input format for these objects is agreed on.

References

[Arn 89] A. Arnold *"Mec: a System for Constructing and Analysing Transition Systems"*, in Acts of the Workshop on Automatic Verification Methods for Finite State Systems, LNCS (1989)

[BFHP 89] B. Backlund, P. Forslund, O. Hagsand, B. Pehrson, "Generation of Graphic Language-oriented Design Environments", 10th IFIP Protocol Specification, Testing and Verification", Twente, North-Hooland (1989)

[BG 87] G. Berry, G. Gonthier, *"The Esterel Synchronous Programming Language: Design, Semantics, Implementation"*, to appear in Comp. Sci. Prog. (1989)

[Bou 85] G. Boudol *"Notes on algebraic calculi of processes"*, Logics and Models of Concurrent Systems, NATO ASI Series F13,K. Apt, Ed.(1985)

[BRSV 89] G. Boudol, V. Roy, R. de Simone, D. Vergamini, *"Process Calculi, from Theory to Practice: Verification Tools"*, in Acts of the Workshop on Automatic Verification Methods for Finite State Systems, LNCS (1989)

[BL 89] G. Boudol, K. Larsen, *"Graphical Versus Logical Specifications"*, INRIA Research Report 1104 (1989).

[BB 89] E. Brinskma, T. Bolognesi, *"Introduction to the ISO Specification Language Lotos "*, The Formal Description Technique Lotos, North-Holland (1989)

[CES 83] E.M. Clarke, E. Emerson, A. Sistla, *"Automatic Verification of Finite-State Concurrent Systems using Temporal Logic Specifications: a practical approach"*, Proc. 10th ACM POPL (1983).

[CPS 89] R. Cleaveland, J. Parrow, B. Streffen, *"A semantics Based Verification Tool for Finite State Systems"*, in Proceedings of the Ninth International Symposium on Protocol Specification, Testing, and Verification, North-Holland (1989).

[Fer 87] J.C. Fernandez, *"Aldebaran: un système de vérification par réduction de processus communicants*, French Thesis, IMAG Grenoble (1987).

[Gla 88] R. van Glabbeek *"The Linear Time - Branching Time Spectrum"*, CWI Report RvG 8801

[GW 89] R. van Glabbeek, W. Weijland, *"Branching Time and Abstraction in Bisimulation Semantics"*, Proceedings 11th IFIP World Comp. Congress, San Francisco (1989).

[GV 89] J.F. Groote, F. Vaandrager, *"Structured Operational Semantics and Bisimulation as a Congruence"*, Proceedings 16th ICALP, Stresa, LNCS (1989).

[GLZ 89] J. Godskesen, K. Larsen, M. Zeeberg, *"TAV (Tools for Automatic Verification) Users Manual"*, University of Aalborg R 89-19 (1989).

A Data Path Verifier for Register Transfer Level Using Temporal Logic Language Tokio

Hiroshi NAKAMURA*   Yuji KUKIMOTO†   Masahiro FUJITA‡
Hidehiko TANAKA†

Institute of Information Sciences and Electronics, The University of Tsukuba, Japan *
Department of Electrical Engineering, The University of Tokyo, Japan †
Fujitsu Laboratories Ltd., Japan ‡

Abstract

A data path verifier for register transfer level is presented in this paper. The verifier checks if all the operations and the data transfers in a behavioral description can be realized on a given data path without any scheduling conflicts. Temporal logic based language *Tokio* is adopted as a behavioral description language in this verifier. In *Tokio*, designers can directly describe concurrent behaviors controlled by more than one finite state machine without unfolding parallelism. The verifier checks for the consistency between a behavior and a structure automatically and lightens the load of designers. The actual LSI chip which consists of 18,000 gates on CMOS gate array has been successfully verified. This verifier is concluded to have the ability to verify practical hardware design.

1   Introduction

Recently, extensive studies have been carried out on the derivation of efficient and error-free data paths in register transfer level assistance. High-level synthesis [5] is one of the solutions to this problem. The approach adopted in high-level synthesis is to synthesize a data path automatically from a given behavioral description. The derived data paths with this approach, however, are not as satisfactory as those which are designed manually yet .

On the other hand, designers initially have the image of a data path to be designed in an actual design process, because they have designed many similar circuits. In addition, they seldom develop hardware which is completely different from the one ever designed. In such situations, it is better to utilize the designers' experience positively instead of synthesizing a data path automatically. It is very important to construct an effective

---

*Address: 1-1-1 Ten'nou-dai, Tsukuba, Ibaraki 305, Japan. E-mail: nakamura@arch3.is.tsukuba.ac.jp
†Address: 7-3-1 Hongo, Bunkyo-ku, Tokyo 113, Japan. E-mail: kuki@mtl.t.u-tokyo.ac.jp
‡Address: 1015 Kamikodanaka, Nakahara-ku, Kawasaki 211, Japan. E-mail: fujita@flab.fujitsu.co.jp

[MV 90] E. Madelaine, D. Vergamini, *"Finiteness Conditions and Structural Construction of Automata for all Process Algebras", this volume (1990)*

[Mil 80] R. Milner *"A Calculus of Communicating Systems", LNCS 92, Springer-Verlag (1980)*

[Mil 83] R. Milner *"Calculi for Synchrony and Asynchrony"*, TCS 25, (1983)

[MdS 87] E. Madelaine, R. de Simone, *"Ecrins, un Laboratoire de Preuve pour les Calculs de Processus"*, (in french), Inria Research Report 672 (1987).

[Res 90] A. Ressouche, *"Ocauto"*, Esterel Technical Documents, CMA-Ecoles des Mines (Sophia-Antipolis) (1990).

[Roy 89] V. Roy, *"Autograph: un Outil d'Analyse Graphique de Processus Parallèles Communicants", French Thesis, Université de Nice (1990)*

[RS 89] V. Roy, R. de Simone, *"An Autograph Primer", I.N.R.I.A. Technical Report 112 , (1989)*

[SV 89] R. de Simone, D. Vergamini *"Aboard Auto", I.N.R.I.A. Technical Report 111 (1989)*

design assistance system based on this approach. However this has not been discussed well so far. Thus we propose a practical assistance system of register transfer level design based on this approach and present a data path verifier, which is the core part of the system.

Figure 1 shows the design flow of the proposed assistance system. The basic idea in this flow is to construct final data paths by modifying initial data paths given by designers.

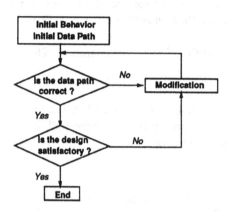

Figure 1: Flow Chart of Practical Design Assistance

At first, a designer gives an initial behavioral description and an initial structure of a data path to be designed. The initial structure is formed through the designer's experience. Now we must verify whether the given behavior can be realized on the given structure. The proposed data path verifier points out when and where scheduling conflicts happen, i.e., which paths and functional units are doubly allocated at which time slot. The structure and the behavior are modified to compensate for that error, and again they are verified. If there is no design error, the performance of the behavior is simulated and the cost of the structure is estimated. If the behavior and the structure satisfy required performance and permitted cost, the design at register transfer level finishes and it proceeds to the lower level design such as logic synthesis. If the cost/performance constraints are not satisfied, the design flow is followed by the stage of modification and the process of verification. This design flow of Figure 1 is justified because the initial data path given by a designer is fairly good in many cases. If the initial data path is fairly good, the revised data path will be very efficient. The data path verifier proposed in this paper plays an important role in this assistance system.

2 An Example

In this section, we explain the semantics of Tokio by showing some simple Tokio programs. After that, taking the example of the circuit which computes square roots, we show how this verifier is effectively used in the process of deriving the proper data path for a pipelined behavior from the data path for a sequential behavior with modification.

2.1 Behavioral Description Language Tokio

Tokio [2][1][4] is a logic programming language based on first-order interval temporal logic [6]. Intuitively, Tokio is regarded as an extension of Prolog with temporal operators. Since Tokio has the notion of time in its own semantics, the various algorithms of hardware can be described flexibly. The essential notion of hardware, such as concurrency and sequentiality, can be specified accurately and simply. Using Tokio, designers can directly describe concurrent behaviors controlled by more than one finite state machine without unfolding parallelism. Parallel behaviors such as pipelined execution can be easily described.

Now we explain the semantics of temporal operators. The expression

head :- p,q.

denotes that the predicates p and q are executed in the same time interval where the predicate head is defined. Concurrency is represented with a *comma* operator.

Sequentiality is expressed as follows.

head :- p && q.

The *chop* operator (&&) specifies the sequential execution of the two predicates p and q. This operator divides the interval where the predicate head is defined into two subintervals as shown in Figure 2. The predicate p is executed in the former interval and q is executed in the latter interval.

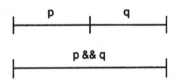

Figure 2: Chop Operator

Conditional branches are described by using *cut* operators(!). The semantics of the following predicates is explained on the right side.

head :- condition1 ,!, p.	*if condition1 = true then execute p*
head :- condition2 ,!, q.	*else if condition2 = true then execute q*
head :- !,r.	*else execute r*

Let us take a simple example shown in Figure 3-(a). The statement "*tmp <= *c + *tmp" represents that the data of register *c is added to the data of register *tmp and that the result of the computation is stored in register *tmp at the next clock. The predicate sub denotes the parallel execution of the computation described above and the decrement of the data of register *c. The predicate main denotes that the computation of the predicate sub is repeated until the data of register *c is equal to zero.

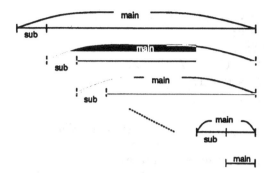

```
main :- *c=0 ,!,*output <= *tmp.
main :- !,sub && main.
sub :- *tmp <= *c + *tmp,
       *c <= *c - 1.
```

(a) Sample Tokio Program   (b) Timing Chart of Sample Program

Figure 3: Sample Tokio Program

As seen from the example, the description in Tokio is based on a top-down approach. This approach is very useful for describing hardware because designers can give behavioral descriptions with high modularity. Using this programming style, we need not unfold concurrency of hardware for every clock cycle. On the other hand, it becomes rather difficult for designers to check the consistency between a behavior and a data path. Our strategy is to adopt a flexible behavioral description language and to verify the consistency automatically.

2.2 Computing Square Roots by Sequential Execution

The example to be taken is the circuit which calculates square roots by using Newton's method. The algorithm of Newton's method is shown in Figure 4-(a). The behavioral description in Tokio and the structure of a data path are shown in Figure 4-(b) and Figure 4-(d) respectively. In this behavioral description, the computation of Newton's method is realized by sequential execution for each input data. In this case, the verification of the data path is not so difficult because there is little concurrency in the description. We have only to verify the data path for each local interval and need not check whether some of these intervals occur concurrently.

2.3 Deriving a Data Path for Pipelined Execution

Now we modify the sequential behavioral description of the circuit computing square roots and derive a pipelined behavioral description from it. It is quite easy to get a pipelined description from the original one by using temporal operators. The pipelined description is shown in Figure 4-(c). We have only to change the top-level predicate main. The second main predicate denotes that stage1 is followed by both main and (stage2 && true). Since main starts the computation for the next input data, this behavior represents a pipelined execution. Figure 5 shows a timing chart for the pipelined execution.

At this point, we must construct a data path for the pipelined behavior, because the data path for sequential computation may not be sufficient. To construct a completely new data path, however, is not efficient. It is better to modify the original data path

```
Y := 0.222222 + 0.888889 * X;
C := 0;
DO UNTIL C = 2 LOOP
  Y := (Y + X / Y) / 2;
  C := C + 1;
ENDDO;
```

(a) Algorithm of Newton's Method

```
main :- *adr = 8,!,true.
main :- !,input && stage1 && stage2 && main.
input :- !, *input1 <= *memory(*adr), *adr <= *adr + 1
     && *reg1 <= 0.222222 + 0.888889 * *input1.
stage1 :-!, *reg2 <= *input1 / *reg1
     &&  *reg2 <= *reg2 + *reg1
     &&  *reg3 <= *reg2 / 2,  *input2 <= *input1.
stage2 :- !, *reg4 <= *input2 / *reg3
     &&  *reg4 <= *reg4 + *reg3
     &&  *output <= *reg4 / 2.
```

(b) Sequential Behavioral Description in Tokio

```
main :- *adr = 8,!,true.
main :- !,input && stage1 && main,(stage2 && true).
```

(c) Pipelined Behavioral Description in Tokio

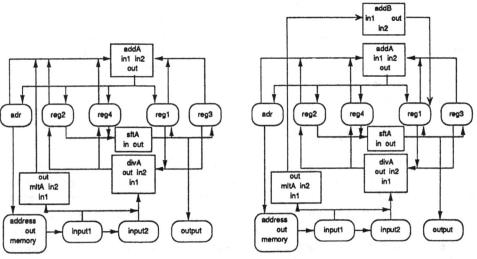

(d) Data Path for Sequential Execution   (e) Data Path for Pipelined Execution

Figure 4: Computation of Square Roots

and to get the proper one. In such a case, we can use this verifier effectively. At first, the verifier checks whether the pipelined behavior can be realized on the original data path. If there are some scheduling conflicts, the behavioral description or the data path is modified considering the output of the verifier in order to avoid the conflicts. This process is repeated until there are no conflicts and the proper data path is constructed.

From the result of the verification, we can say that adder addA is doubly used in the second interval of input and the second interval of stage2 and that these intervals occur concurrently. Thus we add one more adder to the original structure and the final data path is obtained. The data path for the pipelined behavior is shown in Figure 4-(e).

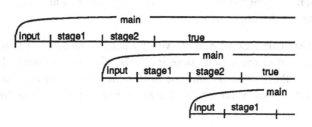

Figure 5: Timing Chart for Pipelined Execution

3 Verification Method

3.1 Overview

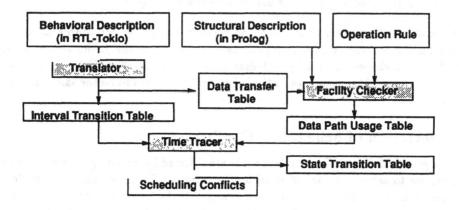

Figure 6: Structure of Data Path Verifier

The data path verifier proposed in this paper checks if all the operations and the data transfers in a behavioral description can be realized on a given data path without any scheduling conflicts. The structure of this verifier is shown in Figure 6. The inputs of this system are a behavioral description in RTL-Tokio (Register Transfer Level Tokio, which is a subset of Tokio), a structural description in Prolog, and operation rules in Prolog. The operation rule is a rule for linking an operation in the behavior with a functional unit in the data path.

The process of the verification is divided into two stages.

Linking a behavior with a data path: To find a set of paths and functional units for each data transfer and operation from the structural description. In this stage, all the data transfers and operations in the behavior are linked with appropriate data path elements in the structure using operation rules, and the link information between them is recorded in a data path usage table. This stage is executed by a *translator* and a *facility checker*.

Detection of scheduling conflicts: To verify whether any paths or functional units are doubly allocated in concurrent time intervals. The concurrency of the behavioral description in RTL-Tokio is unfolded in this process. The state transition table of a control part is also extracted. This stage is executed by a *time tracer*.

3.2 Linking a Behavior with a Data Path:

The link information between the behavior and the structure is derived in accordance with the following steps.

1. To find a set of functional units which realize an operation of each data transfer from the structure.

2. To search for a data path from a source register to the input of the functional unit and that from the output of the unit to a destination register.

The function of each unit is defined by operation rules. At the end of this stage, the link information is recorded in a data path usage table. From this table, we can know which data path element is used by each data transfer and operation. The names of the registers and those of the memories in the behavior are assumed to be the same as those in the structure.

3.3 Detection of Scheduling Conflicts

In this stage, scheduling conflicts in data paths are detected by unfolding the concurrency of a behavior to actual data transfers with a time tracer. The time tracer makes clear what operations and data transfers are done concurrently by traversing all the transitions. After that we check if any paths or functional units are doubly allocated in concurrent time intervals by using a data path usage table. If no path or functional unit is doubly allocated, the given data path turns out to be correct. Otherwise designers modify the behavior and the data path to avoid the detected scheduling conflicts. We have implemented this time tracer in two ways: a forward tracer and a backward tracer.

- *Forward Trace*: At first all the concurrent time intervals are searched for by tracing transitions forward. In the second stage, all these intervals are checked whether they use the same data path element. The scheduling conflicts of data paths are detected in this stage.

- *Backward Trace*: In a backward trace, all the pairs of time intervals which use the same data path element are searched for at first, and they are verified whether they occur concurrently by tracing transitions backward in accordance with an interval transition table.

In this paper, only the algorithm of the forward trace is explained. The algorithm of the backward trace is similar to that of the forward trace except for a direction of tracing.

3.3.1 Unfolding Concurrency of a Behavior

In the first stage, all the concurrent time intervals are searched for with a forward tracer. It traverses all the transitions from an initial state in depth-first. A state in RTL-Tokio description is defined by [*interval-name,clock-number*].

step1: An initial interval I_{init} and an initial clock I_{clock} (usually 0) are selected.
$S_0 = \{[I_{init}, I_{clock}]\}$.

step2: (Unfolding the concurrency of a behavior)
If S_i has predicate calls, all these concurrent intervals are added to S_i and S_i' is obtained. S_i' is a list of the intervals which occur concurrently at the i-th clock.

step3: (Proceeding to the next clock)
S_{i+1} is obtained from S_i' by proceeding to the next clock. The computation of this process is as follows.

1. Increment the clock number I_{clock} for each element of S_i' if the obtained clock number is not larger than the length of that interval.

2. Transfer to the next interval after *chop* operator &&.

Step3 is followed by step2.

Detection of the iteration of a search: Let S_n' be the newly obtained state in step2. If $0 \le \exists i < n$, $S_n' \subseteq S_i'$ holds or S_{n+1} next to S_n' cannot be obtained, the iteration of a search is detected.

3.3.2 Detection of Scheduling Conflicts

All the intervals in each S_i' occur concurrently unless they have exclusive transitive conditions. Using a data path usage table, all the concurrent intervals are checked whether they use the same data path resource or not. If they use the same path or the same functional unit, it results in scheduling conflicts of data paths. In this stage, we check if the data path conflicts can be avoided by using some alternative data paths.

4 Experimental Results and Evaluation

4.1 Network Interface Processor

The largest example to which we have applied this verifier is a network interface processor (NIP) in PIE64 [3]. PIE64 is a parallel inference machine which executes knowledge information processing in parallel, and is under development in our laboratory. This machine consists of 64 inference units and two high speed interconnection networks. The processor manages data transfers and process synchronization between inference units. It has already been designed and is going to be implemented on CMOS gate array. The total number of the gates is about 18,000 including both a data path and a control part. The structure of the processor is divided into four parts. The data path verifier has been applied to the main parts which consist of about 11,000 gates. Since the required performance of this processor is very severe, the behaviors of NIP are 6-stage pipelined.

The verification results of the two parts of the processor are shown in Table 1. The data path verifier is implemented on SUN4/260 using SICStus-Prolog (about 80KLIPS).

	CPU time (sec)				Number of Derived States	Number of State Transitions
	Translator	Facility Checker	Forward Trace	Backward Trace		
Data Transfer Part (Upper Part)	0.80	1.07	2.36	40.1	14	53
Data Transfer Part (Lower Part)	3.23	4.66	183.3	>3600	81	379
Process Synchronization Part	5.97	7.85	1488	>3600	236	1168

Table 1: Verification Results of Network Interface Processor

4.2 Evaluation

Most of the CPU-time is spent in a *time trace* part. Since a forward tracer traverses all the state transitions, its cost is proportional to the number of transitions. The cost of detecting the iteration of the search is proportional to the number of states because traced states are presently recorded with enumeration, Thus the computational complexity for the forward trace is $O(n \times m)$, where n is the number of state transitions and m is the number of states.

In a backward trace, $_NC_2$ pairs of intervals are listed in the worst case, where N is the number of intervals. The worst case occurs when a certain data path element (such as a bus) is used in all the intervals. Though a backward tracer is not so efficient for complete verification from a computational aspect, it is quite useful and efficient for partial verification by selecting one interval and searching for concurrent time intervals with it. Therefore, it is recommended that the important part of design is tested partially using the backward trace at first, and then the whole design are verified using the forward trace.

5 Conclusion

We have presented a data path verifier at register transfer level and proposed a design assistance system based on this verifier. The verifier has been successfully applied to a real LSI chip. Practical hardware design can be assisted with this verifier if an interactive tool for improvement is provided. Our current research aims at the assistance of this improvement stage.

References

[1] T. Aoyagi, M. Fujita, and T. Moto-oka. Temporal Logic Programming Language Tokio. In *Logic Programming Conference '85*, pages 128–137, Springer-Verlag, 1985.

[2] M. Fujita, S. Kono, H. Tanaka, and T. Moto-oka. Aid to Hierarchical and Structured Logic Design Using Temporal Logic and Prolog. In *IEE Proceedings, Vol.133, Pt.E*, pages 283–294, IEE, 1986.

[3] H. Koike and H. Tanaka. Multi-Context Procesing and Data Balancing Mechanism of the Parallel Inference Machine PIE64. In *Fifth Generation Computer Systems*, pages 970–977, ICOT, 1988.

[4] S. Kono, T. Aoyagi, M. Fujita, and H. Tanaka. Implementation of Temporal Logic Programming Language Tokio. In *Logic Programming Conference '85*, pages 138–147, Springer-Verlag, 1985.

[5] M.C. McFarland, A.C. Parker, and R. Camposano. Tutorial on High-Level Synthesis. In *25th Design Automation Conference*, pages 330–336, ACM/IEEE, 1988.

[6] B. Moszkowski. A Temporal Logic for Multi-Level Reasoning about Hardware. In *CHDL '83*, IFIP, 1983.

The use of Model Checking in ATPG for sequential circuits

P. Camurati   M. Gilli   P. Prinetto   M. Sonza Reorda

Dipartimento di Automatica e Informatica
Politecnico di Torino
Turin Italy

Abstract

Some design environments may prevent Design for Testability techniques from reducing testing to a combinational problem: ATPG for sequential devices remains a challenging field. Random and deterministic structure-oriented techniques are the state-of-the-art, but there is a growing interest in methods where the function implemented by the circuit is known. This paper shows how a test pattern may be generated while trying to disprove the equivalence of a good and a faulty machine. The algorithms are derived from Graph Theory and Model Checking. An example is analyzed to discuss the applicability and the cost of such an approach.

1   Introduction

Despite Design for Testability [WiPa82], test pattern generation for synchronous sequential circuits is still needed for some design environments, e.g., those based on Partial Scan [Moto90]. A major class of devices is represented by Finite State Machines (FSMs) that are in extensive use as building blocks in a variety of applications, ranging from VLSI devices to network controllers. As reported in [Wolf90], many ASICs are "*control-dominated*", i.e., they are best modeled in terms of FSMs, rather than in terms of data path and control part. Moreover, following methodologies such as "*Macro Testing*" [BEGP86], complex devices may be partitioned for testing purposes, creating macros composed by one or more FSMs that can be tested individually.

Researchers investigated different approaches to ATPG for synchronous sequential circuits. The extension of fault simulation techniques to test generation for sequential devices has been the object of research [ACAg89], but reduced CPU time and high fault coverage for very complex circuits still cannot be guaranteed. Other works focused on extended combinational deterministic algorithms: starting from Huffmann's model that transforms sequential behavior over time frames into combinational behavior on iterated structures, the basic method extends the D-algorithm [PuRo71]. Many enhancements to it have been proposed, limiting the number of copies [Fuji85], introducing a 9-valued algebra [Muth76], searching for a path to be sensitized [Marl86], [Chen88], exploiting the benefits of techniques used in combinational ATPG [ScAu89].

A common feature of most ATPGs is that they are *structure-oriented*, i.e., the only knowledge they have of the device is its topology. Some efforts have shown a growing interest on algorithms where a certain amount of reasoning is performed or the *function* implemented by the circuit is known. [HuSe89] describes an application of Temporal Logic [ReUr71]: its operators are used both to describe future behavior and to justify the past one and a reasoning process yields a test pattern. The knowledge of the state-transition table is used in [MDNS88] to justify backward the values needed to control an observable fault until the initial state and the primary inputs are reached. The need to know the function restricts the application domain to medium-sized circuits, since the function must be extracted from the structure. Whenever the same approach is embedded within a synthesis system, this limitation is overcome easily [ChJo90], since the state transition table is available immediately.

This paper presents a method to generate test patterns for synchronous sequential circuits working in fundamental mode [McCl86]. The devices are modeled as *Finite State Machines* (FSMs) and may be Mealy or Moore machines, indifferently. This method is based on a strong interaction between Graph Theory and Model Checking [BCDM86]. The basic hypothesis is that the function realized by the device, represented by its automaton, may be extracted from its structure. Once the fault, of the traditional single stuck-at type, is inserted, a faulty automaton is created. The fault detection condition is expressed in theoretical terms within the framework of the product machine [PoMC64] that might in principle be used, although experience shows that it becomes unmanageable for other than trivial circuits. Without any loss of information, it is possible to refer to a Deterministic Finite Automaton (DFA) that is considerably smaller. Other approaches refer to the Error Latency Model (ELM) state table [ShMC76], although their objective is not test pattern generation, rather error latency estimation. The DFA is a Moore Machine and it is a suitable input for the Model Checker of the MCB system [CESi86]. The use of the DFA overcomes the Model Checker's limitation to Moore FSMs, at least for testing applications. The test detection condition is easily stated as a CTL formula and the counterexample facility of MCB is used to compute the test pattern.

Section 2 describes the theoretical framework for FSM modeling, stating the test generation problem in terms of the product machine and of the DFA and showing where Model Checking intervenes. Section 3 illustrates these concepts by means of a simple example. Conclusions are drawn in Section 4, justifying the use of this approach, defining its applicability limits, and pointing to future work.

2   Theoretical framework

Definition 1

A Finite State Machine $\mathcal{M}$ is defined by a 6-tuple [Koha70]:

$$\mathcal{M} = (I, O, S, \delta, \lambda, s^0)$$

where:

I is a finite, nonempty set of input values

O is a finite, nonempty set of output values

S is a finite, nonempty set of states

$\delta : I \times S \rightarrow S$ is the state transition function

$\lambda : I \times S \rightarrow O$ is the output function

s^0 is the initial state. □

A *Mealy* machine is such that $\lambda : I \times S \rightarrow O$.
A *Moore* machine is such that $\lambda : S \rightarrow O$.

Definition 2

Two completely specified FSMs $\mathcal{M}_1$ and $\mathcal{M}_2$ operating on the same input and output sets are *equivalent* iff, for all input sequences, all the elements of the output sequences are equal. □

The test generation problem for a FSM may be stated as follows: given a good machine $\mathcal{M}_1$ and a single stuck-at-i (i ϵ {0, 1}) fault $\mathcal{F}$, generate a faulty machine $\mathcal{M}_2$, operating on the same input, output, and state sets, where the fault $\mathcal{F}$ changes the δ and/or λ functions. If one can demonstrate that the two FSMs are not equivalent, i.e., that there is an input sequence such that the k-th elements ($k \geq 1$) of the two output sequences differ, then a test pattern has been found. If equivalence can be proven, then fault $\mathcal{F}$ is undetectable.

Some ancillary definitions are necessary before stating the equivalence condition of two FSMs in Theorem 1.

Definition 3

Given two FSMs $\mathcal{M}_1 = (I, O, S_1, \delta_1, \lambda_1, s_1^0)$ and $\mathcal{M}_2 = (I, O, S_2, \delta_2, \lambda_2, s_2^0)$, the *product machine* $\mathcal{M}_{12}$ is defined as a 6-tuple:

$$\mathcal{M}_{12} = (I, O_{12}, S_{12}, \delta_{12}, \lambda_{12}, s_{12}^0)$$

where:

$O_{12} = O \times O$

$S_{12} = S_1 \times S_2$

$\delta_{12} : S_1 \times S_2 \times I \rightarrow S_1 \times S_2 : \quad \delta_{12}(s_1, s_2, i) = (\delta_1(s_1, i), \delta_2(s_2, i))$

$\lambda_{12} : S_1 \times S_2 \times I \rightarrow O \times O : \quad \lambda_{12}(s_1, s_2, i) = (\lambda_1(s_1, i), \lambda_2(s_2, i))$

where:

$s_1 \in S_1; \ s_2 \in S_2; \ i \in I$

$s_{12}^0 = (s_1^0, s_2^0)$ □

Given two machines $\mathcal{M}_1$ and $\mathcal{M}_2$, the state set S_{12} of their product machine $\mathcal{M}_{12}$ may be partitioned into two subsets Q_{12} and Z_{12} such that $S_{12} = Q_{12} \cup Z_{12}$ and $Q_{12} \cap Z_{12} = \emptyset$. The two subsets are defined as:

$$Q_{12} = \{ q_{12} = (q_1, q_2) \in S_{12} : \forall\, i \in I \quad \lambda_1(q_1, i) = \lambda_2(q_2, i) \}$$
$$Z_{12} = \{ z_{12} = (z_1, z_2) \in S_{12} : \exists\, i \in I \quad \lambda_1(z_1, i) \neq \lambda_2(z_2, i) \}$$

The following theorem, whose proof is quite obvious and is thus omitted, expresses the equivalence condition for two FSMs.

Theorem 1

Two FSMs $\mathcal{M}_1$ and $\mathcal{M}_2$ are equivalent iff in the product machine $\mathcal{M}_{12}$ the subset Z_{12} is either unreachable from the initial state s_{12}^0 or empty.   □

Proving that the subset Z_{12} is either unreachable from s_{12}^0 or empty is very heavy computationally. It is possible to reduce complexity, without affecting the validity of the method, by noting that only the states of Q_{12} are needed to prove equivalence and that all the states of Z_{12} can be collapsed into a single failure state, indicating that the FSMs are not equivalent. This allows to transform the product machine into a DFA.

Definition 4

A Deterministic Finite Automaton (DFA) is defined by a 5-tuple [Koha70]:

$$DFA = (I,\ S,\ \delta,\ s^0,\ F)$$

where:

I is a finite, nonempty set of input values

S is a finite, nonempty set of states

$\delta : I \times S \rightarrow S$ is the state transition function,

s^0 is the initial state

F is the set of final states f.   □

A DFA, associated to the product machine $\mathcal{M}_{12}$, whose final state indicates a failure in proving equivalence, is defined as follows:

Definition 5

Given two FSMs $\mathcal{M}_1$ and $\mathcal{M}_2$ and their product machine $\mathcal{M}_{12}$, if $s_{12}^0 \in Q_{12}$, the DFA associated with $\mathcal{M}_{12}$ is the 5-tuple:

$$DFA(\mathcal{M}_{12}) = (I,\ Q_{12} \cup F' \cup f,\ \delta_{q12},\ s_{12}^0,\ f)$$

where:

f is the final failure state

$F' = \{ z_{12} \in Z_{12} : \exists i \in I \quad \delta_{12}(q_{12}, i) = z_{12}, \quad q_{12} \in Q_{12} \}$

δ_{q12}:

   if $(t \in Q_{12})$ then $\delta_{q12}(t, i) = \delta_{12}(t, i)$

   if $(t \in F' \wedge (\lambda_{12}(t, i) = (o_1, o_2)) \wedge (o_1 \neq o_2))$

      then $\delta_{q12}(t, i) = f$

      else undefined.   □

By using Definition 4 and Theorem 1, Theorem 2 may be obtained.

Theorem 2

Given two FSMs $\mathcal{M}_1$ and $\mathcal{M}_2$, representing the good and the faulty machines, respectively, every string of the language recognized by DFA($\mathcal{M}_{12}$) is a test pattern.   □

Obviously, if the final state f is not reachable, the two machines $\mathcal{M}_1$ and $\mathcal{M}_2$ are equivalent and therefore fault $\mathcal{F}$ is undetectable.

The construction of the DFA($\mathcal{M}_{12}$) associated to the product machine of the two FSMs leads to the determination of all the test patterns for the fault; sometimes it is sufficient to find only one pattern, possibly the shortest one. In order to do this we use MCB [CESi86]. MCB has as an input the graph of any Moore FSM and a formula of Branching Time Temporal Logic called CTL, describing the property to be verified. We must reduce the DFA of Definition 4 5 to a structure that can be accepted by MCB. The DFA is a Moore FSM in which the output associated with all the states except the final state is 0, while the output associated with the final state is 1. It is important to note that we reduce the problem of the equivalence of two Mealy Machines to that of a simple DFA, which is a trivial Moore Machine; thus we are able to use MCB in finding a test pattern for a generic FSM. The formula to be verified by the MCB asserts that the output of the DFA has to be different from 1. Since the logic is branching in time, the **AG** operator is used to assert that the formula is true for all possible evolutions in the future, i.e., for every path and at every node on the path.

The equivalence formula, if O is the output of the DFA, is:

$$\textbf{AG}(\bar{}\ O).$$

3 An example

In this section we analyze the example of Fig. 1, originally presented in [MDNS88].

The functional description of the FSM expresses the inputs to the D-FFs in terms of the state variables $x_0\,x_1$ and of the input i:

$$\delta_1((x_0, x_1), i) = ((\bar{i} \cdot x_0) + i \cdot x_1), \overline{(i \oplus x_1)})$$

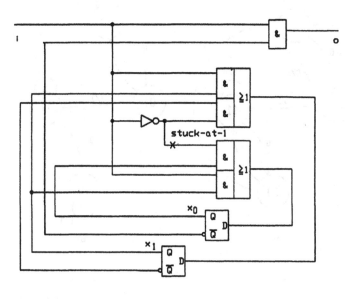

Figure 1: An example

The primary output depends on the state variables $x_0\, x_1$ and on the primary input i:

$$\lambda_1((x_0, x_1), i) = i \cdot \overline{x_0}$$

Fig. 2(a) shows the automaton $\mathcal{M}_1$ of the good machine. Let us consider the stuck-at-1 fault indicated in Fig. 1. The functions corresponding to the faulty FSM are:

$$\delta_2((x_0, x_1), i) = (x_0 + i \cdot x_1,\ \overline{i \oplus x_1})$$
$$\lambda_2((x_0, x_1), i) = i \cdot \overline{x_0}$$

In this particular case the fault's effect is limited to the δ function. The automaton $\mathcal{M}_2$ of the faulty FSM is shown in Fig. 2(b).

The DFA($\mathcal{M}_{12}$) of Fig. 3 is easily built according to Definition 5.

Only 6 states of this automaton are reachable from the initial state; the remaining 4 are unreachable and are not useful to determine test patterns. Two of the 6 states (F_1' and f) are only introduced to formally define the DFA, thus only 4 states are really needed, compared with 16 states for the product machine. The shortest testing sequence is 01011.

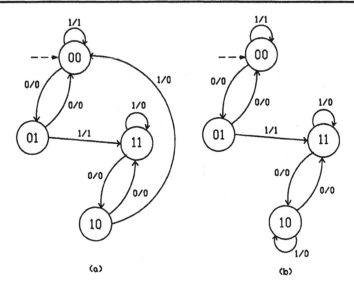

Figure 2: The good and faulty automata for the example of Fig. 1

If we use MCB with Moore FSM corresponding to the DFA of Fig. 3 as an input, the CTL formula to be disproved is:

$$\mathbf{AG}\ (\tilde{\ }\ O).$$

The MCB finds that the formula is false and the counterexample sequence coincides with the test pattern.

4   Conclusions

This paper presented an approach to ATPG for synchronous circuits modeled as Moore or Mealy FSMs. The graph representing state and output transitions is first extracted from the structural description of the circuit, then the effect of the single stuck-at fault is injected, resulting in a faulty graph. The test pattern generation algorithm is based on techniques used to prove the equivalence of FSMs by considering the product machine as a mere conceptual framework. Transforming the product machine's graph into a DFA without unreachable states reduces memory requirements. The advantage of the DFA with respect to the product machine resides in the drastic reduction of the number of states, which in the best case are almost linear in the number of states of the good machine. The Model Checking algorithm is used to disprove, when possible, a CTL formula and MCB's counterexample facility returns the test pattern. A fault

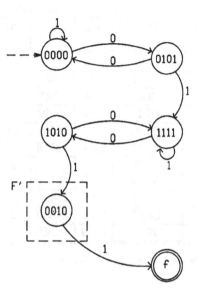

Figure 3: The DFA

simulator [CGSR89] serves the purpose of fault dropping. Preliminary experimental data, on an international benchmark set [BBKo89] (Tab. 1) indicate that the good and faulty machine's graphs are considerably smaller than their upper bound and that the CPU time to create them is reasonable. The main limit is now imposed by MCB, which is able to deal only with a reduced number of states. Results on fault coverage are encouraging: on the s298 circuit 100% coverage for detectable faults is reached.

The novelty of our research resides in modeling both the good and the faulty circuit as graphs, so that the fault detection condition can easily be stated. The use of the DFA and efficient graph simplification algorithms make this approach more efficient. Although the current version refers to the single stuck-at, the algorithm is independent of the fault model, since the latter intervenes only during the generation of the faulty graph. It is thus possible to investigate the use of other models, namely functional ones, whose goodness may be assessed giving a figure of merit in terms of single stuck-at fault coverage.

The programs to extract graphs, to build and reduce the DFA, the combinational ATPG, and the fault simulator have been implemented in separate packages, amounting to a total of 4000 C-code lines and together with MCB, they run on a SUN Sparcstation-1.

This work has been partially supported by the EEC under contract ESPRIT BRA

Circuit	Good Machine			Faulty Machines (reachable states)			
	total states	reachable states	CPU time (s)	avg	max	min	CPU time (s)
s27	8	6	0.05	4	7	3	0.05
s208	256	17	7.81	17	256	1	12.81
s298	16384	218	0.82	189	347	2	0.96
s344	32768	2625	400.17	3216	19693	11	775
s349	32768	2625	405.81	3171	19693	11	768
s382	2M	8865	64.61	5999	13858	2	52.01
s386	64	13	1.37	11	22	2	1.51
s400	2M	8865	66.56	6027	15790	2	53.54
s444	2M	8865	42.70	5929	15790	2	38.71
s526	2M	8868	38.07	6884	14580	2	41.45
s526n	2M	8868	37.47	6879	14580	2	41.22

Table 1: Graphs statistics

3216 "CHARME" and by the Italian National Research Council "Progetto Strategico Collaudo".

References

[ACAg89] V.D. Agrawal, K.T. Cheng, P. Agrawal: "A directed search method for test generation using a concurrent fault simulator," IEEE Transactions on Computer-Aided Design, Vol. 8, n. 2, February 1989, pp. 131-138

[BBKo89] F. Brglez, D. Bryan, K. Koźmiński: "Combinational profiles of sequential benchmark circuits," ISCAS'89: IEEE International Symposium on Circuits And Systems, Portland, OR (USA), May 1989, pp. 1929-1934

[BCDM86] M. Browne, E.M. Clarke, D. Dill, B. Mishra: "Automatic verification of sequential circuits using temporal logic," IEEE Transactions on Computers, Vol. C-35, n. 12, December 1986, pp. 1035-1044

[BEGP86] F.P.M. Beenker, K.J.E. van Eerdewijk, R.B.W. Geritzen, F.F. Peacock, M. van der Star: "Macro Testing: Unifying IC and Board Test," IEEE Design & Test of Computers, December 1986, pp. 26-32

[CESi86] E.M. Clarke, E.A. Emerson, A.P. Sistla: "Automatic verification of finite-state concurrent systems using temporal logic specifications," ACM Transactions on Programming Languages and Systems, Vol. 8, n. 2, April 1986, pp. 244-263

[CGSR89] G. Cabodi, S. Gai, M. Sonza Reorda: "Partitioning Techniques in Multiprocessor Simulators," ESM-89: European Simulation Multiconference, Rome (Italy), June 1989, pp. 311-317

[Chen88] W.T. Cheng: "The Back Algorithm for sequential test generation,"

ICCD'88: IEEE International Conference on Computer Design, Rye Brook, NY (USA), October 1988, pp. 66-69

[ChJo90] K-T. Cheng, J-Y. Jou: "Functional test generation for Finite State Machines," ITC'90: International Test Conference 1990, Washington, DC (USA), September 1990, pp. 162-168

[Fuji85] H. Fujiwara: "Logic testing and design for testability," The MIT Press, Cambridge, MA (USA), 1975

[HuSe89] R.V. Hudli, S.C. Seth: "Temporal Logic based test generation for sequential circuits," IFIP TC 10/WG 10.2 Working Conference on CAD systems using AI techniques, Tokyo (Japan), June 1989, pp. 91-98

[Koha70] Z. Kohavi: "Switching and finite automata theory," Computer Science Series, Mc Graw Hill, New York, NY (USA), 1970

[Marl86] R. Marlett: "An effective test generation system for sequential circuits," DAC-23: 23th IEEE/ACM Design Automation Conference, Las Vegas, NV (USA), June 1986, pp. 250-256

[McCl86] E.J. McCluskey: "Logic Design Principles with Emphasis on Testable Semicustom Circuits," Prentice-Hall, Englewood Cliffs, NJ (USA), 1986

[MDNS88] H.K.T. Ma, S. Devadas, A.R. Newton, A. Sangiovanni-Vincentelli: "Test generation for sequential circuits," IEEE Transactions on Computer-Aided Design, Vol. 7, n. 10, October 1988, pp. 1081-1093

[Moto90] A. Motohara: "Design for Testability of ASICs in Japan," IEEE 13th Annual Workshop on Design for Testability, Vail, CO (USA), April 1990, *(Oral presentation; no proceedings available)*

[Muth76] P. Muth: "A nine-valued circuits model for test generation," IEEE Transactions on Computers, Vol. C-25, n. 6, June 1976, pp. 630-636

[PoMC64] J.F. Poage, E.J. McCluskey: "Derivation of optimum test sequences for sequential machines," 5th Annual Symposium on Switching Theory and Logical Design, 1964

[PuRo71] G.R. Putzolu, J.P. Roth: "A heuristic algorithm for the testing of asynchronous circuits," IEEE Transactions on Computers, Vol. C-20, n. 6, June 1971, pp. 639-647

[ReUr71] N. Rescher, A. Urquart: "Temporal Logic," Springer-Verlag Library of Exact Philosophy N. 3, Springer-Verlag, Berlin (FRG), 1971

[ScAu89] M.H. Schulz, E. Auth: "ESSENTIAL: an efficient self-learning test pattern generation algorithm for sequential circuits," ITC'89: International Test Conference 1989, Washington, DC (USA), September 1989, pp. 28-37

[ShMC76] J.J. Shedletsky, E.J. McCluskey: "The error latency of a fault in a sequential digital circuit," IEEE Transactions on Computers, Vol. C-25, n. 6, June 1976, pp. 655-659

[WiPa82] T.W. Williams, K.P. Parker: "Design for Testability - a survey," IEEE Transactions on Computers, Vol. C-31, n. 1, January 1982, pp. 2-15

[Wolf90] W. Wolf: "The FSM network model for behavioral synthesis of control-dominated machines," DAC-27: 27th IEEE/ACM Design Automation Conference, Orlando, FL (USA), June 1990

Compositional Design and Verification of Communication Protocols, using Labelled Petri Nets

JEAN CHRISTOPHE LLORET

VERILOG, 150 rue Nicolas Vauquelin, 31081 Toulouse cedex,

PIERRE AZÉMA, FRANÇOIS VERNADAT

LAAS-CNRS, 7 Avenue Colonel Roche, F-31077 Toulouse cedex

1 Introduction

This paper proposes a methodology to specify and verify telecommunication protocols by means of Labelled Predicate Transition nets (LPrT). In this paper, the following two principles are used as guidelines.

Modular Specification The structured system decomposition is based upon communication primitives. The rendez-vous communication paradigm of ISO language LOTOS is extended to multi-gate rendez-vous. Several input/output events may appear on a single transition: Petri Net transitions are labelled by sets of communicating events.

Incremental Description A single system is described with respect to several levels of abstraction. Each new abstraction level supplies a more detailed model and new properties are to be verified. Communication by multi-rendez-vous is a first means of abstraction. A second means concerns the data part. To facilitate data abstraction in a LPrT net, logic programming (Prolog) is used as a declarative and prototyping language.

The main contributions of this paper concern the support of former principles.

Multi-rendez-vous is introduced in a stepwise approach, from basic semantical models, that is Labelled transition systems (LTS) and Labelled Petri Nets (LPN), to labelled Predicate Transition nets. The labelled Predicate Transition nets provide the highest abstraction level and are the user interface model. This stepwise definition presents the following characteristics:

didactical: Synchronization aspects are studied in the context of labelled Petri Nets. Communication with value passing is only introduced for Labelled Predicate transition nets.

analytical: Verification techniques are based on the analysis of the LPrT model behaviour. A specific technique, so-called projection or service computation, is derived from observational equivalence [Mil80]). The global service results from the composition of sub-net services. A modular design of labelled PrT nets entails a modular verification.

The composition of Labelled Petri Nets is described in Section 2. A value passing mechanism and the parameterization of Petri Nets are illustrated in Section 3 by means of a specific application.

2 Composition of multi-event Labelled Petri Nets

This section introduces Labelled Petri Nets and multi-rendez-vous operator $|_{LPN}$ as the core language (no value passing) and the basic composition operator, respectively.

The expressive power of this composition operator results from the use of multi-events actions which enable to specify rendez-vous among several (more than two) transitions, on the same or distinct interaction points. This multiple-rendez-vous is an abstraction with respect to implemented communication mechanisms. In the context of the advocated progressive modelling, multi-rendez-vous enables the design of abstract and very compact models.

2.1 Multi-events actions

Communication actions are defined according to the following principles:
events: An event is the most elementary communication unit. Let α be the set of gates, let V be the set of interactions. An event is a couple (g, v) which consists of gate g and interaction v. Event (g, v) is denoted $g(v)$. Let $gate$ be the function which returns the gate of an event $(gate(g(v)) = g)$ and let $\mathcal{E}$ be the set of events.
1. **Multi-event action:** *an action describes several communication events that are performed synchronously on different gates.* A transition is labelled by an action. Because an action refers to a set of events, the expressiveness is increased with respect to the reference languages CCS [Mil80] and LOTOS in which an action is either a single (observable) event or internal action i.
2. **A single event per gate** that is gate is an unshared resource. A service access point is dedicated to a single entity. Consequently, *two distinct events on the same gate do not belong to the same action.*
Formally, an action A is a subset of events $(A \subseteq \mathcal{E})$ such that,
$\forall e_1, e_2 \in A, e_1 \neq e_2 \Rightarrow gate(e_1) \neq gate(e_2)$.
Action $\emptyset$ is denoted i (no event = internal action).

Definition 2.1 Labelled Transition system labelling
Let $proc$ be a labelled transition system; let $\mathcal{T}$ be $proc$ transition set. Let $\alpha_{proc} \subseteq \alpha$ be a subset of gates, let V be the set of interactions.
An event e, connected to a gate in α_{proc}, is a couple $(gate, interaction) \subset \alpha_{proc} \times V$. e is denoted $gate(interaction)$.
$proc$ labelling is couple $(\alpha_{proc}, l_{proc})$ where l_{proc} is the label function. The domain of label function l_{proc} is transition set $\mathcal{T}$. The range of function l is the subset of actions constituted by events connected in α_{proc}.
In the sequel, $\mathcal{L}_{proc}$ denotes couple $(\alpha_{proc}, l_{proc})$.

2.2 Multi-rendez-vous Composition of Labelled Transition Systems

Labelled Transition Systems supply an operational semantics to Labelled Petri Nets. The multi-rendez-vous of labelled transition systems is the interpretation of multi-rendez-vous of Petri nets.

Definition 2.2 *Labelled Transition System*

A labelled transition system *proc* is a 5-tuple $(S, T, -t \rightarrow_{t \in T}, s_0, \mathcal{L})$ where:
- S set of states.
- T set of transitions.
- $-t \rightarrow \subset S \times S$ state change performed by transition t.
- $s_0 \in S$, initial state.
- $\mathcal{L}$ is a labelling as defined above.

Two labelled Transition systems $proc_1$ and $proc_2$ are composed relative to their common gates.

LTS common gates: the set of common gates $\alpha_{proc_1} \cap \alpha_{proc_2}$ is denoted $\alpha_\cap$.

events to be synchronized: events to be synchronized of label $l(t)$ are events whose gate is a shared gate between $proc_1$ and $proc_2$. Notation is $sync_{\alpha_\cap}(l(t))$:

$sync_{\alpha_\cap}(l(t)) = \{e \in l(t) \mid gate(e) \in \alpha_\cap\}$.

Definition 2.3 *LTS composition operator* $|_{LTS}$ Let $proc_i = (S_i, T_i, -t_i \rightarrow_{t_i \in T_i}, s_{i,0}, \mathcal{L}_i)_{i=1,2}$ be two LTS.

Composed LTS $proc_1 |_{LTS} proc_2$ is $(S_1 \mid S_2, T_1 \mid T_2, -t \rightarrow_{t \in T_1 \mid T_2}, s_{1,0} \mid s_{2,0}, \mathcal{L}_1 \mid \mathcal{L}_2)$ where:

$\mathcal{L}_1 \mid \mathcal{L}_2$ is $proc_1 \mid proc_2$ labelling defined by set of gates $(\alpha_{proc_1} \cup \alpha_{proc_2})$ and labelling function l defined on domain $T_1 \mid T_2$.

Function l is defined together with composed states and composed transitions by the following derivation rules (s_i, s_i' and t_i are states and a transition of $proc_i$ respectively):

Independant execution

$$(1) \quad \frac{s_1 - t_1 : l_1 \rightarrow s_1'}{s_1 \mid s_2 - t_1 : l_1 \rightarrow s_1' \mid s_2} \quad (sync_{\alpha_\cap}(l_1) = \emptyset)$$

$$(2) \quad \frac{s_2 - t_2 : l_2 \rightarrow s_2'}{s_1 \mid s_2 - t_2 : l_2 \rightarrow s_1 \mid s_2'} \quad (sync_{\alpha_\cap}(l_2) = \emptyset)$$

Synchronization

$$(3) \quad \frac{s_1 - t_1 : l_1 \rightarrow s_1', \ s_2 - t_2 : l_2 \rightarrow s_2'}{s_1 \mid s_2 - t_1 \mid t_2 : (l_1 \cup l_2) \rightarrow s_1' \mid s_2'} \quad (sync_{\alpha_\cap}(l_1) = sync_{\alpha_\cap}(l_2))$$

2.3 Labelled Petri Nets

Firstly, Labelled Petri Nets are defined as Place/Transition nets of capacity one, associated with a labelling as defined above.

The behaviour of a labelled Petri Nets is then introduced as a labelled transition system. The marking graph and the step graph are two possible candidates for describing a net behaviour. In the step graph, parallel executions are explicitly taken into account and thus a more precise representation of the net behaviour is given. Labelled Petri Net operator $|_{LPN}$ is interpreted on step graphs by Labelled Transition System operator $|_{LTS}$ in such a way that diagram 1 commutes.

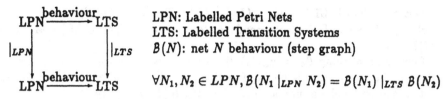

$$\forall N_1, N_2 \in LPN, \mathcal{B}(N_1 \,|_{LPN} N_2) = \mathcal{B}(N_1) \,|_{LTS} \mathcal{B}(N_2)$$

Figure 1: Semantics of Labelled Petri Nets composition "$|_{LPN}$"

Definition 2.4 *Labelled Petri Net*

A labelled Petri Net is a 6-tuple $N = (P, \mathcal{T}, pre, post, M_0, \mathcal{L}_N)$ where:
- P is a set of places;
- $\mathcal{T}$ is a set of transitions, $P \cap \mathcal{T} = \emptyset$
- $pre, post : \mathcal{T} \rightarrow \mathcal{P}(P)$ are two mappings which connect a transition to two sets of places called preconditions and postconditions respectively.
- M_0 is a set of places called initial marking.
- $\mathcal{L}_N$ is the net labelling defined by couple (α_N, l_N) as introduced previously.

Example: net N_1, Fig 2, depicts concurrent transitions t_1 and t_2 (places are circles, transition boxes inscribed by a transition name; if a transition box is connected to a gate, the edge is annoted by the value of the corresponding transition event; the preconditions and postconditions of a transition are the input and output places respectively; places with a token belong to the initial marking).

Definition 2.5 *Firing rule* The transition firing rule proposed for "augmented condition/event nets" in [PD87] is adopted. But the definition of parallel transitions differs because an interaction point is a non shared resource: with respect to labelled Petri nets, two transitions connected to a common interaction point cannot be fired in parallel.

Let M be a marking (or set of places) and T a set of transitions (set T is called *step*). T is M enabled (notation $M[T >$) iff
- T transitions are *pairwise independent* that is two distinct transitions $t_1, t_2 \in T$ share neither a place (*place independance*: $(pre(t_1) \cup post(t_1)) \cap (pre(t_2) \cup post(t_2)) = \emptyset$) nor a gate (*gate independance*: $gate(l_N(t_1)) \cap gate(l_N(t_2)) = \emptyset$, where $gate(l_N(t)) = \{gate(e) \mid e \in l_N(t)\}$).
- preconditions of T transitions are fulfilled ($pre(T) \subset M$, with $pre(T) = \bigcup_{t \in T} pre(t)$) and postconditions which are not preconditions are not fulfilled ($M \cap (post(T) \setminus pre(T)) = \emptyset$).

Marking M' results from firing step T in marking M (notation $M[T > M'$) iff step T is enabled in M and $M' = (M \setminus pre(T)) \cup post(T)$.

The behaviour of a LPN is a LTS such that a state is a reachable marking and a LTS transition is a firable step.

Definition 2.6 *LPN Behaviour*

Let $N = (P, \mathcal{T}, pre, post, M_0, \mathcal{L}_N)$ be a LPN. Reachable markings and firable steps of net N are respectively the smaller sets $M_0[>$ and $M_0[$ defined by:
- $M_0 \in [M_0 >$
- If $M \in [M_0 >$ and T enabled in M with $M[T > M'$ then $M' \in M_0[>$ and $T \in M_0[$.

Net behaviour is LTS, $\mathcal{B}(N) = (M_0[>, M_0[, -T \to_{T \in M_0[}, M_0, \mathcal{L}_{\mathcal{B}})$ where:
- change of state in $\mathcal{B}(N)$ corresponds to step execution:$M, M' \in M_0[>, M - T \to M' \Leftrightarrow M[T > M'$.
- Behaviour labelling $\mathcal{L}_{\mathcal{B}} = (\alpha_{\mathcal{B}}, l_{\mathcal{B}})$ directly follows net labelling: the gate set of $\mathcal{L}_{\mathcal{B}}$ is $\mathcal{L}_N$ one, i.e. $\alpha_{\mathcal{B}} = \alpha_N$. Labelling function $l_{\mathcal{B}}$ is the canonical extension to sets of the net transition labelling function l_N: $\forall T \in M_0[, l_{\mathcal{B}}(T) = \bigcup_{t \in T} l_N(t)$.

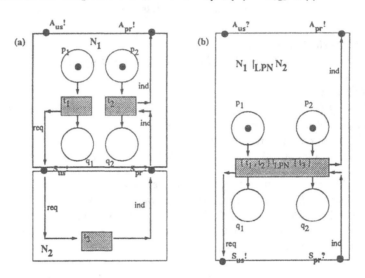

Figure 2: Parallel composition of Labelled Petri Nets

2.4 Multi-rendez-vous Composition of Labelled Petri Nets

Operator $|_{LPN}$ is defined in order to simulate composition of behaviours. The three following characteristics of net multi-rendez-vous are worth noticing:

1. *Parallelism between transitions is taken into account* When composing two labelled Petri nets, two transitions of a net, which fire in parallel, may be merged with a transition of the other net (on Fig 2 transition $\{t_1, t_2\} |_{LPN} \{t_3\}$ results from the merging of transition sets $\{t_1, t_2\}$ and $\{t_3\}$).

2. *Minimal mergings* The possibility of running synchronously (transitions are in parallel) is distinguished from a necessary synchronization (transitions are merged). In the composition of two nets, all synchronization possibilities are preserved but the number of mergings is minimal.

3. The derivation of a composed net $N_1 |_{LPN} N_2$ does not require computation of behaviours $\mathcal{B}(N_1)$ and $\mathcal{B}(N_2)$.

Let $N_i = (P_i, T_i, pre_i, post_i, M_{0,i}, \mathcal{L}_i)_{i=1,2}$ be two LPN with labelling $\mathcal{L}_i = (\alpha_i, l_i)_{i=1,2}$ and let α_n be the common gate alphabet.

Preliminary definitions:

Transition to be merged: Transition t of net N_1 or N_2 is to be merged iff there is at least one event to be synchronized in t label (recall that an event to be synchronized is

an event whose gate belongs to common gate set α_n).
Mergeable Transition sets: Two transition sets $T_1 \subseteq T_1$ and $T_2 \subseteq T_2$ are merge-able iff T_1 (resp. T_2) is a set of transitions *to be merged*, pairwise *independent* and the set of events to be synchronized of T_1 transitions equals the one of T_2 transitions (i.e. $sync_{\alpha_n}(l_1(T_1)) = sync_{\alpha_n}(l_2(T_2))$).
Minimal mergeable Transition sets: T_1 and T_2 are minimal mergeable transition sets iff T_1 are mergeable transition sets and if it is not possible to partition T_1, T_2 merging in smaller ones that is: for all $T_1' \subset T_1$ and $T_2' \subset T_2$, T_1' and T_2' are not mergeable.

Definition 2.7 *LPN composition operator* $|_{LPN}$
  Composed net $N_1 \,|_{LPN}\, N_2$ is $(P_1 \cup P_2, T_1 \,|_{LPN}\, T_2, pre, post, M_{1,0} \cup M_{2,0}, \mathcal{L}_1 \,|_{LPN}\, \mathcal{L}_2)$ where:
• $T_1 \,|_{LPN}\, T_2$ transitions are, on the one hand, T_1 and T_2 transitions which are not to be merged and, on the other, merged transitions $T_1 \,|_{LPN}\, T_2$ where T_1, T_2 are minimal mergeable transition sets.
• Preconditions and postconditions of a nonmerged transition are unchanged in com-posed net. Pre and postconditions of merged transition $T_1 \,|_{LPN}\, T_2$ are defined as the set union of the respective pre and postconditions: $pre(T_1 \,|_{LPN}\, T_2) = pre_1(T_1) \cup pre_2(T_2)$ and similarly for postconditions.
• New labelling $\mathcal{L}_1 \,|_{LPN}\, \mathcal{L}_2$: gate sets of the composed nets are added; the labelling of a nonmerged transition is unchanged; labelling of merged transition $T_1 \,|_{LPN}\, T_2$ is $l_1(T_1) \cup l_2(T_2)$.
  Example: figure 2b depicts composed net $N_1 \,|_{LPN}\, N_2$.

Proposition 2.1 *Properties of multi-rendez-vous composition*
  Operators $|_{LTS}$ and $|_{LPN}$ are commutative and associative (up to state and transition bijective renaming).
  The behaviour of a composed net is the composition of behaviours: $B(N_1 \,|_{LPN}\, N_2) = B(N_1) \,|_{LTS}\, B(N_2)$ (up to state and transition bijective renaming).

3 Application

This section introduces Labelled Predicate Nets by means of an example. LPrT nets are a parameterized version of LPN; they are also PrT nets [Gen88] extended with labels, and featuring, in particular, direct execution in Prolog.
  A remote reading mechanism, the so-called Telereport [HRJ89] application layer, is first modelled then analysed.

3.1 Models

Application entities A_{us} and A_{pr} cooperate through a nonperfect session service S (see Fig 3) in order to provide a remote reading service to a user process.
Application user service. The interface between the user process and entity A_{us} is of particular interest since it determines the external service provided to the user process: user request $req(C)$ is parameterized by requested data code C and is issued on Service Access Point, $A_{us}?$; confirmation $conf(Mess)$ is issued on SAP $A_{us}!$ and can

be of 3 types: *Mess* value is either "*error*": transmission by session service has failed, or "*nak*": no data of code C is available to the provider process, or C read value Val_C. Entity A_{us} is in charge of recovery if a transmission error occurs in the session service. Let NR_{Max} be the maximum number of consecutive recoveries with respect to the same read request. To study the correctness of the recovery mechanism a specific model of session service has been designed: session service may lose consecutively at most NE_{Max} messages. The provided service depends on relative values of parameters NR_{Max} and NE_{Max}.

Configuration. Three configuration classes are distinguished: (1) Perfect session service (no error $NE_{max} = 0$); (2) Faulty session service but less errors than recoveries ($NE_{max} \leq NR_{max}$); (3) Unreliable service: more errors than recoveries ($NE_{max} > NR_{max}$). Furthermore, the requested code may either be available to the provider (expected confirmation with value) or not (expected confirmation: "*nak*"). The set of potential configurations, and the database facts are the following:

	$conf_0$	$conf_1$	$conf_2$	database
NE_{Max}	0	succ(0)	succ(succ(0))	$MaxErr(NE_{max})$.
NR_{max}		succ(0)		$MaxRecover(succ(0))$.
code c_1		c_1 available → value v_1		$codeType(c_1).\ codeVal(c_1,v_1)$.
code c_2		c_2 not available → nak		$codeType(c_2)$.

Session Service Net (see Fig 3, net S)
A request $req(C)$ or a response $resp(Mess)$ may be conveyed without error from user entity A_{us} to provider entity A_{pr} respectively and vice-versa. Primitive req becomes ind and $resp$ becomes $conf$. The corresponding normal transitions are $transmitReq$ and $transmitResp$. Request $req(C)$ or response $resp(Mess)$ may either be lost or incorrectly transmitted. This corresponds to transitions $loseReq$ and $loseResp$. The number of consecutive errors before a reset signal (transition $reset$) is saved by variable K and bounded by logical condition $infMaxErr(K)$. Predicate $infMaxErr(K)$ is defined by the following clause:
```
/* infMaxErr(K) is true if K < NE_max */
infMaxErr(K) :-    maxErr(NE_max),   /* database fact */
                   inf(K,NE_Max).    /* true if K < NE_max */
```
The occurrence of an error results in sending message $conf(error)$ to user entity A_{us} (transition *error*).

User entity of Application protocol (see Fig 3, net A_{us})
The normal transition sequence is (1) *init* receipt of user process request $req(C)$, (2) *send* request transmission to session service, (3) *end* confirmation $conf(Mess)$ is received and returned to the user process, if $Mess \neq error$, that is $Mess$ is a code value or "*nak*"). When a session service error is detected, via primitive $conf(error)$), recovery may take place, i.e. user process request is repeated. Transition *recover* is enabled if less than NR_{max} errors have occured; this enabling condition, $infMaxRecover(K)$, is defined by the following clause:
```
infMaxRecover(K) :-    MaxRecover(NR_max)
                       inf(K,NR_max).
```

In case of confirmation $conf(error)$ and if the maximum number of recoveries is exceeded, transition *end* fires. Signal *reset* is sent to session service.

Provider entity of Application protocol (see Fig 3, net A_{pr})
Provider entity is initially ready to receive read indication $ind(C)$: transition *indication* is enable; procedure $readCode(C, ValOrNak)$ of transition *response* performs a read action; when C is available ($codeVal(C, Val)$ in database) the value of Val is substituted for variable $ValOrNak$; if C is not available variable $ValOrNak$ takes value nak, negative acknowledge. Procedure $readCode$ is defined by the following net clause:

$readCode(C, Val) : -$ $codeVal(C, Val).$ code available
$readCode(C, nak) : -$ $\backslash + codeVal(C, Val).$ code not available

3.2 Verification

An "easy to use" verification technique is to compute the service provided by the protocol and to compare it with the expected one.

The global service provided by a behaviour (LTS) is a reduced LTS, minimal with respect to the state number, which preserves properties of observed communication actions; with respect to gate set G, a local service is derived from the behaviour after the hiding of events which occur on gates outside G. The standard observational equivalence [Mil80] is used to compute the service; observable actions sequences and deadlocks are preserved.

Service Derivation
Let N be a LPN system composed of three sub-nets: $N = N_1 \mid_{LPN} N_2 \mid_{LPN} N_3$. Net service $S(N)$ is defined as the service provided by the behaviour of N, i.e. $S(B_{LPN}(N))$. There are two basic ways to compute service $S(N)$; a third one may also be derived in a combined manner:

Service composition: individual subnet services are computed first, then they are composed. We have, $S(N) \approx S(N_1) \mid_{LTS} S(N_2) \mid_{LTS} S(N_3)$. Partial verifications may be conducted on individual services, and then reused for the verification of the global service. Furthermore, as services are reduced behaviours, composition of individual services can be more efficient than computation of the global netbehaviour.

LPN composition sub-nets are composed to obtain a global net whose behaviour and service are derived in a second step. This approach is mandatory when individual behaviours are unbounded, even if the global behaviour is bounded. For example, in case of PrT nets, unbounded net may be a fifo queue model.

Combined approach: The former computations produce an equivalent result because diagram 1 commutes and because Milner observational equivalence $\approx$ is a congruence with respect to $\mid_{LTS}$. A combined approach may still be useful to gain a better insight into protocols. For example, $S(N) \approx S(N_1 \mid_{LPN} N_2) \mid_{LTS} S(N_3)$: on the one hand N_3 individual service is computed, and on the other net composition $N_1 \mid_{LPN} N_2$ may be performed.

Telereport Service Local service provided to user process is depicted in Fig 4 (transitions represent application service primitives). As long as errors are relatively few, i.e.,

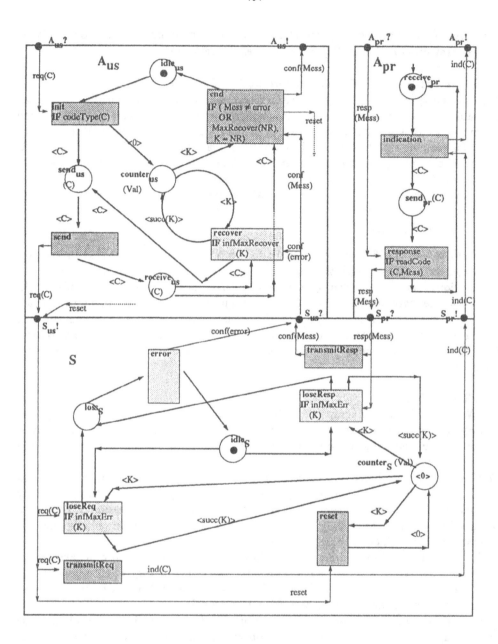

Figure 3: LPrT nets of Application entities A_{us} and A_{pr} and session service S

in configurations $conf_0$, $conf_1$, the provided service is error-free: the recovery mechanism plays the intended role. Configuration $conf_3$ is a degraded configuration: errors may be too numerous, and the application service may not deliver the expected answer (v_1 or nak). Correct status $error$ is however delivered to user process. Confirmation $conf(error)$ follows an internal transition after the last (unrecovered) error.

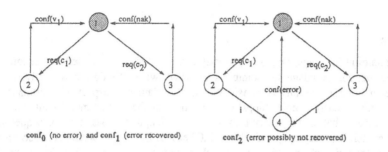

Figure 4: Application service with respect to user process

4 Conclusion

The introduced composition and verification techniques are implemented in the available software PIPN. The tool enforces structured and incremental approaches. It includes a graphical SADT-like editor that defines system architecture. The compiler computes global nets, as a result of subnet composition. State space can be interactively explored by means of the simulator. Finally, the complete behaviour can be computed and then reduced to protocol service. A model checker for CTL logic is also available. A current issue is to combine logic and service verification in a combined approach [PA89].

References

[Gen88] Hartmann J. Genrich. Equivalence transformation of PrT-nets. In *Workshop on Application and Theory of Petri Nets*, pages 229–248, 1988.

[HRJ89] P. Haudebourg, G. Revelaud, and T. Journey. Spécification du protocole de téléreport. Technical Report HR/23-2077 and HR23-1943, Electricité De France, DER, Clamart, 1989.

[Mil80] Robin Milner. *A Calculus of Communicating Systems*, volume 92 of *Lecture Notes in Computer Science*. Springer-Verlag, 1980.

[PA89] F. Vernadat P. Azema, J.C. Lloret. Requirement analysis for communication protocols. In *Workshop On Automatic Verification Methods for Finite State Systems, Grenoble-France*, 1989.

[PD87] U. Montanari P. Degano, R. De Nicola. A distributed operational semantics fot ccs based on condition/event systems. Technical Report 111, Instituto di Elaboratione dell 'Informazione, C.C.R, Pisa, 1987.

Issues Arising in the Analysis of L.0

Linda Ness

Bellcore, Morristown, NJ

linda@bellcore.com

1. Introduction

L.0 is a programming language, which is currently being experimentally used at Bellcore as an executable specification language for communications software. It's design was driven by the special nature of communications systems. The most fundamental aspect of communications systems is that many fairly simple things are happening at the same time, for all time. It is precisely this aspect that is so difficult to specify in languages based on a sequential or asynchronous model. It is easy to specify this in L.0 because L.0 has an abstract concurrent-read-common-concurrent-write shared memory model, a notion of *next*, and because the basic composition operator is *conjunction*. Another fundamental aspect of communications systems is that many of the things happening simultaneously are very similar. Thus, another basic feature of L.0 is quantification, which permits parametrized specifications.

The semantics of the language given originally[1] was a combination of declarative and operational. Now that there is evidence that it is fairly easy to specify communications software in L.0, it is important to try to develop analysis tools. For this it is preferable to have declarative versions of the semantics given in several different staridard paradigms so that analysis tools or algorithms previously developed might be applied to L.0.

One goal of this paper is to present the fundamental semantic constructs of L.0 in the framework of *linear predicate temporal logic*. Temporal logic was chosen because there is a fairly natural fit, and because there has been a great deal of work on its application to the specification and verification of reactive systems[2] [3]. Most of communications software can be viewed as a reactive system. The algorithm for translating L.0 syntax into the constructs given here is not given in this brief paper. It is however given in[4]. Neither is the focus of this paper is not on examples. They may be found in a number of papers[1] [5] [6] [7].

The main reason, that this presentation of the semantics does not immediately lead to the applicability of the various decision procedures and model checkers developed for temporal logic, is that L.0 is based on a subset of predicate (not propositional) temporal logic, and hence permits restricted forms of quantification. Thus, the second goal of this paper is to indicate the uses of quantification in L.0. The simple approach to developing some analysis tools for L.0, is to simply develop them for programs, in which the quantification is restricted to be finite, and the state spaces constructed during execution, are restricted to be contained in a finite universe. This approach is very unsatisfying because quantification is one of the main reasons that L.0 is so expressive. Furthermore, even if this restriction is made, the size of the finite universe, arising in applications of interest is likely to be huge, relative to the size that can be handled by existing tools. Kurshan[8] has developed a formal notion of homomorphism as one approach to handling this problem. An approach would be to develop tools to perform more restricted analysis which would work in presence of less stringents restrictions on quantification.

There has been quite a bit of work on the identification of executable subsets of temporal logic, and the development of executable temporal logic languages, i.e. languages in which a program is a temporal logic formula, and whose execution constructs a model for that formula. The subset of temporal logic exploited by L.0 is very similar to that identified by Dov Gabbay[9] and developed in MetateM. [10] The languages Tempura[11], Lustre[12] and Artic[13] can each be viewed as executable temporal logic languages. Tempura is based on interval temporal logic, rather than propositional, and uses quantification in a more limited way. Unlike L.0 it does not assume any solution to the frame problem. It has been used primarily to specify hardware. The semantics of Lustre can be given in terms of linear temporal logic. However, it appears that quantification is more restricted, since boolean Lustre may be compiled into a finite automaton. Arctic was designed for the specification and implementation of real-time computer systems, and has been applied to music. Its primary data-type is a real-valued function of time.

2. The fundamental constructs of L.0 in terms of temporal logic

Recall that a predicate temporal logic formula is constructed from predicates using the standard boolean operators, the temporal future operators *next* and *until* and universal and existential *quantification*. In addition, temporal past operators may be used, although it has been shown that the past operators do not add expressiveness[14]. The only temporal past operator that will be used here is *previous*.

Here the *until* used will be the weak *until*. In other words it is not required that the second argument of *until* is eventually true on a state. Precisely,

$$S_i \models f \ until \ g \ \text{if}$$

$S_j \models f \text{ for all } j \geq i$
or $(oppE \ k \geq i$ such that $S_k \models g$ and $S_j \models f$ for all j such that $i \leq j < k)$

The temporal future operators *always* and *eventually* may be defined in terms of *until*, using the predicate *false* and negation.

2.1 Predicates on history and on extended history

These are the simplest components of L.0 programs. A *predicate on extended history* is inductively defined to be either:

1. *a predicate or*

2. *a boolean combination of predicates on extended history or*

3. *the past temporal operator previous applied to a predicate on extended history or*

4. *a universally quantified formula of the form* $r => f$ *, where r is a restricting predicate and f is a predicate on extended history or*

5. *an existentially quantified formula of the form r andsign f, where r is a restricting predicate and f is a predicate on extended history.*

A *predicate on history* has the form *previous p*, where *p* is a predicate on extended history. The *restricting predicates* are a subset of the predicates on history. They may be used to ensure the the quantification is finite but unbounded.

The "truth" of a predicate on history on a state in a model, may be deduced by examining only the interpretations for predicates in the previous states, i.e. the *history*. For predicates on extended history, the interpretation for predicates in the current state, may need to be examined as well. In other words, no future interpretations need to be examined. This is the key to executability, since the future states of the model need not have been constructed yet.

2.2 Cause-effect formulas

The second fundamental building block of L.0 programs is the cause-effect formula. Causality is specified using formulas of the form:

$$cause => effect$$

where the *cause* is restricted to be a *predicate on history*, and the effect is any L.0 program. Such a formula is true on a state in a sequence, if either the cause was not true on the history, or the cause was true on the history and the effect is true on the current state. If general formulas were allowed for the cause, the intuition would not correspond to causality, for the truth of the cause might depend on the current and future states.

2.3 A restricted class of until formulas

A restricted class of until formulas is used to specify that several cause-effect formulas apply until the first occurrence of at least one of a set of events, and to also specify the effects of some of those events. Thus, the only until formulas allowed have the form:

- *the binary until operator, applied to a conjunction of cause-effect formulas and a deactivator formula* where a *deactivator formula d* either has form:

$$cause(d) \text{ and } effects(d)$$

where

$$cause(d) = (c_0 \text{ or } c_1 \text{ or } ... \text{ or } c_n)$$

and

$$effects(d) = andsign_{i:\ i\ member\ S}\ (c_i => e_i)$$

Here S *subset* $\{1, ...,n\}$, each of the conjuncts of *effects(d)* is a cause-effect formula, and *cause(d)* is a predicate on history. If none of the events in $cause(d)$ have specified effects, the deactivator specification has the simpler form.

$$d = cause(d)$$

2.4 Universally quantified cause-effect formulas

Universal first order quantification of cause-effect formulas is allowed, providing that the cause contains a conjunct which is a restricting predicate.

2.5 Names of formulas

Formulas may be referred to by name, so that formulas may be structured modularly. Definitions of formula names may be recursive, providing the recursive reference is "over time". This can be expressed in temporal logic using second order quantification.

3. An inductive definition of temporal formulas permitted in L.0

The temporal formulas permitted in L.0 programs are either:

- a predicate

- a conjunction of L.0 programs

- a cause-effect formula where the cause is a predicate on history and the effect is an L.0 program

- a binary until operator, applied to a conjunction of cause-effect formulas and a deactivator formula[1]

- the next operator applied to an L.0 program

- a universally quantified cause-effect formula, where the cause contains a conjunct which is a restricting predicate

- the name of an L.0 program

The original until operator in L.0 had a slightly different semantics in the case of nested untils. However, as in proven in[4], there is an algorithm for mapping these non-standard until formulas to the until formulas of temporal logic, which applies in all reasonable cases.

4. Only safety properties are expressible in L.0

It is not possible to express *eventually* in this subset of temporal logic, because negation of programs is not permitted. The omission of eventually, means that it is not possible to express liveness properties in L.0. However, this omission is not as alarming as it first may seem, for it is possible in L.0 to express pseudo-random sequences, using predicates. This is probably more practical than specifying sequences via the *eventually* operator. Also, the notion of fairness, for which *eventually* is crucial, is not essential to L.0, because L.0 does not base its semantics on the interleaving of atomic events. Instead, the semantics of L.0 is fundamentally the semantics of temporal logic, which is synchronous, and which permits simultaneous occurrence of "events".

4.1 An execution strategy for basic L.0 programs

The subset of temporal logic, consisting of basic L.0 programs is executable, in the sense that there is an algorithm for inductively determining from a basic L.0 program, the obligations it imposes on the current state, and the obligation it imposes on the subsequent sequence of states. This strategy extends to an execution algorithm for general L.0 programs. The decomposition of a basic L.0 program into current obligations and future obligations uses a recursive semantic tautology for basic L.0 until formulas. In fact, an operational semantics for L.0 can be given in terms of a dynamically changing rule set[15].

A fundamental characteristic of the execution strategy is that history, once constructed will never be altered. The key to this it that cause formulas must be predicates on history.

An L.0 program may impose several obligations on the current state, in the sense that a conjunction of predicates must be true on the current state. It is in this sense that L.0 exploits the synchronous semantics of temporal logic. If these obligations conflict, execution of the L.0 program halts. Since there is more than one possible state which satisfies the current obligations, the execution strategy is intrinsically non-deterministic.

5. The current restrictions on the data domain, expressions, and states

Currently the data domain is restricted to be the set of all trees with labeled edges, with the property that all of the child edges of a node have unique labels. The labels are restricted to be from the alphabet of strings of aschii characters. Such a tree is called a namespace. States in L.0 are restricted to be namespaces. Each non-root node of a namespace has a *name*, which is the sequence of labels along the path from the root to the node. Because of the restriction on namespaces, the name of a non-root node in a namespace uniquely identifies it. In general, a *names* is a sequence of labels. Note that namespaces may be equivalently characterized as prefix-closed sets of names. The *suffix-value* of a name on a namespace is the set of strings prefixed by that name in the namespace.

An expression in L.0 programs is either a *name*, a function symbol applied to expressions, a concatenation or union of expressions, or the *suffix value of a name*. The *suffix value* of a name on a namespace is just the set of strings prefixed by the name in that namespace. Geometrically, if the namespace is finite, the value is the tree with labeled edges rooted at the name. The suffix value expression permits simulation of standard variable-value programming, within the more general declarative paradigm. Note that indirection is possible because the suffix value of a namespace is a namespace(which might be a name).

The value of an expression in an L.0 program is a namespace. The algorithm used to interpret expressions is the expected one. The expressions in an L.0 program are only used as arguments of predicates or functions.

6. A restriction on the interpretations of predicate symbols

In an L.0 model, each predicate symbol must be assigned a *predicate definition*. The simplest type of definition is for a predicate symbol of arity 0. Then the definition is either *undefined* or consists of a single *set-theoretic predicate*, which consists of two sets: a *domain*, which is a set of names, and a set of *solutions*, each of which is a subset of the domain[15]. Such a predicate is interpreted as being true on a state, if the intersection of the domain, with the state, is one of the solutions. This is equivalent to a very strong disjunctive normal form, where each element of the domain or its negation occurs in each disjunct. Negation is interpreted as non-existence. There is no restriction that the domain must be finite, here. (With this set-theoretic interpretation, finding a solution of a conjunction of predicates, can be interpreted as finding a global section of a sheaf.)

The definition of an n-ary predicate symbol p, is a mapping from the n-fold cartesian product of *Namespaces(Labels)* to the set of such set-theoretic predicates *cup undefined*.

6.1 Restrictions on the set of predicate symbols permitted

Currently, the predicate definitions allowed are equality of expressions, existence and non-existence of a name, *true* and *false*, and membership in a set of consecutive integers, or in an explicitly given set of names. Furthermore, only limited kinds of equality may be used in effects, and even that in a restricted way: equality of the suffix values of two names, and equality of the suffix value of a name and an expression.

Restricting predicates must contain a conjunction which is either a set membership predicate, or an exists predicate.

Users may provide interpretations for function symbols in "C".

7. A frame assumption

A model for a temporal formula consists of a data domain, an interpretation for expressions, an interpretation for predicate symbols, and a sequence of states. Each of these have been restricted. However, there is one final restriction imposed on a sequence of states, that is to be a model for an L.0 program. This restriction is that successive states are related by one of the replacement rules determined by the predicate, which is the conjunction of predicates, which are to be true on the next state, i.e. the current program obligations. The point is that each predicate (when specific values are supplied for its arguments), has an associated domain and set of solutions. Each solution determines a replacement rule for namespaces, namely replace the intersection of the domain with the namespace by one of its solutions. For a restricted set of predicates (such as are currently in L.0), the result is again prefix-closed, and hence a namespace.

This is nothing but a generalization of the usual frame assumption in sequential programming: namely replace the value of the variable by the new value being assigned. However, this generalization permits the size of the state space to be dynamic. For if the intersection of the domain of a predicate with the previous state space is empty, the set of names making up a solution are added. Conversely, if the intersection of the domain of a predicate with the previous state space is "too large" to be a solution, some of the names are removed to obtain a solution. One of the conveniences in using L.0 as a specification language lies is this concise dynamic allocation and deallocation of "space". To apply analysis algorithms developed for logical formulas, it would be necessary to be able to explicitly specify the frame within an L.0 program, and then consider only the class of programs which explicitly specified it. Fortunately, it appears that this can be done easily, and does not restrict the class of programs, because of the restriction that each predicate must have a specified domain.

8. Uses of quantification in L.0

Universal quantification is extremely powerful. For, as is shown in[4] it permits specification of parameter passing by value, indirection, and the analogue of a set of simultaneous "calls" to the same "procedure". When all three of these uses are combined, one can program in a table driven manner. Thus it is easy to specify a generic non-deterministic finite state machine[1], request handlers that are able to handle a finite but unbounded set of requests each time[16] and specify reconstruction of predicates of particular restricted types[17] In fact using universal quantification, one can easily specify SIMD parallelism.

As seems to be well-known[11], encapsulation can be added by adding existential quantification. Finally, since L.0 permits equality of names, pass by reference can be specified using equality of names and existential quantification. Thus, L.0 augmented by existential quantification, provides permits programmers to program, in a well-structured manner, in temporal logic.

9. Remarks about conjunction and equality

Conjunction permits functional decomposition of specifications. The communication is via shared variables. It seems that well structured L.0 programs are based on a clearly articulated dynamic read-write protocol, among functional components. This permits writing of observer programs which may, for example, filter data, watch for bugs, or write to the screen, to animate the program. Conjunction also makes programs exponentially shorter.

It is interesting to note that when programs are written using encapsulation, it seems possible to restrict the use of equality of names to equality between names local to the parent and child modules. Thus, apart from parameter passing, the uses of equality that seem necessary are standard assignment, and one-way derivations, which deduce that the suffix-value of a name can be the value of a function applied to arguments, which may refer to other current suffix-values.

10. Acknowledgments

L.0 was developed primarily through the joint efforts of Jane Cameron, David Cohen, B. Gopinath, and the author. Prem Uppaluru and Diane Sonnenwald also made contributions to the language, as did a number of the users. B. Gopinath was also the head of the IC* project, during the period when the first version of L.0 was developed. The language implementation was done by David Cohen and Bill Keese (on a sequential machine). The most recent debugger was done by Tim Guinther.

A connection between L.0 and temporal logic has also been recognized by Bob Kurshan, Fred Schneider, Ambuj Singh, and Prem Uppaluru.

REFERENCES

1. E.J. Cameron, D.M. Cohen, L.A. Ness, H.N. Srinidhi, "L.0: A Language for Modeling and Prototyping Communications Software",(to appear in *Proceedings of the Third International Conference on Formal Description Techniques*, Madrid, November 5-8, 1990.)

2. A. Pnueli, "The Temporal Logic of Programs", *Proceedings of the 18th Annual Symposium on Foundations of Computer Science*(1977) pp. 46-57.

3. A. Pnueli, "The Temporal Logic of Programs", *Proceedings of the 18th Annual Symposium on Foundations of Computer Science*(1977) pp. 46-57.

4. L. Ness, "L.O: A Parallel Executable Temporal Logic Language", Bellcore Public Released TM-ARH-014974 September, 1989.

5. E. J. Cameron, N. H. Petschenik, L. Ruston, S. Shah, H. Srinidhi, "From Description to Simulation to Architecture: An Approach to Service-Driven System development", *Proceedings of the First International Conference on Systems Integration*, Morristown, N.J. April 23-26, 1990.

6. D. M. Cohen, T. M. Guinther, L. Ness, "Rapid Prototyping of a Communication Protocol Using a New Parallel Language", *Proceedings of the First International Conference on Systems Integration*, Morristown, N.J. April 23-26, 1990.

7. S. Aggarwal, F.S. Dworak, and P.Obenour, "An Environment for Studying Switching System Software Architecture", *Proceedings of IEEE Global Telecommunications Conference*, 1988.

8. Kurshan, R.P., "Reducibility in Analysis of Coordination", *Discrete Event Systems: Models and Applications*, LNCIS 103(1987), pp. 19-39.

9. D. Gabbay, "Declarative Past and Imperative Future: Executable Temporal Logic for Interactive Systems", in A. Galton, editor, In B. Banieqbal, H. Barringer, and A. Pnueli, editors, *Proceedings of Colloquium on Temporal Logic in Specification*, Altrincham, 1987, pages 402-450. Springer-Verlag, LNCS Volume 398, 1989.

10. H. Barringer, M. Fisher, D. Gabbay, G. Gough, R. Owens, "MetateM: A Framework for Programming in Temporal Logic".

11. B. Moszkowski. *Executing Temporal Logic Programs*. Cambridge University Press, Cambridge. 1987.

12. D. Pilaud, N. Halbwachs, "From a synchronous declarative language to a temporal logic dealing with multiform time", *Proc. Symposium on Formal Techniques in Real Time and Fault Tolerant Systems*, Warwick, Sept 88.

13. R. Dannenberg, "Arctic: Functional Programming for Real-Time Systems", *Proceedings of the Nineteenth Annual Hawaii International Conference on System Sciences*, 1986.

14. O. Lichtenstein, A. Pnueli, L. Zuck, "The Glory of the Past", *Proc. Conf. on Logics of Programs*, Springer-Verlag LNCS #193, 1985, pp. 196-218.

15. E.J.Cameron, D.M.Cohen, B.Gopinath, L.Ness, W.M.Keese, P.Uppaluru, J.R.Vollaro, "The IC* Model of Parallel Computation and Programming Environment," *IEEE Transactions on Software Engineering*, Vol. 14, No 3, March 1988, pp. 317-327.

16. E.J. Cameron, D.M. Cohen, B. Gopinath, L. Ness, "IC*: An Environment for Designing Communications Software", *Proceedings of SETSS '90 7th Int'l Conference on Software Engineering for Telecommunication Switch Systems*, Bournemouth, England, July 3-6, 1989.

17. E.J. Cameron, L. Ness, A. Sheth, "A Universal Executor for Flexible Transactions Which Permits Maximal Parallelism".

AUTOMATED RTL VERIFICATION BASED ON PREDICATE CALCULUS

M. Langevin

Département d'IRO, Université de Montréal, C.P. 6128, Succ. A
Montréal, CANADA, H3C 3J7
e-mail: langevin@iro.umontreal.ca

Abstract

This paper presents a technique for formally verifying synchronous circuits modelled at the Register Transfer Level of abstraction. The circuit's behavior, specification, and hypotheses on its clock sequencing are modelled using a special kind of predicate calculus formulas, named transfer formulas. The proof process consists in applying two general rules of the predicate calculus in a specific order. The automated verification process uses an acyclic graph for representing each transfer formula; in this way, the application of the rules is similar to the manipulation of boolean functions represented by acyclic graphs.

1   Introduction

An error-free design can be assured using good test vectors for simulation, formal verification tools, or silicon compilation. Automated synthesis is the easiest way to design a circuit; unfortunately, no existing tool can handle a wide range of circuit designs. For example, the VTI CAD system can be used to generate the layout of specific components like ROMs, RAMs, PLAs, State Machine, and DataPath, from a high level description [16]; but the compilation of a complete synchronous circuit (datapath and control parts) is not yet available. The validation of these circuits has to be performed with simulation or formal verification tools. Since the complete simulation of a large circuit may be impossible, the designer needs other tools to ascertain the validity of his design; formal verification has been introduced to fill this gap. Much progress has been made in formal verification until now [7]; for example, BULL's PRIAM system has proven the functional correctness of circuits of up to 20000 transistors [11], while the formal verification of a high level description of the RSRE's VIPER processor has been carried out with the HOL system [8].

This paper presents a formal model to prove that the specification of a synchronous circuit is realized by the interconnexion of its components; the circuit specification and structural description are model at the RT Level, but memory component can be latches (asynchronous parts), i.e. the circuit description is more detailed than the circuits with registers only for memory components. Some formalisms have been proposed to perform automated RTL verification [4,15], but these tools do not handle the presence of latches in the design. The systems HOL [6] and VERIFY [3], are accurate formalisms to model the specification and component behavior of the proposed kind of circuit; however, the proofs of correctness in these systems have to be completely or partially human directed, because of the high flexibility of their underlying formalisms. An important feature of a formal verification system is its automation because the circuit designer is not interest to manage the proof; this goal is also pursued in [17,18], where the Boyer-Moore theorem prover is used to verify synchronous design.

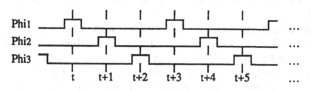117

The proposed method is based on a subset of predicate calculus for modelling circuit behavior. The proof of correctness of a circuit consists in always applying certain logical rules in a specific order, which means that the process can be automated. Direct acyclic graphs (DAGs) are used to represent logical formulas since the manipulation of these formulas is reduces to the graph-based manipulation of boolean functions [5].

2   Level of modelling

The proposed formalism is intended for modelling the behavior of synchronous circuit driven with a multiphase clock, where memory elements could be latches or registers [1]. Carriers are used to represent the nodes of the circuit [2]: there are clock carriers, input carriers, output carriers, and internal carriers for a particular circuit. A carrier represents a bit vector with a certain length.

The circuit behavior is modelled at Register Transfer Level [2]; at this level, the combinational parts are instantaneous. The time granularity used to model the behavior of the circuit components corresponds to the clock phases; a time unit thus corresponds to the interval between the middle point of two adjacent active clock phases. For example, Figure 1 presents the meaning of the time for a circuit which operates in a 3-phase clock mode. This time representation is called the phase-time representation. The clock frequency and sequencing is assumed to assure correct timing.

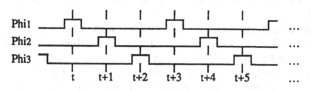

Figure 1

The specification of each circuit's component is a set of transfer functions; a transfer function specifies the possible value that can take a component's output port (an internal or output carrier of the circuit). The components are supposed to be already verified (bottom-up verification), or they will be verified later (top-down verification). Their verification could be performed with other tools of the CAD system: simulators or formal verification tools; in the ideal case, a silicon compiler could be used to generate the components' layout from their specification. A transfer function is defined for any time t, i.e. for any clock phase. A carrier can have memory or not [5]. If a carrier is memoryless, its value at time t+1 depends only on the value of other carriers at time t+1; whereas the value of a memory carrier at time t+1 also depends on the value of carriers at time t (at the previous clock phase).

The specification of a circuit consists in providing transfer functions that specified the possible values for each output carrier and for certain internal carriers; these carriers are the observable carriers of the circuit. The time representation used for these transfer functions corresponds to the clock cycle; this is the cycle-time representation. The circuit's specification is a structural and a temporal abstraction of its behavior [13]. In the specification, the value of an observable carrier V at the end of cycle t+1 (at the last phase of this cycle) must depend only on the values of the input carriers during the cycle and the values of the observable carriers at the end of the previous cycle (the cycle t). These transfer functions must be defined for any clock cycle t. To be able to perform the verification of the circuit, the correspondence between the cycle-time representation and the phase-time representation has to be established. Consider the case of a d-phase clock: the nth phase of cycle $t + 1$ ($1 \leq n \leq$ d) occurs at phase-time d.t + n, i.e. the end of each cycle occurs at a phase-time which is a multiple of d. Hypotheses on the value of clock carriers have to be made when the circuit behavior is specified. A circuit description is the set of its transfer functions which represent the circuit's behavior, the specification, and the hypotheses on the clock sequencing.

3   An example

Let the behavior of a k-bit latch with input D, output Q, and control line 'load' be modelled. The carriers D and Q are described by time function returning bit vectors of length k, while the 'load' carrier is described by a function returning a one bit value. The latch behavior can be modelled with the following first-order formula:

$$load(t+1) * EQU(Q(t+1),D(t+1)) + \overline{load(t+1)} * EQU(Q(t+1),Q(t))$$

where '*' stands for logical 'and', '+' for logical 'or', 'EQU' is the equality predicate, and the variable t represents a clock phase (t is considered to be universally quantified). This formula, called transfer formula, specifies the possible value that can take Q at any phase t+1; if load is true at this phase then the latch is transparent, otherwise the latch keeps its last value.

Two of this latches and one multiplexor are interconnected to form the circuit of Figure 2; after a clock cycle, the value of B is a new value iff 'we' (Write Enable) is high during the second phase. The formulas H1, H2, H3 and H4 constrain the possible values for the clock carriers while the circuit specification is the formula S. To prove that the structure of the circuit meets its specification, it must be shown that the formula S is a logical consequence of the formulas H1, H2, H3, H4, and the formulas F1, F2 and F3, extracted from the circuit structure; F1 and F2 are latch specification while F3 is the multiplexor specification.

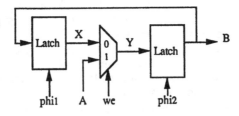

Figure 2

$we(2t+2) * EQU(B(2t+2),A(2t+2)) + \overline{we(2t+2)} * EQU(B(2t+2),B(2t))$	(S)
$EQU(phi_1(2t+1),T)$	(H$_1$)
$EQU(phi_2(2t+1),F)$	(H$_2$)
$EQU(phi_1(2t+2),F)$	(H$_3$)
$EQU(phi_2(2t+2),T)$	(H$_4$)
$phi_1(t+1) * EQU(X(t+1),B(t+1)) + \overline{phi_1(t+1)} * EQU(X(t+1),X(t))$	(F1)
$phi_2(t+1) * EQU(B(t+1),Y(t+1)) + \overline{phi_2(t+1)} * EQU(B(t+1),B(t))$	(F2)

we(t) * EQU(Y(t),A(t)) + $\overline{\text{we(t)}}$ * EQU(Y(t),X(t))   (F3)

The following proof demonstrates that the circuit's behavior meets its specification. It is straithforward for this small example, but could be tedious for larger circuits (note that this proof holds for any width of the datapath). An algorithm to perform automatically this kind of proof is proposed in this paper.

1. phi_2(2t+2) * EQU(B(2t+2),Y(2t+2)) + $\overline{phi_2(2t+2)}$ * EQU(B(2t+2),B(2t+1))   (F2,[t I 2t+1])
2. EQU(B(2t+2),Y(2t+2))   (1,H4)
3. we(2t+2) * EQU(Y(2t+2),A(2t+2)) + $\overline{we(2t+2)}$ * EQU(Y(2t+2),X(2t+2))   (F3,[t I 2t+2])
4. we(2t+2) * EQU(B(2t+2),A(2t+2)) + $\overline{we(2t+2)}$ * EQU(B(2t+2),X(2t+2))   (2,3)
5. phi_1(2t+2) * EQU(X(2t+2),B(2t+2)) + $\overline{phi_1(2t+2)}$ * EQU(X(2t+2),X(2t+1))   (F1,[t I 2t+1])
6. EQU(X(2t+2),X(2t+1))   (5,H3)
7. we(2t+2) * EQU(B(2t+2),A(2t+2)) + $\overline{we(2t+2)}$ * EQU(B(2t+2),X(2t+1))   (4,6)
8. phi_1(2t+1) * EQU(X(2t+1),B(2t+1)) + $\overline{phi_1(2t+1)}$ * EQU(X(2t+1),X(2t))   (F1,[t I 2t])
9. EQU(X(2t+1),B(2t+1))   (8,H1)
10. we(2t+2) * EQU(B(2t+2),A(2t+2)) + $\overline{we(2t+2)}$ * EQU(B(2t+2),B(2t+1))   (7,9)
11. phi_2(2t+1) * EQU(B(2t+1),Y(2t+1)) + $\overline{phi_2(2t+1)}$ * EQU(B(2t+1),B(2t))   (F2,[t I 2t])
12. EQU(B(2t+1),B(2t))   (11,H2)
13. we(2t+2) * EQU(B(2t+2),A(2t+2)) + $\overline{we(2t+2)}$ * EQU(B(2t+2),B(2t))   (10,12)

4 Transfer formulas

In predicate calculus it is possible to use functions that return bit vectors of fixed length. The value taken by a carrier of the circuit at a given time can be modelled as a time function that returns bit vectors. This kind of function is called a signal, as in [10].

Functions can be used also to model high level operators similar to the ones used in RT level HDL's [2]. An operator is modelled as a function that returns a bit vector, and takes as parameters a (possibly empty) list of bit vectors. The operators have only a syntactic meaning, like in [15], i.e. the function of an operator used in the circuit specification has to be realized by a component of the circuit. Hierarchical verification is favored because the functionality of the operator is verified when the corresponding component is verified.

A signal or an operator that returns a bit vector of length 1 can be considered a predicate; it is true if the returned bit is high, and false if the returned bit is low. Special kind of formulas, called transfer formulas, are used to model the transfer functions. The syntax used for the description of a given circuit is similar to the predicate calculus syntax, and corresponds to a subset of first-order predicate calculus (with equality) [12]. The syntax uses temporal expression with the form (c.t+n) to represent a phase-time; the constant c and n are non-negative integers while the variable t can represent a clock phase or a clock cycle. Also, the terms have the type $Quad_n$ defined as $\{0,1,\Phi,\Delta\}^n$; this type represents a vector of n values defined as 0, 1, undefined or high impedance. Here is the syntax:

A) Signals:
    A function V: (time) $\rightarrow$ $Quad_n$, is a signal. Each carrier is represented by a signal.
B) Operators:

1) The function M^k: $(Quad_{n_1} \times ... \times Quad_{n_k}) \to Quad_n$, is associated with each operator M of the circuit,

2) Special operators:

   i)   The constant T: $() \to Quad_1$ models the truth value "true", and the bit high;

   ii) The constant F: $() \to Quad_1$ models the truth value "false", and the bit low;

   iii) The constant UD_n: $() \to Quad_n$ models a vector of n undefined values;

   iv) The constant HZ_n: $() \to Quad_n$ models a vector of n high impedance values;

   v)   The operator BUS_n: $(Quad_n \times Quad_n) \to Quad_n$ models the connection of two sources (defined like the primitive JOIN in [3]).

C) Terms are defined recursively as:

   1) If I is a temporal expression and S is a signal then S(I) is a term,

   2) If $V_1,...,V_n$ are terms and M^n an operator, then $M^n(V_1,...,V_n)$ is a term (the type of V_i has to be the same as the specified type of the ith parameter).

D) Predicates:

   A term P of type $Quad_1$ is considered a predicate, except for HZ_1 and UD_1.

E) Conditions (propositional formulas):

   1) A predicate is a condition,

   2) If G and H are conditions then so are

      i) $\overline{G}$ (logical not)

      ii) $G * H$ (logical and)

      iii) $G + H$ (logical or).

F) Transfers formulas (TFs) for the term U are defined recursively as:

   1) If V is a term then EQU(U,V) is a TF for the term U (this special predicate is called atomic TF (ATF)). It means that the term U takes on the value of the term V; U is the destination while V is the source (the types of U and V have to be the same).

   2) If P is a condition, and F_1 and F_2 are TF for the term U, then $P * F_1 + \overline{P} * F_2$ (IF P THEN F_1 ELSE F_2) is a TF for the term U.

In the following, we will assume that the type restrictions are respected in all TFs. The only variable that could appear in a TF is the time variable t; it is considered to be universally quantified. Therefore, a TF for the term U defines the values that U can take for all t. This is the general form of the TF:

$$cond_1 * EQU(U,U_1) + cond_2 * EQU(U,U_2) + ... + cond_n * EQU(U,U_n)$$

where the $cond_i$ are composed from a set $P_1,...,P_m$ of predicates. This formula specifies that term U takes the value of the term U_i if $cond_i$ is satisfied; U can take one and only one value because all the $cond_i$ are mutually disjoint and their union is a tautology. These formulas can be represented with direct acyclic graphs (DAGs) if the predicates are matched to the nodes of the graph and the atomic TFs EQ(X,Y) are matched to the leaves (Figure 3). The difference of TFs with boolean functions is that more than two values can be transferred in the TFs. A boolean TF is a TF where each U_i is T or F; its representation is the same as for a boolean function. The reduction algorithm for the DAGs that represent boolean functions proposed in [5] can be applied in the same way as for the DAGs that represent TFs. Moreover, as in the case of boolean functions, there exists a canonical form for these DAGs.

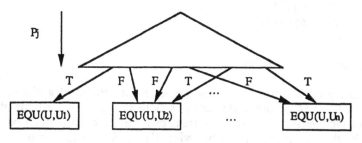

Figure 3

For a given circuit operating in d-phase mode, the description is modelled with specific TFs. For the specification, a TF is defined for each observable carrier at time dt+d (the end of the cycle); its value depends on the value of input carriers at a time between dt+d and dt, and on the value of the observable carriers at the time dt. For the hypotheses on the clock sequencing, a TF is defined for each clock carrier at time dt+n, $1 \leq n \leq d$; its value is either true or false depending on whether or not the clock carrier is active at the nth phase. For the behavior, a TF is defined for each internal and output carrier; if the carrier is a memory carrier, its value at time t+1 depends on the value of the other carriers at time t+1 and t, otherwise, its value at time t depends only on the value of the other carriers at time t.

Example of a circuit description: The circuit NextPC (Figure 4), operating with a 2-phase clock, computes the future value of the PC. At the end of a cycle if the flags' value satisfies the jump condition (JC) then the PC takes on the value of the jump address (JA); otherwise the last value of the PC (the one at the end of the previous cycle) is incremented by 1. The increment operation is modelled with the operator INC while the test condition is modelled with the operator SATISFY; it is assumed that these operators have a verified implementation.

The specification of NextPC is modelled with the transfer formula S while the formulas F1, F2, F3, F4 and F5, extracted from the circuit component, modelled the circuit behavior. The hypothesis on the clock sequencing are the same as for the circuit of Figure 2.

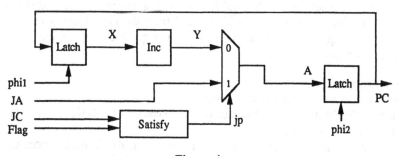

Figure 4

$$\text{SATISFY(JC}(2t+2),\text{Flag}(2t+2)) * \text{EQU(PC}(2t+2),\text{JA}(2t+2)) + \tag{S}$$
$$\overline{\text{SATISFY(JC}(2t+2),\text{Flag}(2t+2))} * \text{EQU(PC}(2t+2),\text{INC(PC}(2t)))$$

$$\text{phi}_2(t+1) * \text{EQU}(\text{PC}(t+1),\text{A}(t+1)) + \overline{\text{phi}_2(t+1)} * \text{EQU}(\text{PC}(t+1),\text{PC}(t)) \qquad (F_1)$$

$$\text{phi}_1(t+1) * \text{EQU}(\text{X}(t+1),\text{PC}(t+1)) + \overline{\text{phi}_1(t+1)} * \text{EQU}(\text{X}(t+1),\text{X}(t)) \qquad (F_2)$$

$$\text{jp}(t) * \text{EQU}(\text{A}(t),\text{JA}(t)) + \overline{\text{jp}(t)} * \text{EQU}(\text{A}(t),\text{Y}(t)) \qquad (F_3)$$

$$\text{EQU}(\text{jp}(t),\text{SATISFY}(\text{JC}(t),\text{Flag}(t))) \qquad (F_4)$$

$$\text{EQU}(\text{Y}(t),\text{INC}(\text{X}(t))) \qquad (F_5)$$

5   Behavioral verification

With the proposed formalism, the proof of correctness of a circuit consists in applying some rules of predicate calculus with the goal of deriving the transfer formulas of the specification from the transfer formulas of the circuit behavior. Let $F_1,...,F_n$ be the TFs of the circuit behavior, $H_1,...,H_m$ be the hypotheses on clock sequencing, and $S_1,...,S_p$ be the TFs of the circuit specification. The goal is to prove that each of the formulas S_i is a theorem when the formulas $F_1,...,F_n,H_1,...,H_m$ are taken as assumptions. A transfer formula S_i specifies the values that an observable carrier V at time dt+d can take as a function of the values of the input carrier between the times dt and dt+d, and the values of the observable carriers at the time dt (the constant d is the number of phases). This kind of TF is called an observable TF, and is defined as:

Def 1:   A term U is observable iff one of the following conditions is true:
       i)   U has the form S(I) where S is an input signal and I a temporal expression,
      ii)   U has the form S(dt) where S is an observable signal,
     iii)   U has the form $M(A_1,...,A_n)$ where M is an n-ary operator and each A_i is observable.
Def 2:   A condition C is observable iff each predicate that occurs in C is observable.
Def 3:   An atomic transfer formula EQU(U,V) is observable iff the term V is observable.
Def 4:   A TF is observable iff each of its conditions and atomic TFs is observable.

An observable TF for a term U specifies the values that U can take as a function of only the values of the observable carrier at time dt and the values of the input carriers at some arbitrary time; the TF S_i is an observable TF for the term V(dt+d). Another TF for V(dt+d) can be extracted from the circuit's behavior since the variable t of the TF for the carrier V can be replaced by any temporal expression. Now, suppose that this TF is transformed in an observable form using logical rules; if this observable formula is identical with S_i then S_i is deduced from the circuit behavior. This is the verification technique proposed in this paper. Two specific rules are used to transform a TF in an observable form:

1)   The substitution rule:
    Suppose that we have the two following TFs:
$$\text{cond}_1 * \text{EQU}(\text{U},\text{U}_1) + \text{cond}_2 * \text{EQU}(\text{U},\text{U}_2) + ... + \text{cond}_n * \text{EQU}(\text{U},\text{U}_n) \qquad (1.1)$$
$$G * \text{EQU}(\text{P},\text{T}) + \overline{G} * \text{EQU}(\text{P},\text{F}) \qquad (1.2)$$
    The substitution of 1.2 in 1.1 consists in replacing all the occurrences of P in the conditions of the TF 1.1 by the condition G of the TF 1.2.

2)   The transition rule:
    Suppose that we have the two following TFs:

$$C_{1,1} * EQU(U,U_1) + ... + C_{1,n} * EQU(U,U_n) \qquad\qquad (2.1)$$
$$C_{2,1} * EQU(V,V_1) + ... + C_{2,m} * EQU(V,V_m) \qquad\qquad (2.2)$$

The transition of 2.1 with 2.2 consists in replacing all the occurrences of V in the terms U_i of 2.1 by its corresponding value specified in 2.2. First, the logical 'and' is applied between the two TFs and the following formula is obtained:

$$C_{2,1} * C_{1,1} * EQU(U,U_1) * EQU(V,V_1) + ... + C_{2,1} * C_{1,n} * EQU(U,U_n) * EQU(V,V_1)$$

$$...$$

$$C_{2,m} * C_{1,1} * EQU(U,U_1) * EQU(V,V_m) + ... + C_{2,m} * C_{1,n} * EQU(U,U_n) * EQU(V,V_m)$$

This transitive rule for equivalence is applied to each product of atomic TFs:

$$EQU(G,H) * EQU(X,Y) => EQU(G,H')$$ where H' is the term H in which the occurrences of X are replaced by Y.

The algorithm used to deduce the observable TF for $V(t')$ ($t' = ct+n$) consists in applying these two rules in a specific order; this algorithm is presented in Figure 5. The initial TF is taken from the list of TFs of the circuit behavior (if the signal represents an output or an internal carrier), or from the list of TFs of the hypotheses on the clock sequencing (if the signal represents a clock carrier). Each non-observable predicate P in the conditions of this TF are replaced by its value specified by the observable TF for P; this transformation is performed with the application of the substitution rule. After, each non-observable source term X of the atomic TF is replaced by its value specified by the observable TF for X; this transformation is performed with the application of the transition rule. The resulting formula is the observable TF for $V(t')$.

The application of the two deduction rules is similar to boolean function graph manipulation when the TFs are represented with a DAG. The substitution rule consists in replacing a predicate P of a TF E by the value specified in the TF D for P. Since the TF D is in the boolean form, this rule can be applied in the same way as the composition of boolean function. This algorithm generates a graph by applying the IF...THEN...ELSE (ITE) function between three graphs: the graph of D, the graph of E where P is restricted to the value true, and the graph of E where the value of P is restricted to false. When the algorithm reaches the leaves of these graphs, the function ITE is applied on three leaves: a leaf that represents a truth value for P, a leaf of E that represents an atomic TF when P is true, and a leaf of E that represents an atomic TF when P is false. Like for boolean functions, the result leaf depends only on the value of the leaf of P.

The transition of a TF D for U with a TF E for V can be performed in the similar way as the application of the logical 'and'. When this function is applied between two boolean functions, the logical 'and' is applied only to the pair of leafs' value. To perform the transition rule, instead of applying the logical 'and', the equality rule is applied to the two leaves $EQU(U,U_i)$ and $EQU(V,V_j)$; the occurrences of the term V in the term U_i are replaced by the term V_j. Unlike the application of the logical 'and', the equality rule is applied in a strict order.

Closed loops in the circuit can be detected when the circuit is verified. During the recursive application of the algorithm to deduce the initial TF for U, if the term U is met, a closed loop in the circuit is detected since the value of the term U depends on the value of the term U [1]. With this algorithm, the way to prove that the specification TF for the term $V(dt+d)$ is a theorem consists in deducing the observable TF for the term

V(dt+d) from the implementation and in showing that it is equivalent to the specified TF. This comparison is carried out by comparing the two DAGs as in [5]. Also, the inclusion of formulas can be used to show the correctness of an incomplete specification.

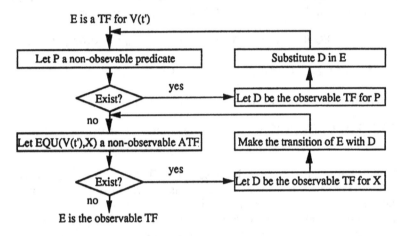

Figure 5

6 Experimental Results

A prototype of the proposed verification technique is under development in Common Lisp. The example of section 4 was successfully verified in about 30 seconds on a Mac Plus. This tools was used in the WORP project [9] at the EPFL, Switzerland; WORP (Watch Oriented RISC Processor) is a RISC processor because it has only one level of programming and each instruction is executed in 4 phases [14]. Its general structure is of the type Harvard, i.e. data and code reside in different memories. It has a 2K of 48-bit instructions saved in a ROM, and 256 of 8-bit words of RAM that can be accessed in a relative or absolute mode.

The circuit is built from the interconnexion of a ROM, a latches bank, an ALU, some working registers, adders, multiplexors, buses, etc. Only a partial verification of this processor has been performed up to now. We specify the possible values that can be stored in the Index Latch of the processor after the execution of any instruction; these values depend on certain fields of the instruction (the ALU code operation, the two source registers for the ALU operation, etc.). A closed loop in the circuit's behavior was detected during this verification. Once this error corrected, the specification has been successfully verified; a trace of roughly 10 pages is produced during the verification and it takes about 7 minutes of real time on a Mac II.

7 Future work

Some improvements to our proposed verification system are planned. First of all, an RT level HDL language could be used to describe the circuit specification, component and hypotheses; the transfer formulas should be deduced directly from the syntax of the HDL (the partial specification of the WORP processor needs 200 lines of edit text and no syntactical analysis is performed). The type validation will be performed directly from the HDL circuit description. Secondly, like the operator BUS, it seems possible to use other predefined operators that have a specific associated meaning. This will be essential in the verification of microcode such as [10]; this requires to associate rewriting rules with the operators, as in [4]. Also to extend the range of verifiable circuits, the pipeline behavior and precharged component will be studied. The study of several circuit examples would help to establish the limitations of our approach.

We try to improve the graph algorithm processing performance by using techniques such as proposed in [11]. Last but not least, the system has to be integrated in a complete CAD system in order to assure the consistency of component models with their real behavior.

Acknowledgements

We want to thank Professor Jacques Zahnd from the EPFL for his help in the formalization of the model, and for his remarks on this research in general. Part of this work was done by the author at the EPFL, Switzerland.

References

[1] F. Anceau, "The Architecture of Microprocessors", Addison-Wesley, 1986.
[2] M. R. Barbacci, "A Comparison of Register Transfer Languages for Describing Computers and Digital Systems", IEEE Trans. on Comp., Vol. C-24, No. 2, February 1975.
[3] H. G. Barrow, "Verify: A Program for Proving Correctness of Digital Hardware Designs", Artificial Intelligence, Vol. 24, 1984.
[4] A. Bartsch, H. Eveking, H.-J. Faerber, M. Kelelatchew, J. Pinder, U. Schellin, "LOVERT -A Logic Verifier of Register-Transfer Level Description", Proceedings of the IMEC IFIP International Workshop on "Applied Formal Methods for Correct VLSI Design", November 1989.
[5] R. E. Bryant, "Graph-Based Algorithms for Boolean Function Manipulation", IEEE Trans. on Comp., Vol. C-35, No. 8, August 1986.
[6] A. Camilleri, M. Gordon, T. Melham, "Hardware Verification Using Higher Order Logic", in From HDL Descriptions to Guaranteed Correct Circuit Designs, ed. D. Borrione, North-Holland, 1987.
[7] P. Camurati, P. Prinetto, "Formal Verification of Hardware Correctness: Introduction and Survey of Current Research", Computer, July 1988.
[8] A. Cohn, "A Proof of Correctness of the VIPER Microprocessor: the First Level", in VLSI Specification, Verification, and Synthesis, eds. G. Birtwistle and P. A. Subrahmanyam, Kluwer Academic Publishers, 1988.
[9] C. Iseli, "WORP: A Watch Oriented RISC Processor", B. Sc. Thesis, Laboratoire de Systèmes Logiques, EPFL, December 1987.
[10] J. Joyce, G. Birtwistle, M. Gordon, "Proving a Computer Correct in Higher Order Logic", Technical Report No. 100, Computer Laboratory, University of Cambridge, December 1986.
[11] J. C. Madre, F. Anceau, M. Currat, J. P. Billon, "Formal Verification in an Industrial Design Environment", to appear in 'The international Journal on Computer Aided VLSI Design'.
[12] Z. Manna, "Mathematical Theory of Computation", McGraw-Hill, 1974.
[13] T. F. Melham, "Abstraction Mechanisms for Hardware Verification", in VLSI Specification, Verification, and Synthesis, eds. G. Birtwistle and P. A. Subrahmanyam, Kluwer Academic Publishers, 1988.
[14] W. Stallings, "Tutorial: Reduced Instruction Set Computers", IEEE Computer Society Press, Order Number 713, 1986.
[15] R. Vemuri, "A Formal Model for Register Transfer Level Structures and Its Applications in Verification and Synthesis", Proceedings of the IMEC IFIP International Workshop on "Applied Formal Methods for Correct VLSI Design", November 1989.
[16] "VTI User's Guide", VLSI Technology, Inc, 1988.
[17] A. Bronstein, C. L. Talcott, "Formal Verification of Pipelines Based on String-Functional Semantics", Proceedings of the IMEC IFIP International Workshop on "Applied Formal Methods for Correct VLSI Design", November 1989.
[18] L. Pierre, "The Formal Proof of Sequential Circuits Described in CASCADE Using the Boyer-Moore Theorem Prover", Proceedings of the IMEC IFIP International Workshop on "Applied Formal Methods for Correct VLSI Design", November 1989.

ON USING PROTEAN TO VERIFY ISO FTAM PROTOCOL

R. Lai(*), K.R. Parker(@), T.S. Dillon(*)

(*)Department of Computer Science and Computer Engineering
La Trobe University
Melbourne, Victoria, Australia

(@)Telecom Research Laboratories
Telecom Australia
Melbourne, Victoria, Australia

Abstract

This paper describes the use of PROTEAN and associated methodology to verify the ISO FTAM (File Transfer, Access and Management) DIS (Draft international standard) protocol and discusses the analysis of its behaviour using PROTEAN facilities. PROTEAN is an automated validation tool developed by Telecom Australia. The formal description technique used is Numerical Petri Nets (NPNs), an extension of Petri Nets. The procedures carried out were based primarily on reachability analysis. The behaviour of the protocol as specified was compared to its service specification. There are two protocol machines specified for FTAM : the basic protocol and the error recovery protocol machines.

1 Introduction

Protocol verification is the demonstration of the correctness, completeness and consistency of the protocol design represented by its formal specification. Techniques for specification and verification of computer network protocols have progressed significantly in the past decade. The success is largely due to the development of the state transition approach for formal specification and to the greater automation of the verification process. Automated protocol verification is the use of computer tools to verify a communication protocol based on its formal specification [2,10].

PROTEAN (PROTocol Emulation and ANalysis) [3], developed by Telecom Australia, is a computer aided tool for analysis of computer communication protocols. It finds faults such as deadlocks, livelocks and maloperations specific to the protocol under test. It is based on a formal description technique called Numerical Petri Nets (NPNs) [11]. An NPN specification is the starting point for verification using PROTEAN. In practice, some faults can be uncovered during the process of creating a precise NPN specification.

ISO 8571 [4] defines a file transfer service and protocol, known as File Transfer, Access and Management (FTAM) [7]. It controls the transfer of whole files or parts of files between end-systems. The protocol is available within the application layer of the OSI (Open Systems Interconnection) reference model [1]. Primitive is an OSI term essentially meaning commands or messages passed between the OSI entities such as the user and the protocol machine.

FTAM supports two services: the Reliable File Service and the User Correctable File Service. For the Reliable File Service, the user states its quality of service requirements, but has no control of error recovery, delegating such considerations to the service provider. For the User Correctable File service, the user has primitives available for error recovery and transfer management. There are 2 protocols specified for FTAM : basic file protocol, supporting the user correctable file service, and error recovery protocol, supporting reliable file service.

This paper describes the use of PROTEAN to verify FTAM DIS protocols specified in NPNs.

2 Numerical Petri Nets

Numerical Petri Nets (NPNs) is a formal description technique that belongs to the family of state transition models. It is an extension of Petri Nets [8]. It was originally developed by Symons [9]. It can be specified algebraically or graphically. The graphical NPN consists of a bipartite directed graph, fixed firing rules and an initial marking. An NPN may have global variables (called P-variables) which transitions can read and write. An NPN has transition enabling conditions and operations which refer not only to the tokens in the input places but may also refer to the global variables. By a marking of a Petri Net we mean a complete specification of the tokens in each of its places and the value of all its global variables.

An NPN, labelled G, can be defined by a quintuple(P,T,Fi,Fo,Mo), where

"P" is the set of all places in the net G, where a place is represented by a circle in the NPN graph and can be used to represent a machine state;

"T" is the set of all transitions in the net G, where a transition is represented by a bar in the NPN graph and can be used to represent an event; each transition includes it's transition conditions and transition operations;

"Fi" are the input firing rules, usually written on the arcs between the input places and their respective transitions; this specifies what tokens are required to be in the input places to enable the transition and what tokens are destroyed in (i.e. removed from) the input places when the transition is fired;

"Fo" are the output firing rules, usually written on the arcs between the transitions and their respective output places; this specifies what tokens are created in (i.e. added to) the output places;

"Mo" is the initial marking of the net, this specifies all the tokens in each place and global variables.

3   Reachability Analysis

A system can be analysed by considering every possible sequence of events and thus every possible "state" of the system. The set of all possible states of the system is known as the Reachability Set, which is unique for a given system NPN with a given initial marking.

In Petri Nets, the relationship between states is given by the Reachability Graph (RG). It may be presented in tabular or graphical form. Reachability analysis is the study of the reachability graph generated for the system being verified.

The RG is used to investigate the properties of the protocol being studied. The main problem with the RG is that it gets very big and unmanageable as the complexity of the system modelled increases, and thus analysis becomes very tedious.

4   PROTEAN

PROTEAN is a user friendly menu-driven system, with on-line help and simple error messages. It has two parts. The first is the NPN Analyzer Program, which handles the NPNs and the generation of the reachability graph. The second is a collection of programs which helps the user detect maloperations and other properties of the protocol using the results from the first stage.

The NPN analyzer program allows NPN subnets to be entered via keyboard or a file. The NPNs are checked for correct syntax. The subnet NPNs can be recalled and combined with other subnets to form larger NPNs. The NPNs can also be modified, listed and displayed graphically.

The NPN is initialised by placing tokens in the places and assigning values to the data variables. A user can then investigate the operation of the net manually or automatically. The manual method allows execution of the net by firing one transition at a time. In the automatic mode the complete reachability set and RG are generated. All deadlocked markings are identified.

PROTEAN contains several programs to investigate properties of the RG. They are RG graphics display, loop detection, livelock detection, reduction of RG and scenario generation. Some illustrations of the use of these are given in section 9.

5   FTAM Model

The operation of the FTAM protocol is modelled by the interaction of two file protocol machines (FPMs). The two FPMs communicate by means of the services available at their lower boundary, in such a way as to provide the service required at their upper boundary.

The file service is defined asymmetrically, with the file service user "A" being the initiator and file service user "B" being the responder. File user "A" can request the user correctable file service or the reliable file service.

The file service and its supporting protocol are concerned with creating a series of stages, a working environment in which the initiator's desired activities can take place. This leads to a set of contexts being established. The period for which some parts of the common state held by the service users is valid is called a "regime". As progressively more shared states are established a nest of corresponding regimes is built up. However, there will, in general, be a time lag between the establishment of a regime at the two ends of the association.

Four types of file regimes are defined: FTAM establishment, file selection, file open, and data transfer regime.

6   Verification Methodology

The procedure used to analyse FTAM follows a methodology which has three main stages.

The FTAM protocol is divided into the basic protocol and the error recovery protocol. For the basic protocol, it is further divided into the basic file protocol, the bulk data transfer protocol, and the basic file protocol under grouping control.

The next phase involves partitioning the protocol into subsystems and creating a complete specification of each of these. This is done by modelling with NPNs to show the systems state changes, data processing and signal flow. The NPNs for each subsystem are joined to form the model of the total system.

Finally the logical operation of the protocol model is then analysed. The possible sequences of events are considered, and the possible states of the system. In particular these are checked to see if the protocol conforms to the requirements of its service specification.

7   Formal Specification of FTAM

The nets are viewed as modelling the operation of the protocol machines. In our approach each service or function provided by the machine is specified via a separate net. We use places to represent ports through which data are sent and received, and global variables to represent states. Predicates are used to govern the conditions of the transition firings.

In this situation, the specification of a single function in a single net has logical advantages. This approach represents the flow of communication primitives closely and makes it easier to follow the behaviour of the entities involved. The data parameters passed by the protocol are modelled by the "attributes" of the tokens.

The specification is based on the Draft International Standard of ISO 8571. The NPN specifications of the whole FTAM protocols are based on the state tables in the Annex of ISO 8571, since the standard states that the Annex is to take precedence over the text.

The detailed formal specifications of the basic file and error recovery protocols are described in [5].

8 FTAM Verification

The NPNs must be first correctly entered into PROTEAN. It is entered in a net form, with each net modelling a service or function of the protocol machines. This allows the behaviour of various sub-nets to be analysed.

There are two methods available in the PROTEAN system to analyse a protocol : single step method and automatic generation. It is not feasible to use single step method to analyse FTAM as there are too many enabled transitions as more nets are added. Using the second method, a RG is automatically generated. The RG can be automatically checked for the presence of any deadlocks. The problem with this method is that the RG gets very large. It then becomes very tedious to analyse the system.

There are many primitives and regimes defined in the protocol. To analyse the protocol by combining all the nets will make the task insurmountable because the RG which would be thus generated will be too huge. For the basic protocol, the analysis was done in 3 phases: basic file protocol, bulk data transfer protocol and basic file protocol under grouping control. The RG was generated for each of the three phases. However, for the bulk data transfer phase, two separate RGs were created with the Cancel Data request included only in one, and the Restart Data request included only in the other.

The issuing of primitives follows the nesting of the regimes. The protocol was examined for behaviour upon receipt of combinations of primitives of interest. The protocol was also checked for situations with error conditions, for example, unsuccessful connection establishment, unsuccessful creation of files, etc. In the NPN model, these are specified by predicates. The net was simulated under all these conditions.

The properties of the protocol were investigated using the Liveness, Language, Scenario and Cycle facilities of PROTEAN. The results of verifying the FTAM basic file and error recovery protocols are described in [5,6].

9 Analysis Using PROTEAN

9.1 Liveness Analysis

The Liveness program determines all of the strongly connected components of the RG. A strongly connected component consists of one or more nodes for each of which a path can be found to every other node in that component. A liveness graph is generated showing how the strongly connected components of the RG are related. Livelocks occur as leaf nodes of this graph with two or more markings (leaf nodes with one marking are deadlocks). If there is only one strongly connected component, the protocol has no livelocks or deadlocks.

Liveness analysis was applied to the RGs generated for the nets created in the specifications of the FTAM protocols. This revealed several leaf nodes which were found to be deadlocks as described in [5,6]. There are no actual livelocks uncovered in the analysis.

9.2 Language Analysis

The reachability graph shows all of the possible sequences of all transitions, ie. the language of all of the transitions. To determine if a protocol meets its service, or to verify other properties, it is useful to determine the language of selected key transitions. The Language program reduces the reachability or language graph to show only the sequences of these user-determined key transitions.

Language analysis was applied to the transitions representing FTAM services to study the behaviour of the FTAM file regimes. It verifies that the protocol does behave according to the FTAM regimes.

For example, a language graph for a net that consists of F-initialize, F-select, F-create, F-deselect, F-delete, and F-terminate is shown in figure 1, where the circles are markings labelled by a number. The transitions which fire to go from numbered one marking to another are as follows:

1	Finirq (F-initialize request)
2	
2	Finicf (F-initialize confirm)
13	
13	Fselrq (F-select request)
15	
13	Fcrerq (F-create request)
14	
14	Fcrecf (F-create confirm)
36	
15	Fselcf (F-select confirm)
36	
36	Fdesrq (F-deselect request)
38	
36	Fdelrq (F-delete request)
37	
37	Fdelcf (F-delete confirm)

38	Fdescf (F-deselect confirm)
59	
59	Fterrq (F-terminate request)
60	
60	Ftercf (F-terminate confirm)
1	

The initial marking is given on the left of the transition name, and the final marking number is below it. The language graph with the legend above shows possible sequences of behaviour. Initially there is no connection, then the FTAM regime is established with the issuing of an F-initialise request and subsequent receipt of an F-initialise confirm. Then the initiator may either issue an F-select request or F-create request, etc.

The language graph is essentially a reduction of the RG with unwanted details suppressed (hence the missing numbers in the the sequence of markings).

9.3 Scenario Generation

The Scenario program finds paths in a RG that match a specified transition or marking sequence. The user may exclude particular nodes or transitions from the sequence and may limit the number of paths found. This facility enabled the tracing of specific behaviour in the FTAM protocol and, in particular, was useful for debugging and tracing the cause of deadlocks.

For example, the deadlock uncovered for the File Open service has the path listed in below.

Path 1: 1 -> (Finirq) -> 2 -> (Ainirq) -> 3 -> (Aassrq) -> 4
    -> (Aassin) -> 5 -> (Binirq) -> 6 -> (Finiin) -> 7
    -> (Finirp) -> 8 -> (Binirp) -> 9 -> (Aassrp) -> 10
    -> (Aasscf) -> 11 -> (Ainirp) -> 12 -> (Finicf) -> 13
    -> (Fselrq) -> 14 -> (Aselrq) -> 15 -> (PdatrqAsel) -> 16
    -> (PdatinBsel) -> 17 -> (Bselrq) -> 18 -> (Fselin) -> 19
    -> (Fselrp) -> 20 -> (Bselrp) -> 21 -> (PdatrqBsel) -> 22
    -> (PdatinAsel) -> 23 -> (Aselrp) -> 24 -> (Fselcf) -> 25
    -> (Fopnrq) -> 26 -> (Aopnrq) -> 27 -> (PdatrqAopn) -> 28
    -> (PdatinBopn) -> 29 -> (Bopnrq) -> 30 -> (Fopnin) -> 31
    -> (Fopnrp) -> 32 -> (Bopnrp) -> 33 -> (PdatrqBopn) -> 34
    -> (PdatinAopn) -> 36 -> (PaltrqBopn) -> 37 -> (Aopnrp) -> 39
    -> (p4Fopncf) -> 41

Marking 41, the final marking in that path, is a deadlock. From this sequence generated by Scenario, the reason for the deadlock can be traced. In this instance the deadlock was due to an error in the FTAM state table, where a negation is missing from action 14.

9.4 Elementary Cycle

The Cycle program lists all of the loops contained in the reachability or language graph. It identifies the largest cycle and may be used for a loop layout of the graph. Cycle is useful in investigating the cyclic behaviour of the protocol.

For each RG, there is at least one cycle. For the example given in section 9.2, there are 4 cycles. (Cycle uses the language graph.) The cycles reveal different paths that can be taken for the FTAM file regimes. They can be obviously seen from the language graph in figure 1, for this simple example, and are as follows:

Cycle 1: 1 2 13 14 36 37 59 60
Cycle 2: 1 2 13 14 36 38 59 60
Cycle 3: 1 2 13 15 36 37 59 60
Cycle 4: 1 2 13 15 36 38 59 60

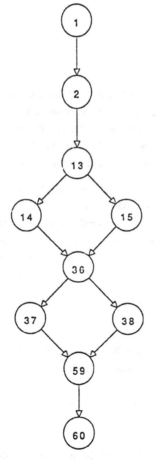

Figure 1.   Language Graph

9.5 Limitations

There are several ways in which facilities currently available in PROTEAN limit the usefulness of it for the analysis of a complex protocol, like FTAM. These include :

- Lack of direct graphical input of NPNs
- Lack of auxiliary net input facilities such as macros
- Lack of refinement facilities for nets
e.g. representation of a subnet at one level as a single transition at a higher level

Other needs, as presented in [3], include handling of infinite reachability set, incorporation of a query language, language comparison, background simulation and improved performance for analysis and graphical layout. Alternative analysis techniques, such as improved reachability analysis, invariant analysis and net reduction, also have great potential for aiding practical analysis. This will greatly improve the range of analysis that will be possible for FTAM, and for other complex protocols.

10 Conclusions

The techniques used to perform a verification of the ISO FTAM DIS protocol using PROTEAN, after subdivision into manageable tasks, have been presented. We have reported results of the analysis of protocol properties performed using PROTEAN, whose utilities proved to be useful in this task. In particular the discovery of deadlocks in the NPN model of FTAM were found, as was an absence of livelocks and conformance to the FTAM service regimes.

While reachability graph generation is fundamentally an exhaustive searching process, we have indicated that with the aid of the PROTEAN automated analysis tool, reachability analysis can be of practical use, even for complex protocols. We thus conclude that reachability-based analysis of Petri-net specifications is a useful approach in the practical verification of complex protocols, and that there is great potential for further tool development.

11 Acknowledgment

We would like to thank Telecom Australia Research Laboratories for the use of their PROTEAN software and to J. Billington for his comments on an early draft of this paper.

References

[1] Bartoli, P.D., "The Application Layer of the Reference Model of Open Systems Interconnection", Proceedings of the IEEE, vol. 71, no. 12, pp. 1404-1407, December, 1983.

[2] Billington, J., Wilbur-Ham, M.C., and Bearman, M.Y., "Automated Protocol Verification", Proceedings of the Fifth International Workshop on Protocol Specification, Testing and Verification, June, 1985.

[3] Billington, J., Wheeler, G.R., and Wilbur-Ham, M.C., "PROTEAN: A High-level Petri Net Tool for the Specification and Verification of Communication Protocols", IEEE Transactions on Software Engineering, Vol. 14, No. 3, pp. 301-316, March 1988.

[4] ISO DIS 8571, "Information Processing Systems - Open Systems Interconnection - File Transfer Access and Management - Parts 1,2,3 and 4", Draft International Standard, ANSI, New York, 1986.

[5] Lai, R., "Formal Specification and Verification of ISO FTAM Protocol", PhD Thesis, La Trobe University, Australia, August, 1989.

[6] Lai, R., Dillon, T.S., Parker, K.R., "Verification Results for ISO FTAM Basic Protocol", Proceedings of Ninth Symposium on Protocol Specification, Testing and Verification, North-Holland, June, 1989.

[7] Lewan, D., and Long, H.G., "The OSI File Service", Proceedings of IEEE, Vol. 71, no. 12, pp. 1414-1419, December, 1983.

[8] Peterson, J.L., "Petri Nets Theory and the Modelling of Systems", Prentice-Hall, Englewood Cliffs, N.J., 1981.

[9] Symons, F.J.W., "Modelling and Analysis of Communication Protocols using Numerical Petri Nets" PhD Thesis, Department of Electrical Engineering Science and Telecommunications, University of Essex, May, 1978.

[10] West, C.H., "An Automated Technique of Communication Protocol Validation", IEEE Transaction on Communications, vol. COM-26, pp.1271-1275, Aug, 1978.

[11] Wheeler, G.R., "Numerical Petri Nets - A Definition", Telecom Australia Research Laboratories Report 7780, Telecom Australia Research Laboratories, May, 1985.

Quantitative Temporal Reasoning[†]

(Extended Abstract)

E. Allen Emerson[1]   *A. K. Mok*[1]   *A. P. Sistla*[2]   *Jai Srinivasan*[1]

1. Department of Computer Sciences,
The University of Texas at Austin,
Austin, TX 78712.

2. GTE Research Laboratories,
Waltham, MA 02254.

Abstract

A substantially large class of programs operate in distributed and real-time environments, and an integral part of their correctness specification requires the expression of time-critical properties that relate the occurrence of events of the system. We focus on the formal specification and reasoning about the correctness of such programs. We propose a system of temporal logic, RTCTL (Real-Time Computation Tree Logic), that allows the melding of qualitative temporal assertions together with real-time constraints to permit specification and reasoning at the twin levels of abstraction: qualitative and quantitative. We show that several practically useful correctness properties of temporal systems, which need to express timing as an essential part of their functionality requirements, can be expressed in RTCTL. We also develop a model-checking algorithm for RTCTL whose complexity is linear in the size of the RTCTL specification formula and in the size of the global state-space graph. Finally, we present an optimal, exponential time tableau-based decision procedure for the satisfiability of RTCTL formulae, which can be used as the basis of a technique to automate the synthesis of real-time programs from specifications.

1   Introduction

Motivated mainly by the virtue of separating concerns, most of the research into the formal specification and reasoning about the correctness of programs has paid little heed to dealing with quantitative temporal properties. In fact, this has proved to be an advantageous abstraction because, in many applications, the correctness properties of a program need to be stated independently of concerns of efficiency, performance, or features (e.g., the speed) of the underlying hardware implementation. Given this, a common characteristic of most temporal or modal logics heretofore proposed for program reasoning (cf. [Pn77], [FL79], [Ab80], [GPSS80], [BHP81], [BMP81], [Wo81], [EC82], [EH82], [Ko82], [EH83], [ES84], [VW84]) is that they provide a formalism for *qualitative* reasoning about change over time. For example, such formalisms allow the expression of assertions such as an event p will *eventually* occur (stated as $F\,p$); note that this assertion places no bound on the time that may elapse before the occurrence of p. Thus, with $p = halt$, $F\,halt$ asserts that a program terminates, and, indeed, such a qualitative sort of correctness temporal property is in fact the strongest one that may be desirable to state of many programs.

On the other hand, there is a substantially large class of programs that operate in distributed and real-time environments (for example, network communication protocols and embedded real-time control systems), an integral part of whose correctness specification requires the expression of time-critical properties that relate the occurrence of events of the system. For example, consider $p = respond$, in the context of a control system. There, we might want

---

[†]This work was supported in part by NSF grant DCR–8511354, ONR URI contract N00014–86–K–0763, and Netherlands NWO grant nf–3/nfb 62–500.

to assert a quantitative correctness property such as $F^{\leq 50}$ *respond*, meaning that a response is guaranteed within *bounded* time, namely, 50 time units. By and large, the specification and verification of such systems has been *ad hoc*. One common technique of coping with such systems (particularly, network communication protocols) has been to abstract out their timing component, use formal techniques that handle qualitative temporal assertions to ensure that their behaviour is correct upto timing constraints, and then incrementally consider quantitative temporal specifications. This method does appear to have advantages in that it allows one to gain considerable intuition for why some of the specifications may be incorrectly conceived, independent of timing constraints, and that it also permits one to manually "fine-tune" time-critical parameters of a system using one's intimate understanding of the system's internals to guide the process.

We remark that methodologies for specifying and reasoning directly about real-time properties have been proposed, but suffer from some limitations. For one thing, such formalisms are often designed with a very specific application in mind and fail to consider or appropriately generalize the kinds of properties expressible, or to characterize the class of properties provably expressible. Also, the ability to describe properties of a system is only one component of a formalism: one also needs develop techniques to effectively, and tractably, reason about such assertions. The few methodologies that do not resort to doing this arbitrarily often use techniques that are *rigid*, in the sense that the correctness of the analysis depends on the specific values of constants in the assertions, and is not easily modified to still be correct for different values of these constants. Moreover, the two-phased capability of being able to reason, first qualitatively, and, then, refining the reasoning to be quantitative, is lost.

In this paper, we provide general techniques to augment systems of temporal logic to handle quantitative assertions. We focus primarily on one system of logic, RTCTL (Real-Time Computation Tree Logic), which extends CTL (Computation Tree Logic, cf. [EC82], [EH82]), a system of logic that has been widely applied to reasoning about program correctness. RTCTL is able to overcome many of the difficulties of the other approaches because it builds on the foundations of temporal logic. For example, it allows the melding of qualitative temporal assertions together with real-time constraints to permit specification and reasoning at the twin levels of abstraction: qualitative and quantitative. It supports not only efficient reasoning at both these levels, but also refinement from the qualitative level down to the quantitative level. Moreover, temporal logic has demonstrably proved to be useful to reason about a variety of discrete systems, and thus, an appropriate extension (such as RTCTL) would naturally allow one to deal with various kinds of applications. And, finally, part of the power of our approach is derived from the fact that standard problems in temporal logic such as *satisfiability* and *model-checking* have been shown to be applicable to automating the construction of, and reasoning about, temporal systems such as concurrent programs (cf. [EC82], [CES83], [MW84], [LP85]), and, furthermore, techniques to effectively tackle these problems are well-established.

Our method is based on the expression of RTCTL assertions in the propositional μ-calculus, extended with natural number ordinal-ranks. We call this new logic Real-μ. In the ordinary μ-calculus ([Ko82]), correctness properties are characterized as extremal fixpoints of predicate transformers similar to those considered by Dijkstra ([Di76]). Here, least fixpoints corresponding to eventualities are annotated with a natural number bound on when they must be fulfilled. We show that several interesting quantitative temporal properties are thereby expressible.

We go on to develop a model-checking algorithm for RTCTL which, like the algorithm for CTL, has complexity linear in the size of the RTCTL specification formula and in the size of the global state-space graph. The key observation that makes this possible is that the model-checking algorithm of [EL88] actually recovers not only whether an eventuality is fulfilled, but also when, based on calculating its rank in the Tarski-Knaster sequence of approximations. Hence, our model-checking algorithm can be generalized to the full language Real-μ in polynomial time complexity.

Next, we focus on the satisfiability problem for RTCTL. We exhibit an exponential time decision procedure for RTCTL, using a *tableau-based* approach (cf. [FL79], [Pr80], [BHP81], [BMP81], [Wo81], [EC82], [EH82], [LPZ85]), and show that it is optimal. The importance of a tableau-based procedure lies in the fact that, unlike the *automata-theoretic* approach (cf. [St81], [WVS83], [VW84], [Em85]), only the tableau-based method has been demonstrably extended to construct a small model of a satisfiable formula. For applications, such as automating synthesis of programs from their specifications, the model corresponds to the global flowgraph of the program, and the ability to generate it is crucial. (Note, therefore, that the alternative approach of deciding RTCTL by translating its formulae to logics—such as the μ-calculus—whose only known decision procedures are automata-theoretic does not suffice.) Thus, our algorithm is a basis for automating the synthesis of programs with timing-constraints, and we expect that the overall synthesis method would be similar to the ones described in [EC82] and [MW84].

The rest of this paper is organized as follows. In the next section, we present the logic RTCTL and some useful assertions expressible in it. Section 3 deals with real-time model-checking, and Section 4 with the satisfiability problem for RTCTL. Finally, Section 5 considers various other quantitative temporal logics derived from CTL.

2 The Logic RTCTL

The system of branching time temporal logic CTL (Computation Tree Logic) has been extensively used to specify and reason about correctness properties of concurrent programs (cf. [EC82], [EH82], [CES83]). One disadvantage of CTL and other extant temporal logics, however, is that they lack the ability to express properties of programs related to real-time. In this section, we define RTCTL (Real-Time CTL), an extension to CTL that permits reasoning about time-critical correctness properties of programs, and give a sample of the kinds of program properties RTCTL can express. We begin, however, with a formal definition of the syntax and semantics of CTL.

Let Σ be an underlying alphabet of atomic propositions P, Q, etc. The set of CTL (Computation Tree Logic) formulae is generated by the following rules:

S1. Each atomic proposition P is a formula.

S2. If p, q are formulae, then so are $p \wedge q$ and $\neg p$.

S3. If p, q are formulae, then so are $A(p\ U\ q)$, $E(p\ U\ q)$, and EXp.

A formula of CTL is interpreted with respect to a temporal structure $M = (S, R, L)$ where S is a set of states, R is a binary relation on S that is total (so each state has at least one successor), and L is a labelling which assigns to each state a set of atomic propositions, those intended to be true at the state. Intuitively, the states of a structure could be thought of as corresponding to the states of a concurrent program, the state transitions of which are specified by the binary relation R. A *fullpath* $x = s_0, s_1, s_2, \ldots$ in M is an infinite sequence of states such that $(s_i, s_{i+1}) \in R$ for each i; intuitively, a fullpath captures the notion of an execution sequence. We write $M, s \models p$ to mean that "formula p is true at state s in structure M". When M is understood we write only $s \models p$. We define $\models$ by induction on formula structure:

S1. $s_0 \models P$ iff P is an element of $L(s_0)$

S2. $s_0 \models p \wedge q$ iff $s_0 \models p$ and $s_0 \models q$
    $s_0 \models \neg p$ iff it is not the case that $s_0 \models p$

S3. $s_0 \models A(p\ U\ q)$ iff for all fullpaths $s_0, s_1, s_2, \ldots$ in M, $\exists i \geq 0$ such that $s_i \models q$ and $\forall j, 0 \leq j < i, s_j \models p$
    $s_0 \models E(p\ U\ q)$ iff for some fullpath $s_0, s_1, s_2, \ldots$ in M, $\exists i \geq 0$ such that $s_i \models q$ and $\forall j, 0 \leq j < i, s_j \models p$
    $s_0 \models EXp$ iff there exists an R-successor t of s_0 such that $t \models p$

The other propositional connectives are defined as abbreviations in the usual way. Other basic modalities of CTL are also defined as abbreviations: AFq abbreviates $A(true\ U\ q)$, EFq abbreviates $E(true\ U\ q)$, AGq abbreviates $\neg EF\neg q$, EGq abbreviates $\neg AF\neg q$, and AXq abbreviates $\neg EX\neg q$.

We now consider some examples of CTL formulae useful to describe qualitative temporal properties of programs. $AF q$, for example, specifies the *inevitability* of q: q must eventually hold along all paths. Thus, $AG(p \Rightarrow AF q)$ says that p *inevitably leads-to* q: q eventually holds along every path stemming from a state at which p is true. Similarly, $EF q$ indicates that q could *potentially* become true: it is true along some one fullpath. Note that none of these modalities allows one to express that q will in fact become true within a certain number, say 10, of state transitions: they merely assert that q will eventually become true.

So we extend CTL to RTCTL. The set of RTCTL (Real-Time Computation Tree Logic) formulae is generated by the rules S1–S3 above together with the rule:

S4. If p, q are formulae and k is any natural number, then so are $A(p\ U^{\leq k}\ q)$ and $E(p\ U^{\leq k}\ q)$.

The temporal structures over which RTCTL formulae are interpreted are the same as CTL structures. The semantics of the new RTCTL modalities are given by:

S4. $s_0 \models A(p\ U^{\leq k}\ q)$ iff for all fullpaths $s_0, s_1, s_2, \ldots$ in M, $\exists i, 0 \leq i \leq k$, such that $s_i \models q$ and $\forall j, 0 \leq j < i, s_j \models p$
    $s_0 \models E(p\ U^{\leq k}\ q)$ iff for some fullpath $s_0, s_1, s_2, \ldots$ in M, $\exists i, 0 \leq i \leq k$, such that $s_i \models q$ and $\forall j, 0 \leq j < i, s_j \models p$

Intuitively, k corresponds to the maximum number of permitted transitions along a path of a structure before the eventuality $p \, U \, q$ holds. We follow the convention that each transition takes unit time for execution (but see the remark near the end of Section 3), so k specifies a time bound.

Some other basic modalities of RTCTL are defined as abbreviations: $AF^{\leq k} q$ abbreviates $A(true \, U^{\leq k} q)$ and $EF^{\leq k} q$ abbreviates $E(true \, U^{\leq k} q)$. We also define the modality $G^{\leq k}$ (for each natural number k) as the dual of $F^{\leq k}$, i.e., $AG^{\leq k} p$ abbreviates $\neg EF^{\leq k} \neg p$, and $EG^{\leq k} p$ abbreviates $\neg AF^{\leq k} \neg p$.

It is worth pointing that the RTCTL modalities elegantly generalize the analogous CTL ones. Specifically, note that $A(p \, U \, q)$ abbreviates $\exists k : A(p \, U^{\leq k} q)$, and, similarly, $E(p \, U \, q)$ abbreviates $\exists k : E(p \, U^{\leq k} q)$. This motivates the following definition: If $A(p \, U \, q)$ is true at a state s of an RTCTL structure, we define the *rank* of $A(p \, U \, q)$ at s as the smallest natural number k such that $A(p \, U^{\leq k} q)$ holds at s. The rank of $AF \, q$, $E(p \, U \, q)$, and $EF \, q$ are defined similarly.

As usual, an RTCTL formula is said to be *valid* if it holds at all states of all structures. From the semantics above, it is easy to verify that the RTCTL formulae $A(p \, U^{\leq k} q) \equiv (q \vee (p \wedge AXA(p \, U^{\leq (k-1)} q)))$ and $E(p \, U^{\leq k} q) \equiv (q \vee (p \wedge EXE(p \, U^{\leq (k-1)} q)))$ are valid for $k \geq 1$. Also $A(p \, U^{\leq 0} q) \equiv q \equiv E(p \, U^{\leq 0} q)$ is valid. These formulae may be regarded as the analogues of the fixpoint characterizations of the CTL modalities AU and EU ([EH82]).

We conclude this section by illustrating how the basic RTCTL modalities could be used to express important correctness properties of programs that must place an explicit bound on the time between events. First, observe that $AF^{\leq k} q$, for example, specifies the *bounded inevitability* of q, i.e., q must hold within k steps along all fullpaths. Thus, the RTCTL formula $AG(p \Rightarrow AF^{\leq k} q)$ specifies that p always leads-to q within a bounded period of time, viz., k time units. This formula is therefore useful to specify, for example, that a system must respond (with the action q) to an environmental stimulus p within k units of time; the importance of specifications of this kind for temporal systems is underscored in [JM87].

As a second example, consider a family of m processes, the schedules of which are required to satisfy the property of k-*bounded fairness*, i.e., each process should be scheduled for execution at least once every k steps of the system. This can be expressed by the RTCTL formula $\bigwedge_{i=1}^{m} AF^{\leq k} P_i \wedge \bigwedge_{i=1}^{m} AG(P_i \Rightarrow AXAF^{\leq (k-1)} P_i)$, where P_i indicates that process i is executed. The first set of conjuncts ensures that each process is in fact executed along the first k steps, and the AG conjuncts ensure that, once executed, a process must be scheduled for execution again within k steps. We may remark, as an aside, that the property that there be at least one k-bounded fair execution sequence is expressed by the formula $E(\bigwedge_{i=1}^{m} F^{\leq k} P_i \wedge \bigwedge_{i=1}^{m} G(P_i \Rightarrow XF^{\leq (k-1)} P_i))$, which does not conform to the syntax of RTCTL.

As a third and final example, consider a system specification that requires that, on sensing an alarm, all normal processes be suspended, and a vigilant mode be entered for at least the next k time units during which only a restricted set of critical activities is performed. The RTCTL formula $AG(alarm \Rightarrow AG^{\leq k} vigilant)$ expresses this requirement.

3   Real-Time Model-Checking

In this section, we present an algorithm for the model-checking problem for RTCTL that is linear in both the size of the structure being checked as well as the length of the input formula. Because of the simplicity of the model-checking problem and the efficiency of its solution, the model-checking approach has found several applications in the automatic verification of temporal systems. So far, model-checking algorithms for several temporal logics (cf. [CES83], [LP85], [EL85], [CG87], [SG87]) have been used to verify a large number of finite-state systems ranging from examples of concurrent programs presented in the academic literature (such as the mutual exclusion example in [OL82]) and network communication protocols ([Sif87]) to VLSI circuits ([Br86]). The capability of RTCTL to allow one to reason quantitatively about time in addition to the qualitative reasoning afforded by CTL can only enhance the utility of this problem to such applications, for timing constraints play a key role in both network protocols and hardware circuits.

The way model-checking is applied to program verification may be summarized as follows. The global state transition graph of a finite-state concurrent system may be viewed as a finite temporal structure, and a correctness specification whose truth is to be determined of the program is expressed as a formula in RTCTL. The model-checking algorithm is used to determine whether the formula is true in the structure, and, thereby, whether the given finite-state program meets a particular correctness specification. It is easily appreciated that this approach is potentially of wide applicability since a large class of concurrent programming problems have finite-state solutions, and the interesting properties of many such systems can be specified in a propositional temporal logic.

Formally, the model-checking problem for RTCTL may be stated as: *Given an RTCTL formula p_0 and a finite temporal structure $M = (S, R, L)$, is there a state $s \in S$ such that $M, s \models p_0$?* (Note that the RTCTL structure is said to be finite if its *size*, $|M|$, defined as $|S| + |R|$, is finite.)

Fig. 1 presents an algorithm that decides this problem. The goal is to determine, for each state s in M, whether $M, s \models p_0$. The algorithm is designed to operate in stages: the first stage processes all subformulae of p_0 of length 1, the second, of length 2, and so on. At the end of the ith stage, each state is labelled with the set of all subformulae of p_0 of length no more than i that are true at the state. As the basis, note that the labelling L of M initially contains the set of atomic propositions (i.e., all subformulae of p_0 of length 1) true at each state of M. To perform the labelling on subsequent iterations, information gathered in earlier iterations is used. For example, a subformula of the form $q \wedge r$, i.e., one whose main connective is $\wedge$, should be added to the labels of precisely those states already labelled with both q and r. Subformulae of the form $\neg q$ are handled in like fashion.

For the modal subformula $A(q \, U^{\leq k} \, r)$, information from the successor states of s as well as that from the state s itself is used. For now assume that the procedure AU_check is always invoked with k instead of $\min(k, |S|)$. Since $A(q \, U^{\leq k} \, r) \equiv r \vee (q \wedge AXA(q \, U^{\leq (k-1)} r))$, $A(q \, U^{\leq k} \, r)$ is initially added to the label of each state already labelled with r. The satisfaction of $A(q \, U^{\leq k} \, r)$ is then propagated outward, by repeatedly adding $A(q \, U^{\leq k} \, r)$ to the label of each state labelled by q and having $A(q \, U^{\leq (k-1)} r)$ in the labels of all successors. It is fairly easy to see that this propagating step need be repeated at most k times, and that states labelled on the ith step actually satisfy $A(q \, U^{\leq i} r)$. Finally, if a state s satisfies $A(q \, U^{\leq i} r)$ for some $i \leq k$, it also satisfies $A(q \, U^{\leq k} \, r)$; hence, the last foreach loop in AU_check adds $A(q \, U^{\leq k} \, r)$ to the labels of such states.

However, if AU_check is invoked with k instead of $\min(k, |S|)$, its complexity would be linear in k. Since k is represented in binary, rather than unary, the complexity of the algorithm would be *exponential* in the length of the binary representation of k, and, hence, in the length of p_0. To overcome this, the invocation to AU_check is made with the minimum of k and $|S|$. To see why this suffices, consider $k > |S|$. Note that for $A(q \, U^{\leq k} \, r)$ to hold at state s, there should not be a fullpath stemming from s along which r never holds. Since M is finite, any such fullpath must contain a loop. Thus, it suffices to check if $A(q \, U^{\leq k} \, r)$ holds along the loop-free initial segments of all paths out of s, and such segments have length at most $|S|$. Hence, it suffices to perform the iteration in the procedure AU_check just $|S|$ times. For much the same reason, to determine if the formula $A(q \, U \, r)$ holds at a state, it suffices to determine if $A(q \, U^{\leq |S|} r)$ holds there. The modalities EU and $EU^{\leq k}$ are handled similarly.

We note that this version of the algorithm can be naively implemented to run in time linear in the length of p_0 and quadratic in the size of the structure M. This is apparent for each of the cases when the main connective of the formula is one of AU, EU, $AU^{\leq k}$ or $EU^{\leq k}$. In the other three cases, the procedure is in fact linear in the size of the structure. However, the techniques of [EL88] are applicable here as well, and can be used to implement the algorithm in time linear in the sizes of both the structure and the formula, i.e., the complexity of the algorithm is $O(|p_0| \times |M|)$. We shall explain this further in the full paper. Thus, we have:

Theorem 1 *The model-checking problem for RTCTL is decidable in time linear in both the size of the input structure as well as the length of the input formula.* □

We should remark that, with minor modifications to the procedures AU_check and EU_check, the above algorithm can as efficiently handle more general temporal structures, ones in which each element of the binary relation R is associated with an integer label that intuitively corresponds to the amount of time taken to "execute" that transition. RTCTL structures as defined in the previous section may be thought of as labelling each element of R with a single unit of time.

As evident from the algorithm, the basic idea behind this mechanical model-checking approach to verification of finite-state systems is to make brute force graph reachability analysis efficient and expressive through the use of temporal logic as an assertion language. Of course, much research in protocol verification—to cite just one area—has attempted to exploit the fact that protocols are frequently finite-state, making exhaustive graph reachability analysis possible. The advantage offered by model-checking seems to be that it provides greater flexibility in formulating specifications through the use of temporal logic as a single, uniform assertion language that can express a wide variety of correctness properties. This makes it possible to reason about, for example, both safety and liveness properties with equal facility. And, now, with RTCTL, quantitative assertions can be handled.

4 Satisfiability for RTCTL

We now turn to the problem of determining the satisfiability of an RTCTL formula. This problem may be stated as: *Given an RTCTL formula f, is there a temporal structure M and a state s of M such that $M, s \models f$?* If f is true at

```
/* Input:   A structure M = (S, R, L) and an RTCTL formula p₀.   */
/* Output: There is a state s ∈ S such that M, s ⊨ p₀.           */

for i := 1 to length(p₀) do begin
    foreach subformula p of p₀ of length i do begin
        case structure of p is of the form
            P, an atomic proposition : /* Nothing to do as states of M already labelled with propositions. */;
            q ∧ r       :  foreach s ∈ S do
                               if q ∈ L(s) and r ∈ L(s) then add q ∧ r to L(s);
            ¬q          :  foreach s ∈ S do
                               if q ∉ L(s) then add ¬q to L(s);
            EX q        :  foreach s ∈ S do
                               if q ∈ L(t) for some R-successor t of s then add EX q to L(s);
            A(q U^≤ᵏ r) :  AU_check (min(k, |S|), q, r);
            A(q U r)    :  AU_check (|S|, q, r);
            E(q U^≤ᵏ r) :  EU_check (min(k, |S|), q, r);
            E(q U r)    :  EU_check (|S|, q, r);
        end; /* case */
    end;    /* foreach */
end;        /* for */
if p₀ ∈ L(s) for some s ∈ S then Output (true)
                              else  Output (false);
```

procedure AU_check (maxrank, q, r);

```
begin
    foreach s ∈ S do
        if r ∈ L(s) then add A(q U^≤⁰ r) to L(s);
    for rank := 1 to maxrank do
        foreach s ∈ S do
            if q ∈ L(s) and A(q U^≤(rank−1) r) ∈ L(t) for every R-successor t of s then
                add A(q U^≤rank r) to L(s);
    foreach s ∈ S do
        if A(q U^≤ʲ r) ∈ L(s) for some j ≤ maxrank then add A(q U^≤maxrank r) to L(s);
end;  /* AU_check */
```

procedure EU_check (maxrank, q, r);

```
begin
    foreach s ∈ S do
        if r ∈ L(s) then add E(q U^≤⁰ r) to L(s);
    for rank := 1 to maxrank do
        foreach s ∈ S do
            if q ∈ L(s) and E(q U^≤(rank−1) r) ∈ L(t) for some R-successor t of s then
                add E(q U^≤rank r) to L(s);
    foreach s ∈ S do
        if E(q U^≤ʲ r) ∈ L(s) for some j ≤ maxrank then add E(q U^≤maxrank r) to L(s);
end;  /* EU_check */
```

Figure 1: A Model-Checking Algorithm for RTCTL.

state s of M, M is said to be a *model* of f. Note that the RTCTL formula f is satisfiable iff $\neg f$ is not valid; hence exhibiting a decision procedure for satisfiability amounts to deciding the validity problem (i.e., determining if a given RTCTL formula is valid) as well.

The satisfiability problem for temporal logics has been shown to have applications to synthesis of concurrent programs from their temporal specifications (cf. [EC82], [MW84], [ESS89], [PR89]). The method determines whether the temporal logic formula expressing the program specifications is satisfiable, and, if so, produces a model of the formula. The model may be viewed as the global flowgraph of a program implementing the specifications, and the program itself can be read off from the model. If the formula is not satisfiable, the specification is *inconsistent*: there is no program that implements it.

As mentioned in the introduction, only the tableau-based approach has been demonstrably extended to produce actual models of satisfiable formulae. Thus, we seek a tableau-based algorithm to decide the satisfiability of RTCTL formulae. A naive way to do this is to translate the given RTCTL formula, f, to an equivalent CTL one, g, by using the fixpoint characterizations of the $AU^{\leq k}$ and $EU^{\leq k}$ modalities to expand each occurrence of these modalities in f. The tableau-based decision procedure for CTL could then be used to determine the satisfiability of g. But the complexity of such an algorithm would be double exponential in $|f|$, as $|g|$ itself would be exponential in $|f|$, and the CTL decision procedure is exponential in the length of its input.

Instead, we outline a direct tableau-based decision procedure for RTCTL, whose complexity is only exponential in the size of its input. Let f be the RTCTL formula whose satisfiability needs to be determined. We first define several useful notions used in the description of the procedure, beginning with the *Fischer-Ladner closure*, $CL(f)$, of an RTCTL formula f (cf. [FL79], [EH82], [LPZ85]). For conciseness of presentation, we assume that f is strictly in the syntax presented, i.e., it does not have any of the abbreviations listed in Section 2. Identifying $\neg\neg p$ with p, and $A(p\,U^{\leq 0}q)$ and $E(p\,U^{\leq 0}q)$ with q for any RTCTL formulae p and q, $CL(f)$ is the smallest set of formulae containing f and satisfying the following eight conditions:

A. $\neg p \in CL(f)$ $\quad\Leftrightarrow\quad p \in CL(f)$,
B. $p \wedge q \in CL(f)$ $\quad\Rightarrow\quad p, q \in CL(f)$,
C. $EX\,p \in CL(f)$ $\quad\Rightarrow\quad p \in CL(f)$,
D. $AX\,p \in CL(f)$ $\quad\Rightarrow\quad p \in CL(f)$,
E. $A(p\,U\,q) \in CL(f)$ $\quad\Rightarrow\quad p, q, AXA(p\,U\,q) \in CL(f)$,
F. $E(p\,U\,q) \in CL(f)$ $\quad\Rightarrow\quad p, q, EXE(p\,U\,q) \in CL(f)$,
G. $A(p\,U^{\leq k}q) \in CL(f) \Rightarrow p, q, AXA(p\,U^{\leq(k-1)}q) \in CL(f)$ for $k \geq 1$, and
H. $E(p\,U^{\leq k}q) \in CL(f) \Rightarrow p, q, EXE(p\,U^{\leq(k-1)}q) \in CL(f)$ for $k \geq 1$.

Note that the size of $CL(f)$ is exponential in $|f|$. We shall call a formula in $CL(f)$ *elementary* if it is of the form $EX\,p$ or $AX\,p$. We define a subset S of $CL(f)$ to be *maximally consistent* iff S satisfies all the following conditions:

1. For each $p \in CL(f)$, $\neg p \in S \Leftrightarrow p \notin S$,
2. $p \wedge q \in S$ $\quad\Leftrightarrow\quad p, q \in S$,
3. $A(p\,U\,q) \in S$ $\quad\Leftrightarrow\quad q \in S$ or $p, AXA(p\,U\,q) \in S$,
4. $E(p\,U\,q) \in S$ $\quad\Leftrightarrow\quad q \in S$ or $p, EXE(p\,U\,q) \in S$,
5. $A(p\,U^{\leq k}q) \in S$ $\quad\Leftrightarrow\quad q \in S$ or $p, AXA(p\,U^{\leq(k-1)}q) \in S$ for $k \geq 1$,
6. $E(p\,U^{\leq k}q) \in S$ $\quad\Leftrightarrow\quad q \in S$ or $p, EXE(p\,U^{\leq(k-1)}q) \in S$ for $k \geq 1$,
7. $A(p\,U^{\leq 0}q) \in S$ $\quad\Leftrightarrow\quad q \in S$,
8. $E(p\,U^{\leq 0}q) \in S$ $\quad\Leftrightarrow\quad q \in S$,
9. $A(p\,U^{\leq k}q) \in S$ $\quad\Rightarrow\quad$ for all $j \geq k$ such that $A(p\,U^{\leq j}q) \in CL(f)$, $A(p\,U^{\leq j}q) \in S$, and
10. $E(p\,U^{\leq k}q) \in S$ $\quad\Rightarrow\quad$ for all $j \geq k$ such that $E(p\,U^{\leq j}q) \in CL(f)$, $E(p\,U^{\leq j}q) \in S$.

We now show that the number of maximally consistent subsets of $CL(f)$ is only exponential in $|f|$. An *eventuality* is any formula of the form $A(p\,U\,q)$, $E(p\,U\,q)$, $A(p\,U^{\leq k}q)$, or $E(p\,U^{\leq k}q)$. We shall call a formula in $CL(f)$ *quantitative* if it is of the form $A(p\,U^{\leq k}q)$, $AXA(p\,U^{\leq k}q)$, $E(p\,U^{\leq k}q)$, or $EXE(p\,U^{\leq k}q)$. We let $\mathcal{H}$ denote the set of quantitative eventualities that appear in f as subformulae. We partition the positive formulae (i.e., formulae not of the form $\neg p$) in $CL(f)$ into $|\mathcal{H}| + 1$ sets: each quantitative eventuality $H = A(p\,U^{\leq k_H}q)$ (respectively, $H = E(p\,U^{\leq k_H}q)$) that appears in f has a corresponding partition, Y_H, which contains all formulae in $CL(f)$ of the form $A(p\,U^{\leq j}q)$ and $AXA(p\,U^{\leq j}q)$ (respectively, $E(p\,U^{\leq j}q)$ and $EXE(p\,U^{\leq j}q)$), where $j \leq k_H$. All other positive formulae in $CL(f)$ are members of a separate partition, Y_0. It is easy to see that $|Y_0|$ is linear in $|f|$, whereas, for any $H \in \mathcal{H}$, $|Y_H|$ is exponential in the number of bits in k_H, and hence, exponential in $|f|$.

Next, we note from Rule 1 above that in constructing any maximally consistent set S, we have two choices for each formula in Y_0: either include it in S or include its negation in S. For the formulae in Y_H, however, Rules 9 and 10 imply that we can effectively choose only one j, viz., the smallest one, which is no more than k_H, such that $A(p\ U^{\leq j}\ q)$ or $E(p\ U^{\leq j}\ q)$ is in S. Also, once this choice of the smallest j is made, Rules 5 and 6 determine the quantitative elementary formulae of H that must appear in S. Thus, the number of distinct maximally consistent sets is of the order of $2^{|Y_0|} \times \prod_{H \in \mathcal{H}}(k_H + 2)$, i.e., of the order of $2^{|f|}$. Note, also, that the number of elements in a maximally consistent set is also exponential in $|f|$.

The decision procedure we outline focusses on establishing a *small-model theorem* for RTCTL, i.e., on showing that each satisfiable RTCTL formula f has a model that is bounded by some small function of its length. The first step in the procedure is to construct the *initial tableau* for f. This tableau, which we denote by T_0, is a directed graph, whose nodes correspond to the maximally consistent sets of $CL(f)$. A node corresponding to the set S is labelled with the formulae in S. We use the elementary formulae in a node to guide us in determining the edges of T_0. An edge is added from node V to node W iff (a) for every formula of the form $AX\ p$ in V, p is in W, and (b) for every formula of the form $\neg EX\ p$ in V, $\neg p$ is in W.

The next step is to prune T_0 by deleting nodes the conjunction of the formulae in whose labels cannot ever label any state of any temporal structure. Despite the fact that RTCTL has more kinds of eventualities than CTL, this step is *identical* to the pruning step for CTL (cf. [EH82]). The main task of the pruning step in the CTL algorithm is to check for each eventuality in the label of each node, whether there is a directed acyclic subgraph of the tableau for that eventuality rooted at that node which certifies fulfillment of that eventuality at that node. For RTCTL, it would appear that such a directed acyclic subgraph would need to be detected for the quantitative eventualities in the label of a node as well; however, this is not required because the local structure of the tableau (i.e., the way the initial tableau is constructed) guarantees that such an acyclic subgraph can always be found. We shall describe the pruning procedure for CTL in the full paper and prove:

Proposition 2 *The above algorithm decides the satisfiability of its input RTCTL formula f correctly and in time* $O(2^{|f|})$. □

Thus, we have a deterministic decision procedure for RTCTL whose complexity is at most exponential in the length of f. Since the problem of determining the satisfiability of CTL formulae is deterministic exponential time complete ([EH82]), and since RTCTL subsumes CTL, our algorithm is optimal. Also, note that the techniques in [EC82] and [EH82] to construct the initial tableau "bottom-up" are applicable to RTCTL as well. Thus the exponential blow-up in $|f|$ need be incurred only in the worst case, rather than in the average case as would be done by the above naive construction of the initial tableau.

5 Other Quantitative Modalities and Temporal Logics

In this section, we briefly consider two other quantitative temporal modalities: $U^{\geq k}$ and $U^{=k}$. Intuitively, $A(p\ U^{\geq k}\ q)$ says that q is true after k or more time instants along each fullpath and p is true till then. Similarly, $A(p\ U^{=k}\ q)$ states that q is true exactly at the kth time instant along all fullpaths and p is true at each of the preceding $k-1$ time instants. More formally, we define the logic CRTCTL (Complete RTCTL) to comprise the formulae generated by the rules S1–S4 in Section 2 together with the rules:

S5. If p, q are formulae and k is any natural number, then so are $A(p\ U^{\geq k}\ q)$ and $E(p\ U^{\geq k}\ q)$, and

S6. If p, q are formulae and k is any natural number, then so are $A(p\ U^{=k}\ q)$ and $E(p\ U^{=k}\ q)$.

We also define two sublogics of CRTCTL: RTCTL$^{\geq}$, whose formulae are obtained by using Rules S1–S3 and S5, and RTCTL$^{=}$, whose formulae are generated by Rules S1–S3 and S6.

The semantics of the new quantitative modalities are given by:

S5. $s_0 \models A(p\ U^{\geq k}\ q)$ iff for all fullpaths $s_0, s_1, s_2, \ldots$ in M, $\exists i, i \geq k$, such that $s_i \models q$ and $\forall j, 0 \leq j < i, s_j \models p$
$s_0 \models E(p\ U^{\geq k}\ q)$ iff for some fullpath $s_0, s_1, s_2, \ldots$ in M, $\exists i, i \geq k$, such that $s_i \models q$ and $\forall j, 0 \leq j < i, s_j \models p$

S6. $s_0 \models A(p\ U^{=k}\ q)$ iff for all fullpaths $s_0, s_1, s_2, \ldots$ in M, $s_k \models q$ and $\forall j, 0 \leq j < k, s_j \models p$
$s_0 \models E(p\ U^{=k}\ q)$ iff for some fullpath $s_0, s_1, s_2, \ldots$ in M, $s_k \models q$ and $\forall j, 0 \leq j < k, s_j \models p$

Other abbreviations of these modalities can be defined as in Section 2. It is worth noting that $AF^{=k}p \equiv AG^{=k}p$ is a validity. So is $A(p\,U^{=k}\,q) \Rightarrow A(p\,U^{\leq k}\,q) \wedge A(p\,U^{\geq k}\,q)$, but the same formula with the implication reversed is not valid. We could also define modalities such as $U^{<k}$ and $U^{>k}$, but as we are dealing with discrete time, these are easily expressed in terms of $U^{\leq k}$ and $U^{\geq k}$ respectively.

From the semantics above, it is easy to verify that the following formulae are valid. First, for each $k \geq 1$: (i) $A(p\,U^{\geq k}\,q) \equiv p \wedge AXA(p\,U^{\geq (k-1)}\,q)$; (ii) $E(p\,U^{\geq k}\,q) \equiv p \wedge EXE(p\,U^{\geq (k-1)}\,q)$; (iii) $A(p\,U^{=k}\,q) \equiv p \wedge AXA(p\,U^{=(k-1)}\,q)$; and (iv) $E(p\,U^{=k}\,q) \equiv p \wedge EXE(p\,U^{=(k-1)}\,q)$. Secondly, for $k = 0$: (v) $A(p\,U^{\geq 0}\,q) \equiv A(p\,U\,q)$; (vi) $E(p\,U^{\geq 0}\,q) \equiv E(p\,U\,q)$; and (vii) $A(p\,U^{=k}\,q) \equiv E(p\,U^{=k}\,q) \equiv q$. These formulae may be regarded as the analogues of the fixpoint characterizations of the CTL modalities AU and EU ([EH82]).

We conclude this section with a summary of results (which we shall prove in the full paper) concerning the logics CRTCTL, RTCTL, RTCTL$^{\geq}$, and RTCTL$^{=}$. First, we note that the expressive power (cf., [GPSS80], [EH83]) of each of these logics is the same as that of CTL as each of them subsumes CTL and the basic quantitative modalities can be expanded into CTL formulae using their fixpoint characterizations. Thus, each of these logics is as expressive as every other.

Secondly, there is a polynomial time model-checking algorithm for each of these logics. In fact, an RTCTL$^{\geq}$ formula, p, can be model-checked over a structure M in time $O(|p| \times |M|^2)$ and an RTCTL$^{=}$ or a CRTCTL formula p can be model-checked in time $O(|p| \times |M|^3)$. The algorithm for RTCTL$^{\geq}$ is similar to that of RTCTL: to test whether a state s in a structure M satisfies $A(q\,U^{\geq k}\,r)$, say, we first rank all states in M that satisfy $A(q\,U\,r)$ with 0, and "radiate" the satisfaction of $A(q\,U^{\geq j}\,r)$, $j \leq k$, outward from these states. To test if s satisfies $A(q\,U^{=k}\,r)$, however, we need to compute all states in M that are k time units away from s. This can be done in time linear in k and polynomial in $|M|$ by an algorithm similar to the one that computes the transitive closure of a directed graph. So, model-checking CRTCTL and RTCTL$^{=}$ is somewhat more computationally intensive.

Finally, the satisfiability problem of RTCTL$^{\geq}$, like that of RTCTL is deterministic EXPTIME-complete; in fact, RTCTL's algorithm can be used with appropriate changes. Surprisingly enough, however, the satisfiability problem of RTCTL$^{=}$ (and, hence, of CRTCTL) is deterministic double exponential time complete. The algorithm outlined for RTCTL can also be modified to handle these logics. The resulting algorithm is double exponential in the number of bits used to represent the integer constants in the input formula, but only single exponential in the length of the remainder of the formula, i.e., in the length of the formula without these integer constants. Thus, this algorithm is likely to be more efficient than translating formulae of these logics to (exponentially longer) formulae in CTL and using CTL's decision procedure to determine their satisfiability.

References

[Ab80]   Abrahamson, K., Decidability and Expressiveness of Logics of Processes, Ph.D. Thesis, Univ. of Washington, 1980.

[Br86]   Browne, M.C., An Improved Algorithm for the Automatic Verification of Finite State Systems Using Temporal Logic, *Proc. Symp. on Logic in Computer Science*, Cambridge, pp. 260–266, 1986.

[BHP81]   Ben-Ari, M., J.Y. Halpern, A. Pnueli, Finite Models for Deterministic Propositional Dynamic Logic, *Proc. 8th Annual International Colloquium on Automata, Languages and Programming*, LNCS#115, Springer-Verlag, pp. 249–263, 1981; a revised version entitled Deterministic Propositional Dynamic Logic: Finite Models, Complexity, and Completeness, appears in Journal of Computer and System Sciences, vol 25, no. 3, pp. 402–417, 1982.

[BMP81]   Ben-Ari, M., Z. Manna, A. Pnueli, The Temporal Logic of Branching Time, *Proc. 8th Annual ACM Symp. on Principles of Programming Languages*, Williamsburg, pp. 164–176, 1981; also appeared in Acta Informatica, vol. 20, no. 3, pp. 207–226, 1983.

[CES83]   Clarke, E.M., E.A. Emerson, A.P. Sistla, Automatic Verification of Finite State Concurrent Systems Using Temporal Logic Specifications: A Practical Approach, *Proc. 10th Annual ACM Symp. on Principles of Programming Languages*, Austin, pp. 117–126, 1983; also appeared in ACM Transactions on Programming Languages and Systems, vol. 8, no. 2, pp. 244–263, 1986.

[CG87]   Clarke, E.M., O. Grumberg, Avoiding the State Explosion Problem in Temporal Model Checking Algorithms, *Proc. of the 6th Annual ACM Symp. on Principles of Distributed Computing*, Vancouver, pp. 294–303, 1987.

[Di76]   Dijkstra, E.W., A Discipline of Programming, Prentice-Hall, 1976.

[Em85]   Emerson, E.A., Automata, Tableaux, and Temporal Logics, *Proc. Conf. on Logics of Programs*, Brooklyn, R. Parikh, editor, LNCS#193, Springer-Verlag, pp. 79–88, 1985.

[EC82]   Emerson, E.A., E.M. Clarke, Using Branching Time Logic to Synthesize Synchronization Skeletons, *Science of Computer Programming*, vol. 2, pp. 241–266, 1982.

[EH82]   Emerson, E.A., J.Y. Halpern, Decision Procedures and Expressiveness in the Temporal Logic of Branching Time, *Proc. of the 14th Annual ACM Symp. on Theory of Computing*, San Francisco, pp. 169–180, 1982; *also appeared in Journal of Computer and System Sciences*, vol 30, no. 1, pp. 1–24, 1985.

[EH83]   Emerson, E.A., J.Y. Halpern, "Sometimes" and "Not Never" Revisited: On Branching versus Linear Time, *Proc. 10th Annual ACM Symp. on Principles of Programming Languages*, Austin, pp. 127–140, 1983; *also appeared in Journal ACM*, vol 33, no. 1, pp. 151–178, 1986.

[EL85]   Emerson, E.A., C.L. Lei, Modalities for Model Checking: Branching Time Logic Strikes Back, *Proc. 12th Annual ACM Symp. on Principles of Programming Languages*, New Orleans, pp. 84–96, 1985; *also appeared in Science of Computer Programming*, vol. 8, pp. 275–306, 1987.

[EL88]   Emerson, E.A., C.L. Lei, Model-Checking in the Propositional Mu-Calculus, *unpublished manuscript*, 1988.

[ES84]   Emerson, E.A., A.P. Sistla, Deciding Full Branching Time Logic, *Information and Control*, vol. 61, no. 3, pp. 175–201, 1984; *also appeared in Proc. of the 16th Annual ACM Symp. on Theory of Computing*, Washington D.C., pp. 14–24, 1984.

[ESS89]   Emerson, E.A., T.H. Sadler, J. Srinivasan, Efficient Temporal Reasoning, *Proc. 16th Annual ACM Symp. on Principles of Programming Languages*, Austin, pp. 166–178, 1989.

[FL79]   Fischer, M.J., R.E. Ladner, Propositional Dynamic Logic of Regular Programs, *Journal of Computer and System Sciences*, vol. 18, pp. 194–211, 1979.

[GPSS80]   Gabbay, D., A. Pnueli, S. Shelah, J. Stavi, On the Temporal Analysis of Fairness, *Proc. 7th Annual ACM Symp. on Principles of Programming Languages*, Las Vegas, pp. 163–173, 1980.

[JM87]   Jahanian, F., A.K. Mok, A Graph-Theoretic Approach for Timing Analysis and its Implementation, *IEEE Transactions on Computers*, vol. C–36, no. 8, pp. 961–975, 1987.

[Ko82]   Kozen, D., Results on the Propositional μ-Calculus, *Proc. 9th Annual International Colloquium on Automata, Languages and Programming*, LNCS#140, Springer-Verlag, pp. 348–359, 1982; *also appeared in Theoretical Computer Science*, vol. 27, no. 3, pp. 333–354, 1983.

[LP85]   Lichtenstein, O., A. Pnueli, Checking That Finite State Concurrent Programs Satisfy Their Linear Specification, *Proc. 12th Annual ACM Symp. on Principles of Programming Languages*, New Orleans, pp. 97–107, 1985.

[LPZ85]   Lichtenstein, O., A. Pnueli, L. Zuck, The Glory of The Past, *Proc. Conf. on Logics of Programs*, Brooklyn, R. Parikh, editor, LNCS#193, Springer-Verlag, pp. 196–218, 1985.

[MW84]   Manna, Z., P. Wolper, Synthesis of Communicating Processes from Temporal Logic Specifications, *ACM Transactions on Programming Languages and Systems*, vol. 6, no. 1, pp. 68–93, 1984.

[OL82]   Owicki, S., L. Lamport, Proving Liveness Properties of Concurrent Programs, *ACM Transactions on Programming Languages and Systems*, vol. 4, no. 3, pp. 455–495, 1982.

[Pn77]   Pnueli, A., The Temporal Logic of Programs, *18th Annual Symp. on Foundations of Computer Science*, Providence, pp. 46–57, 1977.

[PR89]   Pnueli, A., R. Rosner, On the Synthesis of a Reactive Module, *Proc. 16th Annual ACM Symp. on Principles of Programming Languages*, Austin, pp. 179–190, 1989.

[Pr80]   Pratt, V., A Near-Optimal Method For Reasoning About Action, *Journal of Computer and System Sciences*, vol 20, no. 2, pp. 231–254, 1980.

[Sif87]   J. Sifakis, *personal communication*, 1987.

[SC82]   Sistla, A.P., E.M. Clarke, The Complexity of Propositional Linear Temporal Logics, *Proc. of the 14th Annual ACM Symp. on Theory of Computing*, San Francisco, pp. 159–168, 1982; *also appeared in Journal ACM*, vol. 32, no. 3, pp. 733–749, 1985.

[SG87]   Sistla, A.P., S.M. German, Reasoning With Many Processes, *Proc. 2nd Annual Symp. on Logic in Computer Science*, Ithaca, pp. 138–152, 1987.

[St81]   Streett, R.S., Propositional Dynamic Logic of Looping and Converse, Ph.D. Thesis, *MIT LCS Technical Report TR-263*, 1981; *alternatively, see:* Propositional Dynamic Logic of Looping and Converse is Elementarily Decidable, *Information and Control*, vol. 54, no. 2, pp. 121–141, 1982.

[VW84]   Vardi M., P. Wolper, Automata Theoretic Techniques for Modal Logics of Programs, *Proc. of the 16th Annual ACM Symp. on Theory of Computing*, Washington D.C., pp. 446–456, 1984; *also appeared in Journal of Computer and System Sciences*, vol 32, no. 2, pp. 183–221, 1984.

[Wo81]   Wolper, P., Temporal Logic Can Be More Expressive, *22nd Annual Symp. on Foundations of Computer Science*, Nashville, pp. 340–348, 1981; *also appeared in Information and Control*, vol. 56, pp. 72–99, 1983.

[WVS83]   Wolper, P., M. Vardi, A.P. Sistla, Reasoning about Infinite Computation Paths, *24th Annual Symp. on Foundations of Computer Science*, Tucson, pp. 185–194, 1983.

USING PARTIAL-ORDER SEMANTICS
TO AVOID
THE STATE EXPLOSION PROBLEM
IN
ASYNCHRONOUS SYSTEMS

David K. Probst and Hon F. Li
Department of Computer Science
Concordia University
1455 de Maisonneuve Blvd. West
Montreal, Quebec Canada H3G 1M8

ABSTRACT

We avoid state explosion in model checking of delay-insensitive VLSI systems by not using states. Systems are networks of communicating finite-state nonsequential processes with well-behaved nondeterministic choice. A specification strategy based on partial orders allows precise description of the _branching_ and _recurrence_ structure of processes. Process behaviors are modelled by pomsets, but (discrete) sets of pomsets with implicit branching structure are replaced by pomtrees, which have finite presentations by (automaton-like) _behavior_ _machines_. The latter distinguish both concurrency and branching points, and define a finite recurrence structure. Safety and liveness checking are integrated. In contrast to state methods, our methods do not require enumeration or recording of states. We avoid separate consideration of execution sequences that do not differ in their partial order, and ensure termination by recording only a small number of system loop cutpoints -- in the form of system _behavior_ _states_. In spite of the name, behavior states are not states.

Keywords delay-insensitive system, model checking, state explosion, partial-order semantics, branching point, recurrence structure, behavior machine, behavior state.

1. Introduction

There is considerable interest in asynchronous hardware systems, fueled by concerns about clock distribution and component composition in clocked systems [1,3,6,8]. Clearly, there is considerable conceptual overlap between asynchronous hardware systems and asynchronous distributed systems. Here, we focus on hardware systems, although we make our verification assumptions clear, as a first step towards extending the work to other application areas. Most theoretical asynchronous systems research is based on a formal representation strategy that underlies efforts in both verification and synthesis, and can have a major impact on efficiency (for example, on the time needed for verification, or on the size of synthesized objects). In delay-insensitive VLSI systems, system correctness is independent of delays in both asynchronous circuit components and transmission media. Both are specified as asynchronous processes. Delay-insensitive systems are modelled as networks of processes that communicate by direct contact. Since all communication is asynchronous and unbuffered, only local protocols are available to eliminate undesirable inputs. We avoid state explosion in model checking of delay-insensitive VLSI systems by not using states. Our specification strategy allows precise description of the _branching_ and _recurrence_ structure of processes. Processes are modelled by pomtrees, which differ from (discrete) sets of pomsets in making implicit branching structure explicit. Pomtrees have finite presentations by (automaton-like) _behavior_ _machines_ that distinguish both concurrency and branching points, and define a

This research was supported by the Natural Sciences and Engineering Research Council of Canada under grants A3363, A0921 and MEF0040121.

finite recurrence structure. Our slogan is, combine true concurrency with true nondeterminism -- and a finite recurrence structure -- to achieve efficient algorithmic processability. Our model checking strategy is, evaluate graph predicates on a system pomtree rather than state predicates on a system state graph. A small set of loop cutpoints in a finite presentation of the system pomtree is discovered during model checking. The core novelty of our approach is that (i) causality is checked directly, and (ii) behavior states (which are not states) are used only for termination.

2. Abstract specification of asynchronous processes

Abstract specifications define sets of externally-visible infinite computational behaviors, as well as branching points and a finite recurrence structure. Our behaviors are called complete to emphasize that they correspond to some maximal safe use of the process by an environment. In event structure terms, a complete behavior is a maximal conflict-free set of events. A process P has a set of input ports I and a set of output ports O. A process action is a (port, token) pair -- in VLSI, tokens correspond to signal (voltage) transitions. tI (tO) is the set of input (output) actions, and Σ = tI $\cup$ tO is the set of process actions. An event at an input or output port performs a process action defined by the value of the token received or sent. Since the focus in this paper is on systems with cleanly defined control states that are verified separately, tokens are essentially colorless. P's input actions are under the exclusive control of P's environment. P's output actions are under the exclusive control of P. This asymmetry of control is central to the model.

Safety properties (invariance properties) specify what the process is allowed to do; they also specify what the environment is allowed to' do. A safety violation is an occurrence of a proscribed input or output event (in automata-theoretic terms, an undefined transition). After an input safety violation, a process becomes undefined. Liveness properties (inevitability properties) specify what the process is required to do; in our model, they do not specify what the environment is required to do. A liveness violation is a nonoccurrence of a prescribed output event. We group liveness properties into two camps, viz., (1) those related to progress (that is, response), and (2) those related to fairness of conflict resolution. Symmetric (asymmetric) specification of safety (liveness) properties has an interesting structuring effect on model checking (the examination of a special closed network of processes): although safety violations can show up in any process, liveness violations can only show up in the specification.

2.1. Primitive notions

At the level of (infinite) pomtrees, there are two primitive notions: (1) complete behavior (maximal conflict-free set of events), that is, any maximal infinite path through the pomtree, and (2) nondeterministic choice between mutually exclusive sets of process actions, by either process or environment, that is, selection of a particular pomtree branch. Complete behaviors are abstractions of infinite executions of the process that (i) correspond to some maximal safe use, and (ii) contain only necessary temporal precedences between external (interface) events. Concurrent events have no specified temporal relationship. A behavior "contains" any execution state that arises in any execution abstracted by the behavior. Execution state is affected by the performance of process actions in the usual way. P is input nondeterminate when P's environment can choose; P is output nondeterminate when P can choose.

At the level of (finite) behavior machines, there is a third primitive notion: recurrence (looping) in behavior space. Actually, the major precondition for the application of our techniques is the existence of a clean, finite recurrence structure. Behavior machines describe how commands (socket-extended finite pomsets) define transitions between selected pairs of behavior states. The essential use of behavior states is to identify cutpoints in loops of computational behaviors. A behavior state contains all the information in an execution state plus some additional information about how events in the past are temporally related to events in the future. A behavior state is not a state in the sense that recording system execution states is not sufficient to discover a finite presentation of the system pomtree.

2.2. Pomsets and pomtrees

Pomsets and pomset operations have been studied extensively [4]. A labelled partial order (lpo) is a 4-tuple (V, Σ, Γ, μ) consisting of (i) a countable set V of events in a computational behavior, (ii) a finite set Σ of process actions, (iii) a partial order Γ on V that expresses the necessary temporal precedences among the events in V, and (iv) a labelling function $\mu : V \to \Sigma$ mapping each event $v \in V$ to the process action $\sigma \in \Sigma$ it performs. In our model, events at the same port in a given behavior are linearly ordered (concurrent events must be at distinct ports). Since every process has an initial state in which no events have occurred, and since the partial order Γ expresses action enabling (causal dependency among events), Γ must be well-founded (axiom of finite causes). Formally, a **pomset** (partially ordered multiset) is the isomorphism class of an lpo, denoted $[V, \Sigma, \Gamma, \mu]$.

Let p and q be pomsets, and let all partial orders Γ be written as "<". We say that q is a ρ-prefix of p when q is obtainable from p by deleting a subset of the events of p, provided that if event u is deleted and $u < v$, then v is also deleted. We say that α is a π-prefix of p when α is a finite ρ-prefix of p. $\pi(p)$ is the set of π-prefixes of p. We say that q is an **augment** of p when q differs from p only in its partial order, which must be a superset of that of p. If P is a set of pomsets, then $\pi(P)$ is $\cup$ $\pi(p)$, $p \in P$. Let $p = [V, \Gamma, \Sigma, \mu]$ be a pomset and let $V' \subseteq V$. The **projection** onto V' is $p' = [V', \Gamma', \Sigma, \mu']$, where Γ' and μ' are the restrictions of Γ and μ to V'.

We introduce the notion of prefix **envelope**. Let p be a pomset. $°p$ is the set of action labels of initial events of p, that is, the set of $\mu(v)$ of v in p such that $\nexists u$ in p with $u < v$. If $\alpha \in \pi(p)$, then $p - \alpha$ is the projection obtained by deleting the events in α from p. The **envelope** of α in p, denoted $E_p(\alpha)$, is $°(p - \alpha)$. Define $in_p(\alpha) = E_p(\alpha) \cap tI$ and $out_p(\alpha) = E_p(\alpha) \cap tO$. $E_p(\alpha)$ is the set of process **actions** that are concurrently enabled in behavior p after the **events** of partial execution α. We sketch the notion of branch point. Consider the following skeleton presentation of a simple finite pomtree. Let α, β_1 and β_2 be finite pomsets, with $°\beta_1 \cap °\beta_2 = \{ \}$. Lay down pomset α. Now, both concatenate β_1 with set Γ_1 of (α, β_1) precedences, and concatenate β_2 with set Γ_2 of (α, β_2) precedences. The pomtree contains one copy each of α, β_1 and β_2. β_1 and β_2 are the two branches. Let p and q be the two (maximal) paths through the pomtree. By construction, $\alpha \in \pi(p)$, $\pi(q)$ is a maximal (under prefix ordering) common prefix of p and q. Also by construction, $E_p(\alpha) \cap E_q(\alpha) = \{ \}$. We say that α is a **branch point**. If $E_p(\alpha), E_q(\alpha) \subseteq tI (tO)$, then α is an input (output) branch point. Pomtrees are like computation trees except that arcs are maximal determinate behavior segments, and vertices are input or output branch points.

2.3. Determinate processes

Fig. 1 shows a representation of a complete computational behavior p. Process $P = \{p\}$ could be a C-element together with an unconnected wire. The use of pluses and minuses (from rising and falling signal transitions) is partly redundant; once the initial voltage level of a port has been fixed by a reset transition, each new transition at that port is the opposite of the preceding transition. Output port names have been underlined. The complete behavior is obtained by concatenating infinitely many copies of this figure, superimposing adjacent (a^+, b^+, d^+) triples. The figure is intended to illustrate both maximal safe use and necessary temporal precedence. For example, process P is a multithreaded process (it has two threads); the environment may elect to use one thread and not the other. A process with a single thread has essentially only one use. Necessary temporal precedence is explained below.

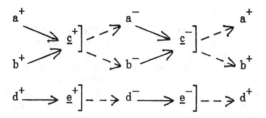

Fig. 1 Representation of a complete computational behavior.

The complete behavior precisely describes when the process (environment) is allowed to perform output (input) actions. In all processes, as illustrated in Fig. 1, the behavior partial order Γ in $p = [V, \Sigma, \Gamma, \mu]$ is the transitive closure Ω^+ of a nontransitive <u>successor</u> relation $\Omega = N \cup \Xi$, where the <u>output protocol</u> N is a relation from input events to output events, and the <u>input protocol</u> Ξ is a relation from output events to input events. In model checking, we use the fact that N (Ξ) is the <u>causal</u> (<u>noncausal</u>) part of Ω, by asymmetry of control. In short, the process (environment) enforces N (Ξ). We also say that the actions in $\text{out}_p(\alpha)$ [$\text{in}_p(\alpha)$] are <u>causally</u> [<u>noncausally</u>] enabled at α.

The semantics of successor arrows is as follows. The process may perform an output action when all of its solid-arrow predecessor events have occurred. The environment may perform an input action when all of its dashed-arrow predecessor events have occurred. Violations of the input (output) protocol are called <u>failures</u> (<u>errors</u>) [6]. After a failure, a process may behave arbitrarily. This means that system pomtrees of (closed) networks of processes are undefined if there is a failure in any component process. Brackets are explained momentarily.

2.4. <u>Delay insensitivity</u>

In general, well-behavedness conditions are motivated by restrictions that allow asynchronous processes to be used as components of delay-insensitive systems [6]. Here, we present a minimal set of assumptions that are used in the model checking approach of this paper. Assumptions that guarantee delay insensitivity are distinguished from more general assumptions -- not shown in this paper -- that guarantee the finite-state character of asynchronous processes, and the bounded-size encoding of behavior states. These general assumptions are what guarantee the existence of a clean, finite recurrence structure in an asynchronous hardware or software system.

Well-behavedness conditions

<u>Rule 1</u> There is no autoconcurrency. Formally, any two events at the same port in $p \in P$ are separated in Ω by at least one event at some other port.

<u>Rule 2</u> There is no specified successor relationship either between two input events or between two output events. Formally, each line in $p \in P$ consists of an infinite sequence of strictly alternating input and output events.

2.5. <u>Progress requirements</u>

We specify progress requirements in a behavior p by bracketing output events. After partial execution $\alpha \in \pi(p)$, a determinate process $P = \{p\}$ is required to perform all bracketed output actions in $\text{out}_p(\alpha)$; this subset is denoted $\text{req}_p(\alpha)$. These actions must be performed eventually, after some arbitrary but finite delay. To simplify model checking, from now on we only consider systems in which $\text{req}_p(\alpha) = \text{out}_p(\alpha)$; this defines <u>restricted</u> process theory. We capture this notion in a rule.

<u>Rule 3</u> In the absence of branching, there is never a choice between performing and not performing an output action. Formally, if $p \in P$, then all output events in p are bracketed.

2.6. Nondeterminate processes

Branching in an asynchronous process arises in either of two ways. There may be free choice (in the Petri net sense) in the environment to apply one of several mutually exclusive access operators, and there may be arbiter choice in the process to respond to one of several concurrent requests to perform a mutually exclusive operation on a shared object. Pomtrees adequately model the branching structure arising from these two sources. The semantics at input (output) branch points is straightforward. An input branch point α defines $in(\alpha) = \{in_j(\alpha) : j \in J\}$, where J indexes the finite input branching at α. This is the family of disjoint sets of input actions that are concurrently enabled at α, respectively, in each of the futures that branch from α. An output branch point α defines $out(\alpha) = \{out_k(\alpha) : k \in K\}$, where K indexes the finite output branching at α. This is the family of disjoint sets of output actions that are concurrently enabled at α, respectively, in each of the futures that branch from α. The semantics at nonbranch points is equally straightforward. If α is within a determinate behavior segment, then there is precisely one enabled set. If α straddles a branch point, then the enabled sets are not disjoint.

Consider an output branch point α in restricted process theory. If P advances to α, then P is required to perform all actions in some out_k within finite time.

Explanation is required for arbiter choice, concerning progress requirements at prefixes of α. Suppose that output action u is required at β and that, before P performs u, P advances to α in which u and v are disjunctively required; P may then perform v and not u. This is an acceptable confusion situation (in the Petri net sense): in an arbiter, a required output action may become disjunctively required after the receipt of new input.

2.7. Behavior machines

Behavior machines are finite presentations of pomtrees. A behavior machine consists of a set of selected behavior states, and a set of commands. A command can be applied in certain behavior states, and produces a new behavior state. Commands are essentially finite pomsets with additional machinery to define the nonsequential serial concatenation -- modulo branching -- of the command to the finite pomtree generated so far. One fact of model checking is that components of systems are subject to nonmaximal safe use, requiring the encoding of behavior states that were not originally selected; for this reason, we encode all behavior states at specification time. Behavior machines specify both safety and progress properties; fairness properties must be specified by a supplementary condition constraining output choice.

Formally, a behavior machine is a set S of selected behavior states (with a distinguished start state), and a set C of commands. The machine describes the possible state transitions (c takes s to t) produced by each command. For each $c \in C$, there is a transition relation $\Delta(c)$ on S, where $(s, t) \in \Delta(c)$ if conceptual execution of command c in state s produces state t.

Fig. 2 shows a behavior machine for a C-element. Each command has the form: behavior state label, socket-extended finite pomset, behavior state label. In constructing behaviors, a new command may be applied provided the command prelabel matches the postlabel of the previous command. Sockets (denoted o) help describe how future behavior follows past behavior. If there is a successor arrow o → u, then o is filled by some predecessor of u according to a well-defined rule. More precisely, if command c can follow partial execution α, then there is an injection (produced by simple labelling) from the set of sockets in c to the set of events in α that can still participate in (new) successor arrows. By convention, there is an imaginary initialization event • with postlabel 0. As a presentational device, we make the vertical placement of symbols in commands significant: a socket is always filled by an action that appears on the same horizontal line. For example, in Fig. 2, the unique socket o can be filled by • (first application of the command) or by c^- (all subsequent applications of the command). Assume that the default vertical placement of • is "middle".

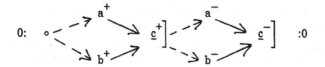

Fig. 2   Behavior machine for a C-element.

Fig. 3 shows a reduced behavior machine for a delay-insensitive arbiter. For conciseness of presentation, we have included several distinct behavior states under label 1. The abuse of notation is intentional. Label 1 in Fig. 3 is an equivalence class of behavior states. The two "commands" on the right in Fig. 3 are equivalence classes of commands. This is explained below. Each of two clients follows a four-cycle protocol, where $\langle A \rangle = \underline{c}^+] \dashrightarrow a^-$ and $\langle B \rangle = \underline{d}^+] \dashrightarrow b^-$ are the critical sections. A two-arrow socket that is always filled by • has been suppressed from command 1 (extreme left). As shown, sockets exist on three horizontal levels. The top ∘ (in command 2) is always filled by a^+ and the bottom ∘ (in command 3) is always filled by b^+. Either middle ∘ can be filled by •, a^- or b^-, perhaps redundantly.

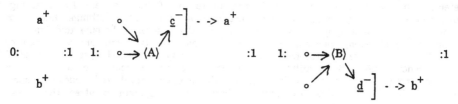

Fig. 3   Reduced behavior machine for a delay-insensitive arbiter.

The reduced behavior machine in Fig. 3 groups three <u>distinct</u> behavior states under label 1, all corresponding to the same execution state. The three may be distinguished as: (1.1) no critical-section entry has occurred (both middle sockets become redundant), (1.2) A's critical-section exit was the most recent (A's middle socket becomes redundant), and (1.3) B's critical-section exit was the most recent (B's middle socket becomes redundant). Behavior states 1.1, 1.2 and 1.3 are output branch points. Behavior states 1.2 and 1.3 form a complete set of loop cutpoints in the full behavior machine; such a set is called a <u>dominator</u> <u>set</u> [2,7]. The full behavior machine is easily obtained. It has four selected behavior states (0, 1.1, 1.2 and 1.3) and seven transitions -- but only five commands (distinct socket-extended pomsets). For example, behavior state 1.2 has a self-loop produced by command 2′ (command 2 minus its middle socket), and a transition to behavior state 1.3 produced by command 3.

The simplest way to encode <u>behavior states</u> is to use names of successor arrows. This is done by recognizing distinct arrows in a behavior machine, and assigning labels. For a finite-state determinate process P = {p}, the behavior state $s(\alpha)$ that corresponds to partial execution $\alpha \in \pi(p)$ is encoded as the set $\Omega(\alpha, p - \alpha)$ of successor arrows with source in α and target in $p - \alpha$. This encoding strategy works equally well for finite-state nondeterminate processes. For example, any behavior state of the arbiter can be encoded by three (small) integers representing the (virtual) successor arrows currently offered by the past to the future. Thus, behavior state 1.2 is encoded as: a^+ (b^+) is currently available to enable the next $\underline{c}^+$ $(\underline{d}^+)$ that appears in any downward path through the pomtree, and a^- is currently available to enable the $\underline{d}^+$ that appears in the immediately adjacent B branch.

We say that two partial executions α, $\beta \in \pi(P)$ are <u>execution equivalent</u>, written α

$\equiv_e \beta$, when their sets of possible futures are equal, that is, $f_P(\alpha) = f_P(\beta)$ [5]. This is standard pomset (or pomtree) equality, based on lpo isomorphism. Each $\equiv_e$ equivalence class of partial executions is an <u>execution</u> <u>state</u> of process P. In the same way, each $\equiv_b$ equivalence class ($\alpha \equiv_b \beta$ when their sets of possible socket-extended futures are equal) is a <u>behavior</u> <u>state</u> of process P. Formally, if α and β are prefixes of infinite paths through pomtree P, then α and β result in the same behavior state of P precisely when the two pomtrees P/α and P/β descending from α and β are equal, where P/α and P/β have been extended to include the $(\alpha, P/\alpha)$ and $(\beta, P/\beta)$ precedences, and isomorphism now requires matching both event labels and arrow labels. In spite of the name, behavior states are not states.

3. Correct implementation

An implementation may exceed the minimum requirements of the specification. For example, it may be more liberal in accepting input and more conservative in producing output, but only if all progress requirements are satisfied. In model checking, a straightforward way to define correctness is to use the mirror of the specification as a conceptual implementation tester [1]. That is, one forms an imaginary closed system by linking mirror mP of specification P to the implementation -- in the general case, a network of processes *Net* --, and then examining certain properties of the resulting system. In the partial order world, the mirror is formed merely by inverting the causal/noncausal interpretation of P's successor arrows. Many things may now be examined. Is there a <u>failure</u> somewhere, causing the system to become undefined? Does the system just stop, violating fundamental liveness? Is some progress requirement violated? Is some conflict resolved unfairly? To study the closed system S - mP|P' -- in the general case, S - mP|*Net* -- by partial order methods, we first regroup <u>all</u> successor arrows (as shown below) to obtain the system causal and noncausal successor relations. Graph predicates are then evaluated on a system pomtree that has two <u>distinct</u> successor relations.

In closed system S, each action is attributed to precisely two processes as a joint action of those two processes. System behaviors are projected onto component alphabets to yield component behaviors. Consider first the trivial case that the implementation is a process. Conceptual execution of S - mP|P' produces <u>corresponding</u> partial executions of specification P and implementation P'. These partial executions correspond by containing the same events, but may not agree on which temporal precedences are necessary. For $\alpha \in \pi(P)$ and $\alpha' \in \pi(P')$, $\alpha \Longleftrightarrow \alpha'$ denotes correspondence. Each α, α' pair is a "doubly-ordered" finite pomset $[V, \Sigma, \Gamma, \Gamma', \mu]$. Corresponding pairs are produced by conceptual execution of S only as long as no safety or liveness violation is detected.

Safety checking occurs on two levels. (i) Any action u that is causally but not noncausally enabled at α, α' is an immediate safety violation. Sets of concurrently causally enabled actions must be concurrently noncausally enabled. (ii) Moreover, while visiting event u, each noncausal arrow (t, u) is checked by a (breadth-first) search for a supporting chain [t, u] of causal arrows. The nonexistence of such a chain is a safety violation. Liveness checking also occurs on two levels. (i) Any external output action v that is noncausally but not causally enabled at α, α' is an immediate liveness violation -- unless v is only disjunctively required (see above). Each set of concurrently causally enabled external output actions must contain some set of concurrently noncausally enabled external output actions. (ii) Moreover, while visiting external output event v, each causal chain [t, v] from an external input event t is checked by a "search" for a matching noncausal arrow (t, v). The nonexistence of such an arrow is a liveness violation.

We give a state-based definition of correctness for S = mP|P'. First, consider determinate P and P'. Correctness of <u>safety</u> properties means: $\forall \alpha, \alpha' : \alpha \Longleftrightarrow \alpha' :$ $in_p(\alpha) \subseteq in_p(\alpha')$ and $out_p(\alpha) \supseteq out_p(\alpha')$. Correctness of <u>progress</u> properties means: $\forall \alpha, \alpha'$ $: \alpha \Longleftrightarrow \alpha' : req_p(\alpha) \subseteq req_p(\alpha')$. In restricted process theory, this reduces to: $out_p(\alpha) =$

$out_{p'}(\alpha') = req_p(\alpha) = req_{p'}(\alpha')$. Next, extend consideration to nondeterminate P and P'. In restricted process theory, correctness now means:

$$\forall \alpha, \alpha' \quad : \alpha \Longleftrightarrow \alpha':$$
$$\forall j \quad \exists j' \quad : in_j(\alpha) \subseteq in_{j'}(\alpha') \wedge \qquad (*)$$
$$\forall k' \quad \exists k \quad : out_k(\alpha) = out_{k'}(\alpha') \qquad (**)$$

These subscripts index sets of actions that are concurrently enabled at α, respectively, in each of the futures that branch from α.

We are now ready for a genuine partial order view that defines correctness without reference to states; in particular, the $\forall \alpha, \alpha'$ quantifier will be dispensed with. Consider a pair α, α', $\alpha \Longleftrightarrow \alpha'$. We regroup the causal and noncausal parts of $\Omega(\alpha)$ and $\Omega'(\alpha')$ to obtain the system causal and noncausal successor relations. We define the <u>system causal</u> successor relation $\tilde{\Omega} = (N' \cup \Xi)^{+-}$, where the superscript denotes transitive closure followed by transitive reduction. Similarly, we define the <u>system noncausal</u> successor relation $\hat{\Omega} = (N \cup \Xi)^{+-}$. When the implementation is a network Net of components P_i, $i \in I$, $\tilde{\Omega} = [(\cup_i N_i) \cup \Xi]^{+-}$, and $\hat{\Omega} = [N \cup (\cup_i \Xi_i)]^{+-}$.

Consider the general case. System pomtree S is undefined if a component fails. Suppose there is no failure (safety violation) in S. Each event v in S is classified as an input event in precisely one (say P_i) of the two processes to which the action performed by event v is attributed. From P_i's specification, there is a well-defined noncausal preset in P_i (say $\Pi_i(v) \subseteq \Sigma_i$) of the P_i action performed by event v. By definition, each event occurs in S because it is causally enabled. In order that P_i not "explode", v must also be noncausally enabled in P_i. This is equivalent to: (i) $\forall u \in \Pi_i(v) : u \in S$, and (ii) for each u, there is a chain of causal arrows [u, v]. This is safety correctness. Some events v that are noncausally enabled in S are, from specification P's point of view, classified as (external) output events. From P's specification, there is a well-defined noncausal preset in mP (say $m\Pi(v) \subseteq m\Sigma$) of the mP action performed by event v. In order that (the nonoccurrence of) event v not be a progress violation, we must have: (i) $v \in S$, and (ii) there exists a chain of causal arrows [u, v] iff $u \in m\Pi(v)$. Condition (ii) means that, after projection of S on mP, precisely $m\Pi(v)$ enables the mP action performed by v. Again, v may be only disjunctively required (see above). This is progress correctness.

4. Model checking

We sketch direct model checking of networks of processes. Given are specification process P and some number of implementation component processes P_i, $i \in I$. Network links map each internal output action of each component process P_i to precisely one internal input action of some other component process P_j [6]. Each internal event is attributed to two component processes. Conceptual links map each action of specification process P, whether input or output, to precisely one action (of the same type) of some component process P_i. Each external event is attributed to the specification process and one component process. External input events are caused by specification process P, while all other events are caused by some component process P_i.

The <u>verification procedure</u> shares superficial structure with the algorithm in [1]. Let D be a complete set of loop cutpoints (dominator set) for P. We enumerate system pomtree S recursively. We maintain (1) a <u>stack</u> to postpone the examination of system pomtree branches (maximal determinate behavior segments), and (2) a <u>table</u> to detect cycles in the enumeration of S. Recording the system behavior state each time P advances to some d ∈ D (at P command completion) leads to discovery of a loop cutpoint for each loop in a finite presentation of system pomtree S.

Fig. 4 shows a typical result of applying a P command to a network, here, in its initial state.

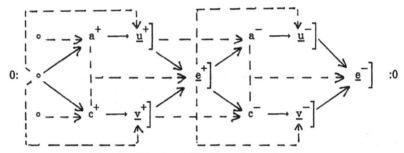

Fig. 4   Direct verification of a network.

Conceptual execution of system S produces n-tuples $<\alpha, \alpha_1, \alpha_2, ... >$ obtained by projection of system behavior prefixes onto processes P and P_i, i ∈ I. In the general case, corresponding partial executions no longer contain the same events. Each $\beta \in \pi(S)$ maps onto a system behavior state, which is the standard vector of behavior states of P and all P_i. Distinguish a "red" problem and a "green" problem. The red problem is, identify each maximal determinate behavior segment of system pomtree S. Branch points of S are scheduled for expansion in the usual way. The green problem is, identify each maximal behavior prefix $\beta \in \pi(S)$ that corresponds to a selected behavior state of specification P. System behavior states of such β's are candidates for entry into the table. The solution to both problems is, conceptually mark the following events in the indicated manner: (1) each initial event of each P command emanating from an input (output) branch point is marked with both a red and a green dot (with a green dot only), (2) each initial event of each P_i command emanating from an output branch point is marked with a red dot only, and (3) each initial event of each P command emanating from a determinate loop cutpoint is marked with a green dot only.

We move forward cleanly to a branch point of S by deferring visiting any red event as long as there are nonred events still enabled. Similarly, we move forward cleanly to a system behavior prefix $\beta \in \pi(S)$ that corresponds to a P rule completion by deferring visiting any green event as long as there are nongreen events still enabled. If the (selected) P behavior state of β has been identified as a loop cutpoint in P, then the system behavior state of β is entered into the table. If this state is already in the table, then β is not extended further.

Let r : $\pi(S) \to \pi(P)$ be the standard projection from system behavior prefixes to specification behavior prefixes. Refinement mapping is nontrivial because r has too many inverse images. The function of green dots is to select only a few "extremal" inverse images of selected behavior states of P -- to create (potential) table entries. For example, some P commands terminate in output-branch loop cutpoints. This is the case for command 1 in the full arbiter behavior machine, with a typical implementation being a ring of DME elements. Command 1 is applied to this implementation by moving downward along all paths of system pomtree S to include any event -- that is causally enabled by a^+ and b^+ -- up to but not including any green event (here, $\underline{c}^+$ and $\underline{d}^+$). With this implementation, command 1 of P corresponds to a finite initial pomtree prefix of S rather than a determinate system behavior segment. In our table, we record the two system behavior states that correspond to the <u>endpoints</u> of the two maximal paths through this finite system subpomtree.

155

Here is the algorithm summary. Recursively generate the system pomtree S by extending all system behavior prefixes $\beta \in \pi(S)$ as long as no safety or liveness violation is detected. Starting from the initial global state, apply P commands to the implementation by generating all causally enabled events in all possible futures up to but not including any green event. As events are generated, evaluate the graph predicates discussed in section 3. At P command completion, if the postlabel t in $s: c : t$ belongs to the dominator set D of P, then record all pairs (t, N), where N is any network behavior state that corresponds to t by the construction in the previous paragraph. Do not apply a command with prelabel t to the implementation in network behavior state N if (t, N) is already in the table. Terminate the algorithm when it is no longer possible to apply any P command.

5. Conclusion

Behavior machines -- finite presentations of pomtrees -- precisely describe the branching and recurrence structure of processes, with benefits for both analysis and synthesis. Partial-order model checking allows us -- whenever there is a moderate amount of true concurrency but not an excessive amount of true nondeterminism -- to avoid the state explosion that frequently haunts verifiers based on state graphs, and limit any combinatorial explosion to that created by the branching structure of processes. Loop cutpoint detection in behaviors appears to require behavior states; such states contain information about how future behavior follows past behavior. Based on an understanding of the state space, the specifier/designer can create a useful separation of concerns for the verifier by structuring processes into (i) determinate behavior segments that are either completed or eliminated during conceptual execution (a pretending atomicity property), and (ii) input or output branch points. We obtain a dramatic performance improvement in model checking, and this even though only the specification process P gets to pretend atomicity.

References

[1] D.L. Dill, "Trace theory for automatic hierarchical verification of speed-independent circuits", Ph. D. Thesis, Department of Computer Science, Carnegie Mellon University, Report CMU-CS-88-119, February 1988. Also MIT Press, 1989.

[2] Z. Manna and A. Pnueli, "Specification and verification of concurrent programs by ∀-automata", Proc. of 14th ACM Symposium on Principles of Programming Languages, January 1987, pp. 1-12.

[3] A.J. Martin, "Compiling communicating processes into delay-insensitive VLSI circuits", Distributed Computing, Vol. 1, No. 4, October 1986, pp. 226-234.

[4] V.R. Pratt, "Modelling concurrency with partial orders", Int. J. of Parallel Prog., Vol. 15, No. 1, February 1986, pp. 33-71.

[5] D.K. Probst and H.F. Li, "Abstract specification of synchronous data types for VLSI and proving the correctness of systolic network implementations", IEEE Trans. on Computers, Vol. C-37, No. 6, June 1988, pp. 710-720.

[6] D.K. Probst and H.F. Li, "Abstract specification, composition and proof of correctness of delay-insensitive circuits and systems", Technical Report, Department of Computer Science, Concordia University, CS-VLSI-88-2, April 1988 (Revised March 1989).

[7] D.K. Probst and H.F. Li, "Partial-order model checking of delay-insensitive systems". In R. Hobson et al. (Eds.), Canadian Conference on VLSI 1989, Proceedings, Vancouver, BC, October 1989, pp. 73-80.

[8] J.v.d. Snepscheut, "Trace theory and VLSI design", Lect. Notes in Comput. Sci. 200, Springer Verlag, 1985.

A STUBBORN ATTACK ON STATE EXPLOSION
(abridged version)

Antti Valmari
Technical Research Centre of Finland
Computer Technology Laboratory
PO Box 201
SF-90571 Oulu
FINLAND

Tel. +358 81 509 111

ABSTRACT

*The paper presents the **LTL preserving stubborn set method** for reducing the amount of work needed in the automatic verification of concurrent systems with respect to linear time temporal logic specifications. The method facilitates the generation of **reduced state spaces** such that the truth values of a collection of linear temporal logic formulas are the same in the ordinary and reduced state spaces. The only restrictions posed by the method are that the collection of formulas must be known before the reduced state space generation is commenced, the use of the temporal operator "next" is prohibited, and the (reduced) state space of the system must be finite. The method cuts down the number of states by utilising the fact that in concurrent systems the nett result of the occurrence of two events is often independent of the order of occurrence.*

1. INTRODUCTION

The automatic verification of temporal properties of finite-state systems has been a topic of intensive research during the recent decade. A typical approach is to generate the state space of the system and then apply a model checking algorithm on it to decide whether the system satisfies given temporal logic formulas [Clarke & 86] [Lichtenstein & 85]. A well known problem of the approach is that the state spaces of systems tend to be very large, rendering the verification of medium-size and large non-trivial systems impossible with a realistic computer. This problem is known as the *state explosion* problem.

Concurrency is a major contributor to state explosion. It introduces a large number of execution sequences which lead from a common start state to a common end state by the same transitions, but the transitions occur in different order causing the sequences to go through different states. This phenomenon has been recognized long ago and the choice of a coarser level of atomicity has been suggested as a partial solution (see [Pnueli 86]). Unfortunately, the power of coarsening the level of atomicity is limited. Consider a system consisting of n processes which execute k steps without interacting with each other and then stop. The system has $(k+1)^n$ states. Each of the processes of the system can be coarsened to a single atomic action. Coarsening reduces the number of states to 2^n which is still exponential in the number of the processes [Valmari 88c]. However, it seems intuitively that to check various properties of the system it would be sufficient to simulate the processes in one arbitrarily chosen order, thus generating only $nk+1$ states.

To our knowledge the first person to suggest a concurrency-based state space reduction method potentially capable of changing a state space from exponential to polynomial in the number of processes was W. Overman [Overman 81]. Overman's work is little known, perhaps because he considered a very restricted case (the terminal states of systems consisting of processes which do not branch or loop), and the algorithm he gave as part of his method for finding certain sets was not efficient enough from the practical point of view. He suggested also a modified method with a faster algorithm, but the modification destroyed the ability of changing exponential state spaces to polynomial.

The problems in Overman's approach were effectively solved by Valmari when he presented the so-called *stubborn set* method [Valmari 88a, 88b]. The stubborn set method has been developed in a series of papers [Valmari 88a, 88b, 88c, 89a, 89b]. Originally the method could be used only to investigate deadlocks but more advanced versions of the method have been gradually developed in order to verify more properties. The method was initially applied to ordinary Petri nets but now it is applicable to a rather general model of concurrency, the *variable/transition* systems. Two profoundly different versions of the method have been distinguished: *weak* and *strong*. The weak theory is more complicated and more difficult to implement, but it leads to better reduction results.

The present paper extends the stubborn set method to almost full linear temporal logic — almost, because the operator "next state" is forbidden. For simplicity, we have chosen to use the strong stubborn set framework, although with minor refinements the results are valid in the weak theory as well. The theory is developed in Chapter 2. Chapter 3 (not in this abridged version) discusses how the theory may be implemented and Chapter 4 contains an example.

This paper is a revised and abridged version of a paper with the same name which appeared in DIMACS Technical Report 90-31, "Workshop on Computer-Aided Verification", Rutgers University, NJ, USA, June 1990, Volume I.

2. DEFINITIONS AND BASIC THEOREMS

2.1 Variable/Transition Systems

To develop the stubborn set method we look at concurrent systems as systems consisting of a finite set V of *variables* and a finite set T of *transitions*. Each variable v has an associated set called *type* and denoted by $type(v)$, and at every instant of time v has a unique *value* belonging to its type. Assuming an ordering of V, the Cartesian product of the types of the variables is the set of *syntactic states* and is denoted by S. The value of variable v at syntactic state s is denoted by $s(v)$. There is a partial *next state* function *next* from $S \times T$ to S which defines when a transition is *enabled* and what is the result of the *occurrence* of an enabled transition. Transition t is enabled in state s, denoted by $en(s,t)$, iff $next(s,t)$ is defined. If $next(s,t) = s'$ we say that t may *occur* at s producing s' and often write $s -t\rightarrow s'$. We often merge the states between successive transition occurrences and write $s_0 -t_1\rightarrow s_1 -t_2\rightarrow \ldots -t_n\rightarrow s_n$ instead of $s_0 -t_1\rightarrow s_1 \wedge s_1 -t_2\rightarrow s_2 \wedge \ldots \wedge s_{n-1} -t_n\rightarrow s_n$. A sequence like this is called a *finite execution sequence* and its *length* is n. The *concatenation* of the execution sequences $\sigma = s_0 -t_1\rightarrow \ldots -t_n\rightarrow s_n$ and $\rho = r_0 -d_1\rightarrow \ldots -d_m\rightarrow r_m$ where $s_n = r_0$ is defined by $\sigma^\circ\rho = s_0 -t_1\rightarrow \ldots -t_n\rightarrow s_n -d_1\rightarrow \ldots -d_m\rightarrow r_m$. "$\rightarrow$" is defined by $s \rightarrow s' \Leftrightarrow \exists t \in T: s -t\rightarrow s'$. "$\rightarrow^*$" is the reflexive transitive closure of "$\rightarrow$". There is a distinguished state s_0 called the *initial state* of the system. A *variable/transition system* or *v/t-system* is the 5-tuple $(V,T,type,next,s_0)$, where the components of the 5-tuple are as just explained.

Section 4.1 contains a non-trivial example of a v/t-system.

The stubborn set method relies on the analysis of certain relationships between transitions. Let us define and explain some necessary concepts.

Definition 2.1 Let $t, t' \in \mathbf{T}$, $s \in \mathbf{S}$, $V \subseteq \mathbf{V}$ and $T \subseteq \mathbf{T}$.

- t is *enabled with respect to* V at s, denoted by $en(s,t,V)$, iff
  $$\exists s' \in \mathbf{S}: en(s',t) \wedge \forall v \in V: s'(v) = s(v).$$

- T is a *write up* set of t with respect to V, iff for every $t' \in \mathbf{T}$ and $s' \in \mathbf{S}$
  $$\neg en(s,t,V) \wedge s -t'\to s' \wedge en(s',t,V) \Rightarrow t' \in T.$$

- t *accords with* t', denoted by $t \leftrightarrow t'$, iff for every $s \in \mathbf{S}$
  $$en(s,t) \wedge en(s,t') \Rightarrow \exists s', s_1, s'_1 \in \mathbf{S}: s -t\to s_1 -t'\to s'_1 \wedge s -t'\to s' -t\to s'_1. \quad \square$$

The intuition behind the definition of $en(s,t,V)$ is perhaps best understood by noticing that if t is *not* enabled w.r.t. V, then it is necessary to modify the value of at least one variable in V to enable t. The definition has the following rather obvious consequence: $en(s,t) \Leftrightarrow en(s,t,\mathbf{V}) \Leftrightarrow \forall V \subseteq \mathbf{V}: en(s,t,V)$.

A write up set of the transition t with respect to the set V of variables is any set of transitions containing at least the transitions which have the potential of modifying the status of t from disabled w.r.t. V to enabled w.r.t. V. We do not require the write up set to be the smallest such set, because minimality is not needed in the theory of stubborn sets and the smallest set may be difficult to find in practice. For instance, if the specification of transition t' contains a writing reference to a variable in V but has a complicated enabling condition evaluating to **false**, t' does not belong to the smallest write up set of t w.r.t. V because it is never enabled. However, it may be difficult to see that t' can be ruled out. Our definition allows the use of a write up set which can be easily computed and is an upwards approximation of the minimal set. Every transition t and subset of variables V has at least one write up set, namely the set of all transitions $\mathbf{T}$. From now on we assume that a unique write up set is defined for every (t,V)-pair. We denote it by $wrup(t,V)$.

The definition of according with can be illustrated graphically:

$$
\begin{array}{ccc}
s -t'\to s' & & s -t'\to s' \\
| & & | \quad\quad | \\
t & \Rightarrow & t \quad\quad t \\
\downarrow & & \downarrow \quad\quad \downarrow \\
s_1 & & s_1 -t'\to s'_1
\end{array}
$$

According with is a static, symmetric commutativity property of transition pairs. Concurrent systems typically contain several pairs of transitions according with each other. Let us call the set of variables tested, read or written by a transition its *reference set*. Two transitions accord with each other if their reference sets are disjoint. The same is true even if there are common variables in the reference sets, as long as the transitions never write to them. Transitions writing to a fifo queue accord with transitions reading from it, unless there are other variables in common. Transitions corresponding to different locations in the code of a sequential process accord with each other independent of to what variables they refer, because they are never simultaneously enabled.

2.2 Stubborn Sets

We are ready to define *semistubborn* sets of transitions:

Definition 2.2 Let $T \subseteq \mathbf{T}$ and $s \in \mathbf{S}$. T is *semistubborn* at s, iff $\forall t \in T$:

(1) $\neg en(s,t) \Rightarrow \exists V \subseteq \mathbf{V}: \neg en(s,t,V) \wedge wrup(t,V) \subseteq T$

(2) $en(s,t) \Rightarrow \forall t' \notin T: t \leftrightarrow t'$ □

Part (1) of the definition guarantees that a disabled transition belonging to a semistubborn set can be enabled only by the transitions in the set. Part (2) says that transitions in a semi-stubborn set accord with outside transitions. It is not difficult to prove that at least the empty set and the set of all transitions **T** are semistubborn at every state, thus semistubborn sets always exist. Furthermore, if a transition outside a semistubborn set occurs, the set remains semistubborn. This is a justification for the peculiar term semi*stubborn*. The significance of semistubborn sets is in the following theorem:

Theorem 2.3 Let $T \subset \mathbf{T}$ and $s_0 \in \mathbf{S}$ such that T is semistubborn at s_0. Let $n \geq 1$ and

$$\sigma = s_0 -t_1\rightarrow s_1 -t_2\rightarrow \dots -t_{n-2}\rightarrow s_{n-2} -t_{n-1}\rightarrow s_{n-1} -t_n\rightarrow s_n$$

be a finite execution sequence such that $t_1, \dots, t_{n-1} \notin T$ and $t_n \in T$. There is the finite execution sequence

$$\sigma' = s_0 -t_n\rightarrow s'_0 -t_1\rightarrow s'_1 -t_2\rightarrow \dots -t_{n-2}\rightarrow s'_{n-2} -t_{n-1}\rightarrow s'_{n-1},$$

where $s'_{n-1} = s_n$. □

The theorem can be illustrated graphically. σ corresponds to the top and right edges of the figure, while σ' corresponds to the left and bottom sides.

$$
\begin{array}{ccccccc}
s_0 & -t_1\rightarrow & s_1 & -t_2\rightarrow & \dots & -t_{n-2}\rightarrow & s_{n-2} & -t_{n-1}\rightarrow & s_{n-1} \\
\downarrow t_n & & & & & & & & \downarrow t_n \\
s'_0 & -t_1\rightarrow & s'_1 & -t_2\rightarrow & \dots & -t_{n-2}\rightarrow & s'_{n-2} & -t_{n-1}\rightarrow & s'_{n-1} \\
& & & & & & & & = s_n
\end{array}
$$

The proof is a straightforward application of the definitions, and can be found in [Valmari 88b, 89a, 89b].

Theorem 2.3 allows us to move the occurrence of a transition belonging to a semistubborn set from future to the current state. However, this is of course of no use if no transition in the semistubborn set is going to occur in the future. This is certainly the case if the set is empty. Therefore we augment the definition by the requirement that there must be an enabled transition in the set:

Definition 2.4 Let $T \subseteq \mathbf{T}$ and $s \in \mathbf{S}$. T is *stubborn* at s, iff T is semistubborn at s, and $\exists t \in T: en(s,t)$. □

A stubborn set exists exactly when there is an enabled transition, because then the set of all transitions **T** is stubborn. However, as will soon become obvious, it is advantageous (but not mandatory) to use stubborn sets containing as few enabled transitions as possible.

2.3 Reduced State Spaces

A variable/transition system defines a labelled directed graph called its *state space* in a natural way:

Definition 2.5 The *state space* of the *v/t-system* $(\mathbf{V},\mathbf{T},type,next,\mathbf{s}_0)$ is the triple $(\mathbf{W},\mathbf{E},\mathbf{T})$, where

- $\mathbf{W} = \{s \in \mathbf{S} \mid \mathbf{s}_0 \rightarrow^* s\}$
- $\mathbf{E} = \{(s,t,s') \in \mathbf{W} \times \mathbf{T} \times \mathbf{W} \mid s -t\rightarrow s'\}$ □

Let *TS* be a function from **S** to the set of the subsets of **T** such that *TS(s)* is stubborn if $\exists\, t \in$ **T**: *en(s,t)*, and *TS(s)* = ∅ otherwise. We call the function *TS* a *stubborn set generator*. The stubborn set method uses a stubborn set generator to generate a *reduced state space* as follows. The generation starts at $\mathbf{s_0}$. Assume *s* has been generated. In ordinary state space generation, every transition enabled at *s* is used to generate the immediate successor states of *s*. In the stubborn set method, only the enabled transitions in *TS(s)* are used. If *TS(s)* contains less enabled transitions than **T**, the number of immediate successors is reduced. This often leads to a reduction in the total number of states. To distinguish between reduced and ordinary state space concepts we use underlining notation as follows:

- $\underline{en}(s,t) \Leftrightarrow en(s,t) \wedge t \in TS(s)$

- $s -t\underline{\rightarrow} s' \Leftrightarrow s -t\rightarrow s' \wedge t \in TS(s)$

- $s \underline{\rightarrow} s' \Leftrightarrow \exists\, t \in TS(s): s -t\rightarrow s'$

- "$\underline{\rightarrow}*$" is the reflexive and transitive closure of "$\underline{\rightarrow}$".

Definition 2.6 Assume a stubborn set generator *TS* is given. The *reduced state space* of the v/t-system (**V**,**T**,*type*,*next*,$\mathbf{s_0}$) is the triple ($\underline{\mathbf{W}},\underline{\mathbf{E}},\mathbf{T}$), where

- $\underline{\mathbf{W}} = \{s \in \mathbf{S} \mid \mathbf{s_0} \underline{\rightarrow}* s\}$

- $\underline{\mathbf{E}} = \{(s,t,s') \in \underline{\mathbf{W}} \times \mathbf{T} \times \underline{\mathbf{W}} \mid s -t\underline{\rightarrow} s'\}$ □

The definition implies that $\underline{\mathbf{W}} \subseteq \mathbf{W}$ and $\underline{\mathbf{E}} \subseteq \mathbf{E}$. The ordinary state space is a special case of a reduced state space, because we may choose *TS(s)* = **T** for states with enabled transitions. However, our goal is to keep the reduced state space small, to save effort in its generation and the model checking afterwards. Perhaps surprisingly, always choosing the stubborn set with the smallest number of enabled transitions does not necessarily lead to the smallest reduced state space [Valmari 88c]. However, it is obvious that if T_1 and T_2 are stubborn and the set of enabled transitions in T_1 is a proper subset of the corresponding set of T_2, then T_1 is preferable. We say that a stubborn set is *optimal* if it is the best possible in this respect. [Valmari 88a, 88b] give a linear and an (under certain reasonable assumptions) quadratic algorithm for finding almost optimal and optimal stubborn sets, respectively. The linear algorithm is particularly attractive because its best case complexity is better than linear; if it finds a stubborn set close to its starting point, it optimises it as much as it can and stops without investigating the rest of the v/t-system. The linear algorithm is briefly described in Chapter 3 of the unabridged version.

2.4 Execution Sequences in Reduced State Spaces

This section is devoted to a construction which, given a finite execution sequence of the system under analysis, finds an execution sequence which is present in the reduced state space and is, roughly speaking, a permutation of an extension of the former. The construction is in the heart of most proofs in the stubborn set theory. It proceeds in steps. Each step appends a transition occurrence to the end of the constructed sequence. The transition occurrence is either picked and removed from the original sequence, in which case we say that an original transition occurrence is *consumed*, or a fresh new transition occurrence is found for the purpose. The construction may be continued until the original sequence is exhausted. It may happen that only fresh transitions are used from some step onwards, in which case the construction may be continued endlessly.

The construction is presented formally below. σ is the original finite execution sequence. The constructed sequence after step *i* is denoted by $\underline{\sigma}_i$ and its last state by $\underline{s}_i$. The execution sequences ρ_i and σ_i correspond to the unconsumed part of the original sequence and the original sequence appended by the occurrences of the fresh transitions used by the construction. *k(i)* is the length of ρ_i, that is, the number of still unconsumed transition occurrences.

It may be helpful to notice that $\underline{\sigma}_i \cdot \rho_i$ exists and its first and last state are the same as the first and last state of σ_i. Furthermore, the transition occurrences of $\underline{\sigma}_i \cdot \rho_i$ are the same as the transition occurrences of σ_i, but not necessarily in the same order.

Construction 2.7 Let TS be a stubborn set generator and $\sigma = s_0 -t_1 \rightarrow \ldots -t_n \rightarrow s_n$ be a finite execution sequence. The states $\underline{s}_i$ and execution sequences $\underline{\sigma}_i$, σ_i and $\rho_i = r_{0,i} -d_{1,i} \rightarrow \ldots -d_{k(i),i} \rightarrow r_{k(i),i}$ are defined recursively as follows. $k(i)$ is defined as the length of ρ_i.

- $\sigma_0 = \rho_0 = \sigma$ and $\underline{\sigma}_0 = s_0 = \underline{s}_0$.

- If $k(i) = 0$, that is, ρ_i consists of one state and no transition occurrences, the construction cannot be continued.

- If $k(i) > 0$, $d_{1,i}$ is enabled at $r_{0,i}$, thus $TS(r_{0,i})$ is stubborn. There are two cases.

  (1) ρ_i contains at least one occurrence of a transition belonging to $TS(r_{0,i})$. We define $k'(i) = k(i)$, $\sigma'_i = \sigma_i$ and $\rho'_i = \rho_i$.

  (2) ρ_i contains no occurrences of transitions belonging to $TS(r_{0,i})$. By Definition 2.4 $TS(r_{0,i})$ contains an enabled transition t'_i. By (2) of Definition 2.2 t'_i is enabled at $r_{1,i}, \ldots, r_{k(i),i}$. We define $k'(i) = k(i)+1$, $d_{k(i)+1,i} = t'_i$, $\sigma'_i = \sigma_i -t'_i \rightarrow r_{k'(i),i}$ and $\rho'_i = \rho_i -t'_i \rightarrow r_{k'(i),i}$.

By construction, ρ'_i contains at least one occurrence of a transition belonging to $TS(r_{0,i})$. Let $1 \le j(i) \le k'(i)$ be chosen such that $d_{1,i} \ldots d_{j(i)-1,i} \notin TS(r_{0,i})$ and $d_{j(i),i} \in TS(r_{0,i})$. By Theorem 2.3 there is the sequence $\rho''_i = r_{0,i} -d_{j(i),i} \rightarrow r'_{0,i} -d_{1,i} \rightarrow \ldots -d_{j(i)-1,i} \rightarrow r_{j(i),i} -d_{j(i)+1,i} \rightarrow \ldots -d_{k'(i),i} \rightarrow r_{k'(i),i}$. By the $(i-1)$th construction step $r_{0,i} = \underline{s}_i$. Hence $d_{j(i),i} \in TS(\underline{s}_i)$ and we may define $\underline{s}_{i+1} = r'_{0,i}$, $\underline{\sigma}_{i+1} = \underline{\sigma}_i -d_{j(i),i} \Rightarrow \underline{s}_{i+1}$ and $\sigma_{i+1} = \sigma'_i$. ρ_{i+1} is defined by $\rho''_i = r_{0,i} -d_{j(i),i} \rightarrow \rho_{i+1}$. After this i:th construction step $r_{0,i+1} = r'_{0,i} = \underline{s}_{i+1}$. $\square$

In the future we will need the fact that in the above construction $d_{x,i} = d_{x,i+1}$ for $x = 1, \ldots, j(i)-1$. Also the order of transition occurrences in $\underline{\sigma}_{i+1} \cdot \rho_{i+1}$ is otherwise the same as the order of transition occurrences in $\underline{\sigma}_i \cdot \rho_i$, but either the occurrence of $d_{j(i),i}$ has moved to a different location, or a new transition occurrence has been introduced.

2.5 Linear Temporal Logic Preservation Theorem

In this section we state and prove the theorem underlying the linear temporal logic (LTL) preserving stubborn set method. LTL formulas state properties of infinite sequences of states. Infinite execution sequences of a v/t-system give naturally rise to infinite sequences of states. We also consider *stopping* execution sequences which are finite and end at a state without enabled transitions. As usual, we extract infinite sequences of states from stopping execution sequences by letting the last state repeat forever. We say that the infinite or stopping execution sequence σ *satisfies* the LTL formula φ iff φ is a true statement about the infinite sequence of states extracted from σ. We say that φ is *valid* in state s in a given state space, iff there is no infinite or stopping execution sequence σ in the state space starting at s such that σ satisfies $\neg \varphi$.

Let $\Phi = \{\varphi_1, \ldots, \varphi_f\}$ be a collection of LTL formulas. We associate with Φ the set $vis(\Phi)$ of *visible transitions*. Transition t is *properly visible*, iff there are syntactic states s and s' such that $s -t \rightarrow s'$ and the truth value of at least one state predicate appearing in at least one formula in Φ is different at s and s'. $vis(\Phi)$ is a set containing at least the properly visible transitions. The reason why we allow in $vis(\Phi)$ the presence of transitions which are not properly visible is the same as the reason of allowing the write up set to be an upwards approximation of the smallest set with the required property. Transition t is *visible* if $t \in vis(\Phi)$, otherwise t is *invisible*. The property we will use in the future is that when an invisible transition occurs the truth values of the state predicates of the formulas in Φ do not change.

The LTL preserving stubborn set method works only with *stuttering-invariant* formulas. Informally, stuttering-invariance means that the truth value of a formula on an infinite sequence of states does not change if one or more or even all of the states of the sequence are multiplicated, where multiplication means the replacement of the state by a positive finite number of copies of the state. Among the common LTL operators, "next state" (O) and "previous state" may introduce formulas which are not stuttering-invariant. Formulas containing no other temporal operators than "henceforth" ($\Box$), "eventually" ($\Diamond$), "until" ($\mathcal{U}$), and their corresponding past operators and the operators derived from them are stuttering-invariant (see [Lamport 83]).

We call the sequence of states $s_0 s_1 \ldots s_n$ an *elementary cycle*, if $n > 0$, $s_0 = s_n$, $s_i \neq s_j$ where $0 \leq i < j < n$ and $s_i \to s_{i+1}$ where $0 \leq i < n$. We can now formulate the theorem underlying the LTL preserving stubborn set method:

Theorem 2.8 Let $(V,T,type,next,\mathbf{s}_0)$ be a v/t-system and $\mathbf{S}$ and $(\mathbf{W},\mathbf{E},\mathbf{T})$ its set of syntactic states and its state space, respectively. Let Φ be a collection of stuttering-invariant LTL formulas such that the domain of the state predicates in the formulas is $\mathbf{S}$. Let TS be a stubborn set generator and $(\underline{\mathbf{W}},\underline{\mathbf{E}},\underline{\mathbf{T}})$ the corresponding reduced state space such that the following hold:

(1) $\underline{\mathbf{W}}$ is finite

(2)   For every $\underline{s} \in \underline{\mathbf{W}}$, either
      (a) $TS(\underline{s})$ contains no enabled visible transitions, or
      (b) $vis(\Phi) \subseteq TS(\underline{s})$

(3)   For every $\underline{s} \in \underline{\mathbf{W}}$, if there is an enabled invisible transition, then $TS(\underline{s})$ contains an enabled invisible transition

(4)   Every elementary cycle of $(\underline{\mathbf{W}},\underline{\mathbf{E}},\underline{\mathbf{T}})$ contains at least one state $\underline{s}$ such that $vis(\Phi) \subseteq TS(\underline{s})$.

Claim: $\varphi \in \Phi$ is valid at $\mathbf{s}_0$ in $(\mathbf{W},\mathbf{E},\mathbf{T})$ if and only if φ is valid at $\mathbf{s}_0$ in $(\underline{\mathbf{W}},\underline{\mathbf{E}},\underline{\mathbf{T}})$. $\Box$

(For proof see the unabridged version of this paper.)

3. IMPLEMENTATION OF THE METHOD

(This abridged version does not contain Chapter 3.)

4. EXAMPLE

To demonstrate the power of the LTL preserving stubborn set method we consider a version of the resource allocator system specified in [Pnueli 86]. Our system consists of a resource allocator and $n \geq 2$ customers which communicate via $2n$ Boolean variables r_i and g_i, initially F. The behaviour of customer i is shown below.

1:	$r_i := T$	/* t_{i1} */
2:	when $g_i \Rightarrow$ goto 3	/* t_{i2} */
3:	$r_i := F$	/* t_{i3} */
4:	when $\neg g_i \Rightarrow$ goto 1	/* t_{i4} */

The customer may use the resource when it is in state 3. The resource allocator behaves as follows.

1:	when $r_i \Rightarrow g_i := T$; goto $2i$	/* d_{i1} */
$2i$:	when $\neg r_i \Rightarrow g_i := F$; goto 1	/* d_{i2} */

The system has $(n+1) \cdot 3^n$ states (for proof see the unabridged version).

4.1 Example System as a V/T-System

The resource allocator system can be seen as a variable/transition system:

$$\mathbf{V} = \{A, C_1, ..., C_n, r_1, ..., r_n, g_1, ..., g_n\}$$

A is the state of the allocator and C_i is the state of the i:th customer

$$\mathbf{T} = \{t_{11}, t_{12}, t_{13}, t_{14}, t_{21}, ..., t_{24}, ..., t_{n1}, ..., t_{n4}, d_{11}, d_{12}, d_{21}, d_{22}, ..., d_{n1}, d_{n2}\}$$

$type(A) = \{1, 21, 22, ..., 2n\}$, $type(C_i) = \{1, 2, 3, 4\}$ and $type(r_i) = type(g_i) = \{F, T\}$

$\mathbf{s}_0(A) = 1 \wedge \forall\ i: \mathbf{s}_0(C_i) = 1 \wedge \mathbf{s}_0(r_i) = \mathbf{s}_0(g_i) = F$

next is too large to be listed here in full. It can be determined from the program code. For instance, $next(s, t_{12})$ is defined if $s(C_1) = 2 \wedge s(g_1) = T$. If $next(s, t_{12}) = s'$ (i.e. $s - t_{12} \rightarrow s'$) then $s'(C_1) = 3 \wedge \forall\ v \neq C_1: s'(v) = s(v)$.

We need an upper approximation to the set $\{(t, t') \in \mathbf{T} \times \mathbf{T} \mid \neg\ t \leftrightarrow t'\}$. If the sets of variables referred to by t and t' are disjoint, then $t \leftrightarrow t'$, because then the occurrence of t does not directly affect the environment of t' and vice versa. Let $ref(v)$ denote the set of transitions referring to a variable v. We conclude that the union of $ref(v)^2$ for every variable v of the system is an upper approximation to the set. We continue by investigating how the approximation can be improved.

We have $ref(A) = \{d_{11}, d_{12}, d_{21}, d_{22}, ..., d_{n1}, d_{n2}\}$. Because of the control structure of the resource allocator, transitions of the form d_{i2} are never enabled simultaneously with any other transition in $ref(A)$. Thus the left hand side of the implication in the definition of "$\leftrightarrow$" is never satisfied and the implication is always true, if $t = d_{i2}$ and $t' \in ref(A) - \{d_{i2}\}$. Consequently, the corresponding transition pairs (t, t') and (t', t) can be eliminated. A transition seldom accords with itself, so we choose not to try to eliminate the pairs (d_{i2}, d_{i2}). $\neg\ d_{i1} \leftrightarrow d_{j1}$ holds for $1 \leq i, j \leq n$, because d_{i1} can disable d_{j1}. As a result, the pairs (d_{i1}, d_{j1}) remain. So we have eliminated all pairs except (d_{i2}, d_{i2}) and (d_{i1}, d_{j1}), where $1 \leq i, j \leq n$. By similar argument all pairs except $\{(t, t) \mid t \in ref(C_i)\}$ can be eliminated from the sets $ref(C_i)^2$, as $ref(C_i) = \{t_{i1}, t_{i2}, t_{i3}, t_{i4}\}$.

As $ref(r_i) = \{t_{i1}, t_{i3}, d_{i1}, d_{i2}\}$, we have investigated $ref(r_i)^2$ except the pairs (t_{ij}, d_{ik}) and (d_{ik}, t_{ij}), where $j \in \{1, 3\}$ and $k \in \{1, 2\}$. Consider the states s enabling both t_{i1} and d_{i1}. We have $s(r_i) = T$ and $s(C_i) = s(A) = 1$. When t_{i1} or d_{i1} occurs the only variables whose values may be changed are r_i, g_i, C_i and A, so we investigate them only. If $s - t_{i1} \rightarrow s_1$ and $s - d_{i1} \rightarrow s'$ then $s_1(r_i) = s'(r_i) = T$, $s_1(g_i) = s(g_i)$, $s'(g_i) = T$, $s_1(C_i) = 2$, $s'(C_i) = 1 = s_1(A)$ and $s'(A) = 2i$. Therefore $en(s_1, d_{i1})$ and $en(s', t_{i1})$. Let $s_1 - d_{i1} \rightarrow s'_1$ and $s' - t_{i1} \rightarrow s''_1$. By computing the values of r_i, g_i, C_i and A in s'_1 and s''_1 we see that $s'_1 = s''_1$. Thus $t_{i1} \leftrightarrow d_{i1}$. By similar argument it can be shown that $t_{i3} \leftrightarrow d_{i2}$. Only the pairs (t_{i1}, d_{i2}), (t_{i3}, d_{i1}) and their inverses and the pairs of the form (t, t) were not eliminated from $ref(r_i)^2$. Investigating $ref(g_i)$ in the similar way leaves as the total only the following pairs left:

$$(t, t),\ (d_{i1}, d_{j1}),\ (t_{i1}, d_{i2}),\ (t_{i2}, d_{i2}),\ (t_{i3}, d_{i1}),\ (t_{i4}, d_{i1}),\ \text{where } 1 \leq i, j \leq n \text{ and } t \in \mathbf{T}.$$

Also the predicates $en(s, t, V)$ and the sets $wrup(t, V)$ for $t \in \mathbf{T}$ and for some $V \subseteq \mathbf{V}$ are needed by the stubborn set method. Consider t_{12}, for example. $en(s, t_{12})$ holds iff $s(C_1) = 2 \wedge s(g_1) = T$. Therefore $en(s, t_{12}, \{C_1\}) \Leftrightarrow s(C_1) = 2$ and $en(s, t_{12}, \{g_1\}) \Leftrightarrow s(g_1) = T$. We may choose $wrup(t_{12}, \{C_1\}) = \{t_{11}\}$, because it is the only transition whose occurrence can assign 2 to C_1. Similarly $wrup(t_{12}, \{g_1\}) = \{d_{11}\}$. Following the same principles we can evaluate $en(s, t, \{v\})$ and define $wrup(t, \{v\})$ for every $t \in \mathbf{T}$ and $v \in \mathbf{V}$. All the so defined $wrup$ sets contain exactly one transition, excluding the sets $wrup(d_{i1}, \{A\})$ which are all equal to $\{d_{12}, d_{22}, ..., d_{n2}\}$.

4.2 Reduced State Space of the Example System

Now we construct a reduced state space of the system. We want to know whether the system guarantees that the resource cannot be used simultaneously by two customers (it does), and whether a customer which has requested a resource eventually uses the resource (not true). Because of the symmetry of the system the first requirement can be specified by the LTL formula $\Box(\,(C_1 \neq 3) \vee (C_2 \neq 3)\,)$, where C_1 and C_2 are the states of customers 1 and 2, respectively. The second requirement can be encoded as $\Box(\,(C_1 = 2) \Rightarrow \Diamond(C_1 = 3)\,)$. The transitions which can modify the truth values of the state predicates $C_1 \neq 3 \vee C_2 \neq 3$, $C_1 = 2$ and $C_1 = 3$ are $t_{11}, t_{12}, t_{13}, t_{22}$ and t_{23}. Thus we choose $vis(\Phi) = \{t_{11}, t_{12}, t_{13}, t_{22}, t_{23}\}$.

In the initial state $\mathbf{s_0}$ the transitions t_{i1} and no other transitions are enabled. If we want to build a stubborn set $TS(\mathbf{s_0})$ around t_{i1} then we have to take d_{i2} into the set because $\neg t_{i1} \leftrightarrow d_{i2}$. d_{i2} is disabled, thus we have to find $V \subseteq \mathbf{V}$ such that $\neg en(\mathbf{s_0}, d_{i2}, V)$ and include $wrup(d_{i2}, V)$ to the set. We can choose $V = \{A\}$ and $wrup(d_{i2}, \{A\}) = \{d_{i1}\}$. $\neg en(\mathbf{s_0}, d_{i1}, \{r_i\})$ holds and $wrup(d_{i1}, \{r_i\})$ $= \{t_{i1}\}$, but t_{i1} is already in the set. We can thus stop with the set $TS(\mathbf{s_0}) = \{t_{i1}, d_{i1}, d_{i2}\}$. It is stubborn and contains exactly one enabled transition, namely t_{i1}. To satisfy Assumption (3) of Theorem 2.8 it is reasonable to choose the i so that $i > 1$, and the algorithm in Chapter 3 indeed does so. So only one transition is fired in $\mathbf{s_0}$, namely t_{i1}. Call the resulting new state s'.

By applying the above reasoning again one can see that if $j \neq i$, $\{t_{j1}, d_{j1}, d_{j2}\}$ is stubborn in s'. In an attempt to avoid visible transitions, the algorithm in Chapter 3 chooses $\{t_{j1}, d_{j1}, d_{j2}\}$ for some $j \geq 2$ and so on until the state s_{12*} such that $s_{12*}(A) = s_{12*}(C_1) = 1$ and $s_{12*}(C_i) = 2$ for $2 \leq i \leq n$ is reached. At this stage we have generated n states.

The enabled transitions at s_{12*} are t_{11} and d_{i1} for $i \geq 2$. Assumption (3) forces us to include at least one of d_{i1}, $i \geq 2$, into $TS(s_{12*})$. Because $\neg d_{i1} \leftrightarrow d_{j1}$ we conclude $\forall\, i \geq 1\colon d_{i1} \in TS(s_{12*})$. Intuitively, this reflects the fact that it is essential which customer gets the resource. d_{11} is disabled and, as before, takes us to t_{11}. That is, also C_1 is given the chance to take the resource. So we have to take all enabled transitions into $TS(s_{12*})$. In essence, the algorithm gives all the other customers the chance to take the resource before t_{11} occurs, because the occurrence of t_{11} modifies the value of the state predicate $C_1 = 2$ in Φ. The algorithm tries to find out what can happen before the state predicate value is modified.

For the continuation of the reduced state space generation see the unabridged version of this paper. Assumption (4) of Theorem 2.8 is satisfied without further action. The total number of states generated is $11n - 6$.

The validity of the formulas $\Box(\,(C_1 \neq 3) \vee (C_2 \neq 3)\,)$ and $\Box(\,(C_1 = 2) \Rightarrow \Diamond(C_1 = 3)\,)$ can now be checked from the reduced state space. The former is true, the latter is not. The reduced state space is linear in the number of customers, while the ordinary state space is exponential.

5. CONCLUDING REMARKS

We showed how to generate reduced state spaces such that the truth values of LTL formulas are preserved, provided that the formulas are given before the reduced state space generation commences and they do not contain the "next state" operator. In the example in Chapter 4 the reduction of the size of the state space is from exponential to linear in the size of the system. This is a very good result. However, it is currently not known how well the LTL preserving stubborn set method performs on the average. One may expect that the size of the reduced state space increases when more and more variables are referred to by the formulas to be preserved.

By the time of the writing of this paper the LTL preserving stubborn set method has not been implemented. However, a related stubborn set state space reduction method has been implemented into a tool called *Toras* [Wheeler & 90]. Toras is being developed in Telecom Australia Research Laboratories. Among other features, it supports an as yet unpublished version of the stubborn set method which preserves the *failure semantics* [Brookes & 84] of systems. The method differs from the one presented in this paper in that it does not need Assumptions (1) and (4) of Theorem 2.8. To give an example of the performance of Toras, a certain version of the n dining philosophers system has 3^n-1 states, and the basic stubborn set method reduces the number to $3n^2-3n+2$ states. For the 100 philosopher system ($\approx 10^{47}$ states) Toras generated the predicted 29 702 states in 20 minutes CPU time on a Sun 3/60 [Wheeler & 90].

ACKNOWLEDGEMENTS

This work has been supported by the Technology Development Centre of Finland (TEKES).

REFERENCES

[Aho & 74] Aho, A. V., Hopcroft, J. E. & Ullman, J. D.: *The Design and Analysis of Computer Algorithms.* Addison-Wesley, Reading, Massachusetts 1974, 470 p.

[Brookes & 84] Brookes, S. D., Hoare, C. A. R. & Roscoe, A. W.: *A Theory of Communicating Sequential Processes.* Journal of the ACM 31 (3) 1984, pp. 560–599.

[Clarke & 86] Clarke, E. M., Emerson, E. A. & Sistla, A. P.: *Automatic Verification of Finite-State Concurrent Systems using Temporal Logic Specifications.* ACM Transactions on Programming Languages and Systems 8 (2) 1986 pp. 244–263.

[Lamport 83] Lamport, L.: *What Good is Temporal Logic?* Information Processing '83, North-Holland pp. 657–668.

[Lichtenstein & 85] Lichtenstein, O. & Pnueli, A.: *Checking that Finite State Concurrent Programs Satisfy their Linear Specification.* Proceedings of the Twelfth ACM Symposium on the Principles of Programming Languages, January 1985 pp. 97–107.

[Overman 81] Overman, W. T.: *Verification of Concurrent Systems: Function and Timing.* Ph.D. Dissertation, University of California Los Angeles 1981, 174 p.

[Pnueli 86] Pnueli, A.: *Applications of Temporal Logic to the Specification and Verification of Reactive Systems: A Survey of Current Trends.* In: Current Trends in Concurrency, Lecture Notes in Computer Science 224, Springer 1986 pp. 510–584.

[Valmari 88a] Valmari, A.: *Error Detection by Reduced Reachability Graph Generation.* Proceedings of the Ninth European Workshop on Application and Theory of Petri Nets, Venice, Italy 1988 pp. 95–112.

[Valmari 88b] Valmari, A.: *Heuristics for Lazy State Generation Speeds up Analysis of Concurrent Systems.* Proceedings of the Finnish Artificial Intelligence Symposium STeP-88, Helsinki 1988 Vol. 2 pp. 640–650.

[Valmari 88c] Valmari, A.: *State Space Generation: Efficiency and Practicality.* Ph.D. Thesis, Tampere University of Technology Publications 55, 1988, 169 p.

[Valmari 89a] Valmari, A.: *Eliminating Redundant Interleavings during Concurrent Program Verification.* Proceedings of Parallel Architectures and Languages Europe '89 Vol 2, Lecture Notes in Computer Science 366, Springer 1989 pp. 89–103.

[Valmari 89b] Valmari, A.: *Stubborn Sets for Reduced State Space Generation.* Proceedings of the Tenth International Conference on Application and Theory of Petri Nets, Bonn, FRG 1989 Vol. 2 pp. 1–22. A revised version to appear in Advances in Petri Nets 90, Lecture Notes in Computer Science, Springer.

[Wheeler & 90] Wheeler, G. R., Valmari, A. & Billington, J.: *Baby Toras Eats Philosophers but Thinks about Solitaire.* Proceedings of the Fifth Australian Software Engineering Conference, Sydney, NSW, Australia, 1990 pp. 283–288.

USING OPTIMAL SIMULATIONS TO REDUCE REACHABILITY GRAPHS

Ryszard Janicki
Department of Computer Science and Systems
McMaster University
Hamilton, Ontario, Canada, L8S 4K1

Maciej Koutny
Computing Laboratory
The University of Newcastle upon Tyne
Newcastle upon Tyne NE1 7RU, U.K.

ABSTRACT

We here discuss an approach which uses the optimal simulation - a kind of reachability relation - to enable reasoning about important dynamic properties of a concurrent system. The optimal simulation usually involves only a very small subset of the possible behaviours generated by the system, yet provides a sufficient information to reason about a number of interesting system's properties (such as deadlock-freeness and liveness). In this paper we show how the optimal simulation might be used to generate a reachability graph which is usually much smaller than the standard reachability graph of the system; however, both graphs essentially convey the same information about its dynamic behaviour.

INTRODUCTION

High complexity of the design of concurrent programs, such as inherently concurrent communication protocols, made apparent the need for appropriate formal specification methods, and specialised verification techniques enhanced by computer-aided tools for automated analysis of concurrent programs. Examples of the verification techniques include algebraic transformations of CSP and CCS [Hoa85,HM85]; temporal logic model checkers [CG87, CES86]; and invariant methods developed for Petri nets [MS82].

The process of verification of dynamic properties of concurrent systems often involves some kind of reasoning about the complete state-space of the system, e.g. proving deadlock-freeness requires showing that it is not possible to reach a state in which no transition is enabled. Reasoning about the complete state-space of concurrent systems has one serious drawback which is a combinatorial explosion of the state-space. Even a simple concurrent system can generate many hundreds or thousands of states. Moreover, the higher the degree of concurrency of the system is (the degree of concurrency is roughly the number of sequential subsystems) the faster its state-space becomes unmanageable. To cope with this problem a number of sophisticated techniques have been developed, such as *induction* [Kel76] which employs invariants to prove that a property is true in all the states of the system, and *reduced state-space analysis* [Jen87,Val89,God90] in which reasoning about the complete state-space is replaced by the analysis of its reduced representation [MR87].

In [JK89] and [JK89a] we discussed a possibility of defining a fully expressive reachability relation on the system's histories which would be a 'small' subset of the complete reachability relation. We defined such a reduced reachability relation and called it the *optimal simulation*. It enables reasoning about a number of dynamic properties of a concurrent system, and at the same time requires a minimal computational effort. Optimal simulation has been defined in a very general trace-based setting which makes it applicable to different models for concurrency, such as Petri nets, CCS, CSP, or automata-based models.

The reachability graphs of finite-state systems can be regarded as *finite representations of reachability relations*. Since the optimal simulation provides the same information about relevant dynamic properties of the system as the full reachability relation [JK89,JK89a], the reachability graph of optimal simulation (i.e. its finite representation) and the full reachability graph may be considered as equivalent. The optimal simulation is always a subset of the full reachability, but of course it does not mean that the reachability graph of optimal simulation is always (much) smaller than the full reachability graph. However, we do claim that in the case of concurrent systems exhibiting high degree of concurrency (i.e. those with many sequential components), the reachability graph of optimal simulation is much smaller than the full reachability graph. Thus it is advantageous to use optimal simulation as a tool to *reduce the size of reachability graphs of concurrent systems*.

Unfortunately, as opposed to the full reachability graph, in the general case it is not clear how to generate the reachability graph for optimal simulation in an efficient way. In some sense this is a negative side-effect of the above mentioned generality of optimal simulation. In this paper we will outline how such a graph (as we claim, reduced in the majority of cases) can be constructed for Petri nets which can be decomposed onto finite state machines.

We will not prove here that optimal simulation is indeed behaviourally equivalent to the full reachability, nor justify the construction of the optimal simulation relation. These issues have been dealt with in [JK89] which is widely available (Lecture Notes in Computer Science 366), and in [JK89a] which can be sent on request. Proofs of technical results presented in this paper can be found in [JK89,JK90].

Note that our approach is based on the assumption that concurrent behaviours (histories) can be modelled by causal partial orders. We will represent those partial orders by certain equivalence classes of step sequences, generalising the notion of traces of [Maz86].

We would like to point out that the major methodological difference between our approach to minimise reachability graphs and those developed in [Jen87,Val89,God90] is that we do not try to minimise (or even deal with) the full reachability graph in an explicit way. All what we are trying to do is to build a reachability graph which represents the optimal simulation relation (a subset of full reachability). We then make a claim, based on the general properties of optimal simulation, that in majority of cases, such a graph is much smaller than the original full reachability graph.

1 MOTIVATION

Execution paths generated by Petri nets can be represented by *step sequences* - each step being a finite set of transitions executed simultaneously. Consider the Petri net in Fig. 1.1. Its behaviour might be briefly described in the following way: All step sequences must begin with transition a. After that one can simultaneously execute transitions b and c, or execute b followed by c, or execute c followed by b. The net generates three step sequences leading to a deadlock, $n_1 = \{a\}\{b,c\}$, $n_2 = \{a\}\{b\}\{c\}$ and $n_3 = \{a\}\{c\}\{b\}$. Suppose now that we were about to find

168

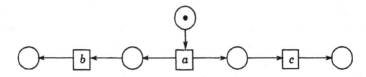

Fig. 1.1

all the deadlocks of the net by following as few step sequences as possible and by selecting possibly shortest step sequences. An exhaustive search would include n_1, n_2 and n_3. But we may observe that all these lead to the same deadlocked marking, and that n_1 is shorter than both n_2 and n_3. Hence an efficient search should include just one path, n_1.

The above example is an instance of the following general problem: *Is there a way of executing a net which is both expressive and efficient?* By an expressive execution we mean one providing enough information to verify relevant properties of the system, e.g. liveness or termination, whereas by an *efficient* execution we mean one which requires minimal computational effort, e.g., by avoiding execution paths providing redundant information. Referring to our example, one may observe that n_1 has a straightforward operational interpretation as it follows the rule: always choose a *maximal* set of independent transitions to be executed next, a rule which characterises *maximally concurrent* execution. Employing maximal concurrency is an attractive idea, both conceptually and from the point of view of implementation. Unfortunately, there are cases in which maximally concurrent execution is not sufficiently expressive (see [JLKD86] for necessary and sufficient condition where it is). To show this we take the net in Fig. 1.2. The maximally concurrent execution can find only one deadlocked marking of the net, by following step sequence $\rho_1=\{a,b\}\{d\}$. The other deadlocked marking, which might be reached by following $\rho_2=\{b\}\{c\}$, is left undetected.

In [JK89,JK89a] we have defined, by generalising maximally concurrent execution, the *optimal simulation* which is both expressive and efficient way of executing the net for verification purposes. Fig. 1.2 shows both the full reachability graph of the net and the reachability graph of the optimal simulation. The latter one is smaller, but because in this case only two transitions a and b can be fired concurrently, the difference in size is not significant. However, in the case of net in Fig. 1.3, the reachability graph of the optimal simulation is isomorphic to that in Fig. 1.2, while (as one may easily check) the full reachability graph would hardly fit on a single page.

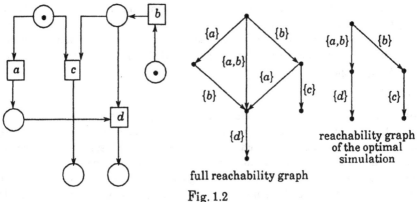

full reachability graph

reachability graph of the optimal simulation

Fig. 1.2

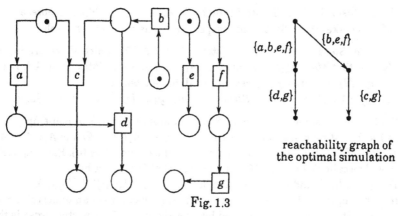

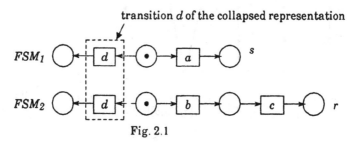

Fig. 1.3

2 SMD NETS AND THE OPTIMAL SIMULATION

In our discussion we will use state machine decomposable (SMD) nets which model non-sequential systems composed of a number of sequential subsystems which synchronise by means of common transitions. SMD can provide semantical basis for more complex Petri nets [Rei85] and other models like COSY [LSC81], CCS [HM85], and CSP [Hoa85] (see [Tau89]).

In the SMD nets, a sequential subsystem is represented by a *finite state machine* which is a triple (S_i,T_i,F_i) such that S_i and T_i are disjoint finite sets of *places* and *transitions*, and $F_i \subseteq S_i \times T_i \cup T_i \times S_i$ is the *flow* relation such that for every $t \in T_i$ there is exactly one s and exactly one p satisfying $(s,t) \in F_i$ and $(t,p) \in F_i$.

An *SMD-net* is a tuple $N=(FSM_1,...,FSM_n,M_{init})$ such that each $FSM_i=(S_i,T_i,F_i)$ is a finite-state machine, $S_i \cap S_j = \varnothing$ for $i \neq j$, and M_{init} is the *initial* marking. (Marking is a set of places which has exactly one place in common with each S_i.) In what follows we will assume that N is fixed and denote: $S=S_1 \cup ... \cup S_n$, $T=T_1 \cup ... \cup T_n$ and $F=F_1 \cup ... \cup F_n$.

For a set of transitions A we denote: $A^\bullet=\{s \mid \exists t \in A. (t,s) \in F\}$ and $^\bullet A=\{s \mid \exists t \in A. (s,t) \in F\}$.

Fig. 2.1 shows an SMD net. As usual, places are represented by circles, transitions by boxes, the flow relation by arcs, and marking by tokens.

Let $ind \subseteq T \times T$ be the set of all pairs of transitions (a,b) such that there is no T_i comprising both a and b. Such a and b are interpreted as *independent*, and only independent transitions can be executed simultaneously.

Let *Ind* be the set of *steps*, each step being a non-empty set of mutually independent transitions, i.e. $Ind=\{A \subseteq T \mid A \neq \varnothing \wedge \forall a,b \in A. a=b \vee (a,b) \in ind\}$.

For the net of Fig. 2.1 we have $ind=\{(a,b),(b,a),(a,c),(c,a)\}$ and $Ind=\{\{a\},\{b\},\{c\},\{d\},\{a,b\},\{a,c\}\}$. We could have defined N as $N=(S,T,F,M_{init})$ for which there are finite state machines (S_i,T_i,F_i) such that $S=S_1 \cup ... \cup S_n$, $T=T_1 \cup ... \cup T_n$, $F=F_1 \cup ... \cup F_n$, and M_{init} is a marking

transition d of the collapsed representation

Fig. 2.1

with exactly one token in each S_i (see [Tau89]); however, the 'collapsed' representation - $N=(S,T,F,M_{init})$ - makes the definition of independent transitions less readable.

A *step sequence*, $\sigma \in Steps$, is a sequence of steps $\sigma=A_1...A_k$ for which there are markings $M_0,M_1,...,M_k$ such that $M_0=M_{init}$ and for all i, ${}^\bullet A_i \subseteq M_{i-1}$ and $M_i=(M_{i-1}-{}^\bullet A_i)\cup A_i{}^\bullet$. Later we will denote $mar_\sigma=M_k$. The empty step sequence will be denoted by λ.
For the net of Fig. 2.1 we have $\{a,b\}\{c\}\in Steps$ and $mar_{\{a,b\}\{c\}}=\{s,r\}$, but $\{a,b\}\{d\}\notin Steps$.

Let $\simeq$ be the least equivalence relation on *Steps* containing all pairs of non-empty step sequences (σ,ω) such that $\sigma=\sigma_1A\sigma_2$ and $\omega=\sigma_1A_1A_2\sigma_2$, where $A_1\cap A_2=\emptyset$ and $A_1\cup A_2=A$. The equivalence class of $\simeq$ containing step sequence σ will be denoted by $[\sigma]$. Each equivalence class H of $\simeq$ will be called a *history*, $H\in Hist$. For the net of Fig. 2.1 we have
$[\{a,b\}]=\{\{a\}\{b\},\{b\}\{a\},\{a,b\}\}$ and $Hist=\{[\lambda],[\{d\}],[\{a\}],[\{b\}],[\{a,b\}],[\{b\}\{c\}],[\{a\}\{b\}\{c\}]\}$.
Step sequences belonging to H can be seen as different realisations of an underlying concurrent history which itself may be represented by a partial order. This partial order is the intersection of all the partial orders induced by the step sequences in H. This is illustrated in Fig. 2.2 for history $H=[\{a\}\{b\}\{c\}]$ of the net of Fig. 2.1. Note that step sequence $\sigma=\{a\}\{b,c\}\{a\}$ induces a partial order in which the first occurrence of a precedes the occurrences of b and c, and the occurrences of b and c are un-ordered and both precede the second occurrence of a. Every partial order induced by step sequences has the following property: its disorder relation is transitive. Such orders are sometimes called *stratified partial orders*.

The concepts of $\simeq$ and a history $H\in Hist$ are natural generalisations of similar concepts from the theory of *partial commutative monoids* [CF69,Maz86,Zie89]. If we restricted step sequences to just sequences (interleavings) in the definition of *Hist* then we would get exactly the classical notion of Mazurkiewicz traces [Maz86] (the name 'trace' is sometimes used [Hoa85] to mean 'sequence', so we write 'Mazurkiewicz trace' to avoid any confusion). Our representation of a history as a partial order is a natural generalisation of the result of [Szp30] on the representation of partial orders by the set of their linearisations. We represent partial orders by the set of their *stratifications*. The basic advantage of the approach in [Maz86] is that a causal partial order may be represented by just one interleaving. In our case, every history may be represented by just one step sequence. In particular, we may choose the shortest one, i.e. maximally concurrent. This idea will be developed further when the definition of *canonical* step sequence - a very fundamental concept of our approach - will be given.

For every history H, *enabled(H)* is the set of all steps A such that σA is a step sequence for at least one $\sigma \in H$. It turns out that if $\sigma,\omega \in H$ and $A\in enabled(H)$ then $mar_\sigma=mar_\omega$, $\sigma A\in Steps$

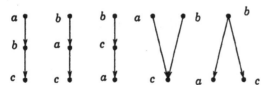

partial orders induced by step sequences belonging to history   partial order underlying
$H=\{\{a\}\{b\}\{c\}, \{b\}\{a\}\{c\}, \{b\}\{c\}\{a\}, \{a,b\}\{c\}, \{b\}\{a,c\}\}$ history H

Fig. 2.2

and $\sigma A \simeq \omega A$. Hence we can define mar_H to be the marking mar_σ, and $H\circ[A]$ to be the history $[\sigma A]$.

In [JK89,JK89a] we introduced the notion of a *simulation* which is a kind of reachability relation on the histories representing a possible mode of executing the net. In this paper we will deal only with two simulations, the *full* and *optimal* ones. The full simulation is simply defined as $FULL = \{(G,H) \in Hist \times Hist \mid \exists A. \ H = G\circ[A]\}$. $FULL$ represents the dynamic behaviour of N in a complete way. Its advantage is relatively straightforward definition and natural interpretation, its disadvantage is the size of its reachability graph. Even for small nets the graph grows beyond any manageable size, making the formal verification of the net's properties extremely difficult. It was our goal in [JK89,JK89a] to find possibly smallest simulation which could be used for the verification of relevant net properties. As a solution we proposed the *optimal* simulation, OPT. There are three reasons why OPT can be regarded as the optimal simulation:

(1) There are a number of behavioural properties which are common to $FULL$ and OPT. For example, $FULL$ and OPT generate the same sets of deadlocked markings. It is also possible to verify liveness using OPT. Indeed, as we claimed in [JK89a], $FULL$ and OPT essentially capture *the same behavioural properties of the net* (a concurrent system, in general).

(2) OPT involves a minimal set of histories, i.e. each proper subset of OPT is less expressive than $FULL$, and it may not be used, e.g., to verify the deadlock-freeness.

(3) The information about the net is generated in OPT using the shortest step sequences. For instance, each deadlocked marking will be generated by following the shortest step sequence leading to it.

Moreover, there is no other simulation which would satisfy (1)-(3). OPT is defined as follows.

A step sequence $\sigma = A_1...A_k$ is *canonical* if for all $i \geq 2$ and $a \in A_i$ there is $b \in A_{i-1}$ such that $(a,b) \notin ind$. Intuitively, in canonical step sequence the execution of transitions is never delayed (no transition can be moved from A_i to A_{i-1}). It can be shown [CF69,JLKD86] that every history H contains exactly one canonical step sequence, $can(H)$. To define OPT we first introduce an auxiliary reachability relation on histories: $CAN = \{(G,H) \mid can(H) = can(G)A\}$. We also define $Hist_{max}$ to be the set of all histories H whose canonical step sequence ends with a maximal step, i.e. if $can(H) = \sigma A$ and $A \subseteq B \in enabled([\sigma])$ then $A = B$.

The *optimal* simulation OPT is defined as the smallest subset of CAN such that for every $H \in Hist_{max}$ there are histories $H_1,...,H_m$ satisfying: $H_1 = [\lambda]$, $H_m = H$, and $(H_i,H_{i+1}) \in OPT$, for all $i < m$. We also denote $Hist_{opt} = \{[\lambda]\} \cup \{H \mid (G,H) \in OPT\}$. For the net of Fig. 2.1 we have $Hist_{max} = \{[\{d\}],[\{a,b\}],[\{a,b\}\{c\}]\}$ and $OPT = \{([\lambda],[\{d\}]),([\lambda],[\{a,b\}]),([\{a,b\}],[\{a,b\}\{c\}])\}$.

3 REACHABILITY GRAPH OF THE OPTIMAL SIMULATION

A reachability graph of the full simulation can be defined as: $RG_{FULL} = (V, Arcs, M_{init})$, where $V = \{mar_H \mid H \in Hist\}$ is the set of nodes; $Arcs = \{(mar_H, A, mar_{H\circ[A]}) \mid A \in enabled(H)\}$ is the set of arcs; and M_{init} is the initial node.

The above definition is not very useful in the case of the optimal simulation. The reason is that even if G and H two are histories in $Hist_{opt}$ satisfying $mar_G = mar_H$ and $(H,H\circ[A]) \in OPT$, then it does not necessarily follow that $(G,G\circ[A]) \in OPT$ (see [JK90] for an example). Hence the construction used to define RG_{FULL} in which histories leading to the same marking were assigned the same node of the graph would not work. To guarantee that $(H,H\circ[A]) \in OPT \Leftrightarrow (G,G\circ[A]) \in OPT$ we strengthen the condition $mar_G = mar_H$, as follows.

Let $V \subseteq CAN$ comprise all pairs of histories $(G,G \circ[A])$ such that there is a non-empty set $I \subseteq \{1,...,n\}$ satisfying the following (below $R = \bigcup_{i \in I} S_i$, $U = \bigcup_{i \in I} T_i$ and $V = \bigcup_{i \notin I} T_i$):

(3.1) $A \cap U = \emptyset$.

(3.2) There is $t \in U\text{-}V$ such that $\{t\} \in enabled(G)$.

(3.3) There is no $u \in U \cap V$ such that $\cdot u \cap R \subseteq mar_G$.

The last definition is illustrated in Fig. 3.1. The idea behind V is that in all continuations of $G \circ[A]$ in CAN the tokens in the subnets FSM_i, for $i \in I$, will be 'frozen'. Hence, by (3.2), no such continuation can yield a step which is maximal.

For every history H, let Γ_H be the set of all A such that $(H,H \circ[A]) \in CAN\text{-}V$. It can be shown that if $(H,H \circ[A]) \in OPT$ then $A \in \Gamma_H$. Hence when generating the reachability graph of OPT instead of taking $enabled(H)$ as the potential next steps for a history H we can restrict ourselves to the (usually much smaller) set Γ_H. What is, however, more important, Γ_H can be used to identify histories with identical continuations in OPT:

Let $\sim$ be a relation on histories such that $G \sim H$ if $mar_G = mar_H$ and $\Gamma_G = \Gamma_H$. It can be shown that if $G,H \in Hist_{opt}$ and $G \sim H$ then $(G,G \circ[A]) \in OPT$ implies $(H,H \circ[A]) \in OPT$ and $G \circ[A] \sim H \circ[A]$. Hence a reachability graph of the optimal simulation $RG_{OPT} = (V,Arcs,v_{init})$ can be defined in the following way:

(1) $V = \{(mar_H,\Gamma_H) \mid H \in Hist_{opt}\}$.

(2) $Arcs = \{((mar_H,\Gamma_H),A,(mar_{H\{A\}},\Gamma_{H\{A\}})) \mid (H,H \circ[A]) \in OPT\}$.

(3) $v_{init} = (M_{init},\Gamma_{[\lambda]})$.

The above definition is an operational one, i.e. it can be used to describe an efficient algorithm constructing RG_{OPT}.

Generating reachability graph RG_{FULL} is usually done in a loop which checks the already generated nodes and steps 'enabled' at those nodes (nodes are labelled with markings). If there exists a node labelled by mar_H and a step A which have not yet been tried, the algorithm generates marking $M = mar_{H\{A\}}$. It then adds a new node labelled by M and an arc to the graph if M has not yet been a label; otherwise it draws an arc to the node labelled by M.

The algorithm generating RG_{OPT} follows in principle the same pattern. There is, however, one essential difference. An arc cannot be accepted as belonging to RG_{OPT} before another arc, labelled with a maximal step, is found which can be reached from that arc. Hence one first generates an auxiliary reachability graph in a similar way as it is done for RG_{FULL} and then prunes the arcs from which an arc labelled with a maximal step cannot be reached, obtaining RG_{OPT}. A formal description of this algorithm can be found in [JK90].

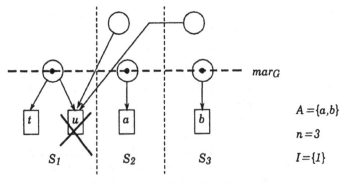

$A = \{a,b\}$

$n = 3$

$I = \{1\}$

Fig. 3.1: u is excluded by (3.3)

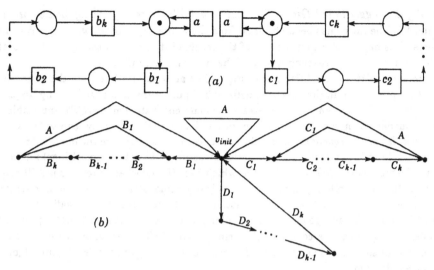

Fig. 3.2

To illustrate the last definition we take the net N_k of Fig. 3.2(a). Fig. 3.2(b) shows RG_{OPT}, where $A=\{a\}$, $B_i=\{b_i\}$, $C_i=\{c_i\}$ and $D_i=\{b_i,c_i\}$. RG_{OPT} has $3k$ vertices and $3k+5$ arcs which compares favourably with RG_{FULL} with k^2 vertices and $3k^2+1$ arcs.

CONCLUDING REMARKS

In this paper we presented the idea behind an algorithm generating reachability graph of the optimal simulation - a way of executing a system directly generalising the maximally concurrent execution [JK89]. Together with the results obtained in [JK89] and [JK89a], this gives a strong indication that the graph RG_{OPT} would in general case be much smaller than the full reachability graph. Furthermore, the higher the degree of concurrency the system exhibits, the more can be gained by using RG_{OPT} instead of RG_{FULL}. There can, however, be situations where RG_{OPT} does not have an apparent advantage over RG_{FULL}. For example, RG_{OPT} of a net can be bigger than RG_{FULL} (see [JK90]). Another problem can be identified by taking the net of Fig. 3.2. If we remove transition b_k and join b_{k-1} with the place holding a token, then although RG_{OPT} for the modified net will be smaller than RG_{FULL}, it will have $o(k^2)$ nodes and arcs. There are two points to be made which show that problems of such kind are less serious than it might look at the first glance.

Let MAX be the maximally concurrent simulation [JLKD86,JK89], and let RG_{MAX} be its reachability graph. (Formally, MAX is the maximal subset of CAN which only involves histories from $Hist_{max}$.) It is not difficult to see that $RG_{MAX} \subseteq RG_{FULL}$ and $RG_{MAX} \subseteq RG_{OPT}$, as well as $MAX \subseteq OPT \subseteq FULL$. Furthermore, OPT is a minimal subset of $FULL$ containing the same behavioural information as $FULL$ does and, intuitively, OPT is only 'slightly' bigger than MAX. As the result, RG_{OPT} is only 'slightly' bigger than RG_{MAX}. On the other hand, $RG_{MAX} \subseteq RG_{FULL}$, and the difference between RG_{MAX} and RG_{FULL} depends strongly on the degree of concurrency exhibited by the net. If the net contains only a few concurrent transitions then the difference between RG_{MAX} and RG_{FULL} is rather small, and in such a case RG_{OPT} might be bigger than RG_{FULL}. When the net contains many concurrent transitions the difference betwe n RG_{MAX} and RG_{FULL} increases dramatically, while RG_{OPT} is still *only 'slightly' bigger than* RG_{MAX}. This can be well illustrated by taking the nets from Fig. 1.2 and 1.3. The latter has been obtained from the former by adding two simple concurrent subnets. This had no effect on the size of the reachability graph of the optimal simulation, while the size of the full reachability graph has increased significantly.

The problems with the modified net of Fig. 3.2(a) are essentially due to the generality of *OPT*. The optimal simulation has been defined in a pure transition-based setting. In particular, the lack of any reference to the states of a concurrent system makes *OPT* applicable to almost all models for non-sequential computation, but sometimes it may lead to less efficient solution as far as the reachability graphs are concerned. Referring to the modified net of Fig. 3.2(a), by taking into account the particular structure of the states (markings) for this net, we may further reduce RG_{OPT} to a graph which has $o(k)$ nodes and arcs and is still equivalent to the full reachability graph. What this clearly demonstrates is that for specific system models, e.g. for those which support the notion of state, it is possible to modify RG_{OPT} in a way which takes advantage of some particular properties of that model. RG_{OPT} should not therefore be regarded as a complete blueprint for an efficient reduction of the reachability graphs, but in some cases as particularly suitable starting point for developing algorithms for such a reduction.

Although in this paper we consider only nets which can be decomposed onto finite state machines, our approach can be extended to other kinds of nets and models. In [JK90], where all the proofs are given, we use asynchronous automata of [Zie89] as a model of concurrent system. Moreover, [Tau89] enables a translation of our results to CCS and TCSP. In fact, if the behaviour of a concurrent system can be adequately modelled in terms of Mazurkiewicz traces [Maz86], then the approach presented above can always be applied.

Our final comment is that the fusion of our approach with one of the approaches which deal explicitly with reachability graphs, as those of [Val89] and [God90], is likely to lead to highly efficient algorithms for reduced reachability graph generation.

ACKNOWLEDGMENT
We would like to thank Antti Valmari for his comments on the modified net of Fig.3.2. The work of the first author was supported by a grant from NSERC No. OGP 0036539. The work of the second author was supported by ESPRIT BRA 3148 Project DEMON.

REFERENCES

[CF69]   Cartier P., Foata D., *Problemes combinatoires de communication et rearrangements*, Lecture Notes in Mathematics 85, Springer 1969.

[CG87]   Clarke E.M., Grümberg O., *Research on Automatic Verification of Finite-State Concurrent Systems*, Ann. Rev. Comp. Sci. 2(1987), 269-290.

[CES86]   Clarke E.M., Emerson E.A., Sistla A.P., *Automatic Verification of Finite-State Systems using Temporal Logic Specifications*, ACM Transactions on Programming Languages and Systems 8(1986), 244-263.

[God90]   Godefroid P., *Using Partial Orders to Improve Automatic Verification Methods*, Proc. of CAV'90, this volume.

[HM85]   Hennessy M. and Milner R., *Algebraic Laws for Nondeterminism and Concurrency*, JACM 32(1985), 136-161.

[Hoa85]   Hoare C.A.R., *Communicating Sequential Processes*, Prentice-Hall, 1985.

[JLKD86]   Janicki R., Lauer P.E., Koutny M., Devillers R., *Concurrent and Maximally Concurrent Evolution of Non-Sequential Systems*, Theoretical Computer Science 43(1986), 213-238.

[JK89]   Janicki R., Koutny M., *Towards a Theory of Simulation for Verification of Concurrent Systems*, Lecture Notes in Computer Science 366, Springer 1989, 73-88.

[JK89a]   Janicki R., Koutny M., *Optimal Simulation for Verification of Concurrent Systems*, Technical Report No. 89-05, McMaster University,Hamilton, Ontario, 1989.

[JK90]   Janicki R., Koutny M., *On Some Implementation of Optimal Simulations*, Technical Report No. 90-07, McMaster University,Hamilton, Ontario, 1990 (also to appear in the ACM/AMS DIMACS series).

[Jen87]   Jensen K., *Coloured Petri Nets*, LNCS 254, Springer 1987, pp. 248-299.

[Kel76]   Keller R.M., *Formal Verification of Concurrent Programs*, CACM 19(7), 1976, 371-384.

[LSC81]   Lauer P.E., Shields M.W., Cotronis J.Y., *Formal Behavioural Specification of Concurrent Systems without Globality Assumptions*, Lecture Notes in Computer Science 107, Springer 1981, 115-151.

[MS82]   Martinez J., Silva M., *A Simple and Fast Algorithm to Obtain All Invariants of a Generalized Petri Net*, Informatik-Fachberichte 52, Springer 1982, 301-310.

[Maz86]   Mazurkiewicz A., *Trace Theory*, Lecture Notes in Computer Science 255, Springer 1986, 297-324.

[MR87]   Morgan E.T, Razouk R.R., *Interactive State-Space Analysis of Concurrent Systems*, IEEE Transactions on Software Engineering 13(10), 1987.

[Rei85]   Reisig W., *Petri Nets*, Springer 1985.

[Szp30]   Szpilrajn-Marczewski E., *Sur l'extension de l'ordre partial*, Fundamenta Mathematicae 16 (1930), pp. 386-389.

[Tau89]   Tauber D., *Finite Representations of CCS and TCSP Programs by Automata and Petri Nets*, Lecture Notes in Computer Science 369, Springer 1989.

[Val89]   Valmari A., *Stubborn Sets for Reduced State Space Generation*, Proc.of the 10th International Conference on Application and Theory of Petri Nets, Bonn, June, 1989.

[Zie89]   Zielonka W., *Safe Executions of Recognizable Trace Languages by Asynchronous Automata*, Lecture Notes in Computer Science 363, Springer 1989.

Using Partial Orders to Improve Automatic Verification Methods (Extended Abstract)*

Patrice Godefroid

Université de Liège, Institut Montefiore, B28
4000 Liège Sart-Tilman, Belgium
Email: godefroid@montefiore.ulg.ac.be

Abstract

In this paper, we present a verification method for concurrent finite-state systems that attempts to avoid the part of the combinatorial explosion due to the modeling of concurrency by interleavings. The behavior of a system is described in terms of partial orders (more precisely in terms of Mazurkiewicz's traces) rather than in terms of interleavings. We introduce the notion of "trace automaton" which generates only one linearization per partial order. Then we show how to use trace automata to prove program correctness.

1 Introduction

Finite-state methods are quite widely used for concurrent program verification. Indeed, they have several advantages: they are simple and easy to understand and they can be fully automated. Unfortunately, these methods also have some serious drawbacks. They are not always applicable and, when they are applicable, they are often limited by combinatorial explosion.

The frustrating fact is that a lot of this combinatorial explosion is unnecessary: it is due to the modeling of concurrency by interleavings. For example, the concurrent composition of two n-state processes having completely independent activities is represented by an n^2-state process.

Of course, it has been recognized for some time that concurrency and nondeterminism are not the same thing. This observation has inspired a fairly large body of work on "partial order" models of concurrency [Lam78] [Maz86] [Pra86] [Win86]. With very few exceptions, work in this area is limited to rather abstract semantical models.

In this paper, we take a very pragmatic point of view towards partial order models. Our goal is to develop verification methods for concurrent finite-state systems that avoid the part of the combinatorial explosion due to the modeling of concurrency by interleaving. We present a framework in which this can be done successfully.

To define a verification method, four elements are necessary: a representation of programs, a representation of properties, a semantics according to which we compare programs and properties, and an algorithm for doing this comparison [Wol89].

For representing programs and properties, we chose one-safe place/transition-nets (P/T-nets) [Rei85] [Roz86]. This is a well-known formalism and it fits very well both with the interleaving and the partial order approaches. As semantic model, we use Mazurkiewicz's traces [Maz86]. However, we use this model in such a way that the results of our verification are identical to what one would obtain with an interleaving model of concurrency. Precisely, we verify that the language of firing sequences of the

---

*This research is supported by the European Community ESPRIT BRA project SPEC (3096).

one-safe P/T-net N_I representing the implementation is included in the language of firing sequences of the net N_S representing the specification.

Our verification method works in several steps. First, we build an automaton from the net representing the implementation. However, this automaton does not represent all interleavings. It only represents one interleaving for each Mazurkiewicz trace in the semantics of the implementation. This automaton can be very much smaller than the one representing all interleavings. We then compare this automaton to the net for the specification taking into account the dependency relation of the Mazurkiewicz trace semantics.

2 One-safe P/T-nets

Definition 2.1 *A one-safe place/transition-net (P/T-net), is a quadruple $N = (S, T, F, M_0)$ where S and T are finite, disjoint, nonempty sets of respectively* places *(local states) and* transitions; $F \subseteq S \times T \cup T \times S$ *is a* flow relation *such that $dom(F) \cup cod(F) = S \cup T$ (no isolated elements); and $M_0 \subseteq S$ is the* initial marking.

P/T-nets are represented graphically using boxes (or straight lines) to represent transitions, circles to represent places, and arrows to represent the flow relation. In such a representation, circles corresponding to places in the initial marking are marked with dots (tokens).

For each $x \in S \cup T$, the sets $\cdot x = \{y : (y, x) \in F\}$, $x\cdot = \{y : (x, y) \in F\}$, and $\cdot x\cdot = \cdot x \cup x\cdot$ are called respectively the *preset*, the *postset* and the *proximity* of x.

A *marking* M (or *global state*) of such a net N is a subset of S.

Definition 2.2 *Let M be a marking of N. A transition $t \in T$ is* M-firable[1] *iff $(\cdot t \subseteq M) \wedge (t\cdot \cap (M \setminus \cdot t) = \emptyset)$.*

Thus each place of each marking will contain at most one token. An M-firable transition $t \in T$ may fire and yield a *successor marking* M' of M which is such that $M' = (M \setminus \cdot t) \cup t\cdot$.

If t fires from M to M' we write $M [t > M'$. A *reachable marking* of N is a marking M such that $\exists t_1, \ldots, t_n \in T$: $M_0 [t_1 > M_1 [t_2 > M_2 [\ldots [t_n > M_n = M$. The set of all reachable markings of N will be denoted by $mark(N)$.

The sequence $t_1, t_2, \ldots, t_n$ is called a *firing sequence* of N. The set of all firing sequences of N will be called the *firing sequence language* of N. This language is prefix closed.

Definition 2.3 *A one-safe P/T-net N is called* contact-free *iff for all $M \in mark(N)$ and for all $t \in T$: $(\cdot t \subseteq M) \Rightarrow (t\cdot \cap (M \setminus \cdot t) = \emptyset)$.*

3 Automata

A finite-state deterministic automaton is a quadruple (S, Σ, δ, s_0) where S is a finite set of states; Σ is an alphabet; $\delta : S \times \Sigma \to S$ is a deterministic transition function; and s_0 is the starting state.

The language generated by such an automaton is the set of words $w = a_1 a_2 \ldots a_n$ such that there exist $s_i = \delta(s_{i-1}, a_i)$, for all $i : 1 \leq i \leq n$. An automaton can be represented by a directed graph. The nodes of this graph represent the states of S while the edges represent the transition function and are labeled with elements of Σ.

It is easy to define an automaton that generates the firing sequence language of a given one-safe P/T-net N: $S = mark(N)$; $\Sigma = T$, the set of transitions of N; $\delta(M, t) = M'$ iff $M, M' \in mark(N)$ and $M [t > M'$; and $s_0 = M_0$, the initial marking of N. Clearly, the states of this automaton correspond to the reachable markings of N.

The language L accepted by such an automaton will be referred to as the *sequential behavior* of N. The words (sequences) of this language can be viewed as sequential observations of the behavior of N,

---

[1] Note that a C/E-system [Rei85] (or an Elementary Net system [Roz86]) is more restrictive than a one-safe P/T-net since the requirements for the firing of a transition t in a C/E-system are $(\cdot t \subseteq M) \wedge (t\cdot \cap M = \emptyset)$.

i.e. observations made by observers able to see only a single event occurrence at a time. The ordering of symbols in these words reflects not only the (objective) causal ordering of event occurrences (transitions of the net), but also a (subjective) observational ordering resulting from a specific view of concurrent actions: whenever there are concurrent transitions in the net N, the corresponding automaton introduces an "artificial" nondeterminism whose nondeterministic choices correspond to the possible interleavings of these concurrent transitions. Therefore, the structure of such an automaton alone does not make it possible to decide whether the difference in ordering is caused by a conflict resolution (a nondeterministic decision), or by different observations of concurrency (interleavings). In order to extract the causal ordering of event occurrences, we will use the notion of *trace* [Maz86].

4 Traces

First, we define the notion of *concurrent alphabet*.

Definition 4.1 *A concurrent alphabet is a pair $\Sigma = (A, D)$ where A is a finite set of actions, called the alphabet of Σ, and D is a binary, symmetrical and reflexive, relation in A, called the dependency in Σ.*

$D(A) = (A, A^2)$ stands for the concurrent alphabet of total dependency on A, and $I_\Sigma = A^2 \setminus D$ stands for the *independency* in Σ.

Definition 4.2 *Let Σ be a concurrent alphabet; A^* represents the set of all finite sequences (words) of symbols in A, $\cdot$ stands for the concatenation operation, and the empty word is noted ϵ. We define the relation $\equiv_\Sigma$ as the least congruence in the monoid $[A^*; \cdot, \epsilon]$ such that $(a, b) \in I_\Sigma \Rightarrow ab \equiv_\Sigma ba$.*

The relation $\equiv_\Sigma$ is referred to as the *trace equivalence over Σ*.

Definition 4.3 *Equivalence classes of $\equiv_\Sigma$ are called* traces over Σ.

A trace characterized by a word w and a concurrent alphabet Σ is denoted by $[w]_\Sigma$.

Thus a trace over a concurrent alphabet $\Sigma = (A, D)$ represents a set of words defined over A only differing by the order of adjacent symbols which are independent according to D. For instance, if a and b are two symbols of A which are independent according to D, the trace $[ab]_\Sigma$ represents the two words ab and ba. A trace is an equivalence class of words. A *trace language* is a set of traces over a given concurrent alphabet.

Let us return to the one-safe P/T-net N. We define the *dependency* in N as the relation $D_N \subseteq T \times T$ such that:

$$(t_1, t_2) \in D_N \Leftrightarrow {}^\cdot t_1 \cap {}^\cdot t_2 \neq \emptyset. \tag{1}$$

The complement of D_N is called the *independency* in N. If two independent actions occur next to each other in a firing sequence, the order of their occurrences is irrelevant (since they occur concurrently in this execution). Let $\Sigma_N = (T, D_N)$ be the concurrent alphabet associated with N and let L be the firing sequence language of N. We define the *trace behavior* of N as the set of equivalence classes of L defined by the relation $\equiv_{\Sigma_N}$. These equivalence classes are called *firing traces* of N. Such a class (trace) corresponds to a partial order (i.e. a set of causality relations) and represents all its linearizations (words).

To describe the behavior of a one-safe P/T-net by means of traces rather than sequences, we will need the dependency D_N of N (which can be deduced from the statical structure of N as defined by (1)) and *only one* linearization for each trace. Consider a language L' representing one arbitrary linearization (word) for each possible trace of the net. Let L' be such that

$$L = \bigcup_{w' \in L'} Pref(lin([w']_{\Sigma_N}))$$

where $lin([w]_{\Sigma_N})$ denotes the set of linearizations (words) of the trace (equivalence class) $[w]_{\Sigma_N}$ and $Pref(w)$ denotes the prefixes of w. Clearly, L' is a regular language. So, *the behavior of a one-safe*

P/T-net is fully characterized by the dependency D_N and an automaton which generates exactly L'. Let us call this automaton a trace automaton for N.

To construct such an automaton, we do not need to compute all the reachable markings of N: whenever several independent transitions are firable, we fire only one of these transitions in order to generate only one interleaving (linearization) of these transitions.

In the next section, we present an algorithm to construct a trace automaton for a given contact-free one-safe P/T-net N.

5 Constructing the Trace Automaton

The algorithm presented in Figure 1 is a classical depth-first search of the reachable markings of the net N with some important modifications.

The algorithm uses a *Stack* to hold the configurations that remain to be examined. Each configuration is composed by a marking M and two additional information: a *"Sleep set"* and a *"NDinfo"*. A *Sleep* set is a subset of M-firable transitions. A "sleeping" transition will never be fired in the remainder of the search starting from the current marking M. Thus *Sleep* denotes a set of transitions which are firable but which will not be fired. A *NDinfo* is an information which identifies the nondeterministic branch of the current marking M. Different possibilities for solving a nondeterminism lead to different nondeterministic branches. The root of all nondeterministic branches is the initial marking. A (hash) table H is also used to store the states of the automaton that have been explored. As usual, these states will correspond to reachable markings of the net N. A *Sleep* set, a *NDinfo* and a *"succ"* set are associated to each state. The *succ* set contains the transitions leaving that state in the trace automaton. The states reachable from a given state of the trace automaton are obtained by firing, in the net, the transitions of the *succ* set associated with that state.

Since we suppose that N is contact-free, we do not have to check the requirement $t^{\cdot} \cap (M \setminus {}^{\cdot}t) = \emptyset$ before firing a transition t from a marking M. Dealing with contact-free one-safe P/T-nets is not a restriction. Indeed for every one-safe P/T-net there exists an equivalent contact-free one-safe P/T-net (i.e. a net that has the same firing sequence language) [Rei85]. Transitions $t_1, t_2, \ldots, t_n$ are referred to as being *in conflict* iff $({}^{\cdot}t_1 \cap {}^{\cdot}t_2 \cap \ldots \cap {}^{\cdot}t_n) \neq \emptyset$ (their occurrences are mutually exclusive and lead to different nondeterministic branches). A transition t that is not in conflict with another transition is *conflict-free*.

The first modification w.r.t. the classical algorithm is that we do not systematically fire all firable transitions from a given marking: we only choose some of them since we want to construct an automaton that generates only one interleaving whenever several independent transitions are firable. Our choices are motivated by the following two principles: we want to minimize the number of states of the trace automaton; we have to consider all possible behaviors of the net: each firing sequence of the net must be represented by some trace generated by the trace automaton.

Remember that each word w generated by the trace automaton defines a trace $[w]_{\Sigma_N}$. This means that all the linearizations of this trace and all the prefixes of these linearizations are firing sequences of the net N.

In order to minimize the number of states to be constructed, we define a priority scheme to choose amongst the firable transitions those that are to be fired. The highest priority (priority 1) is given to conflict-free transitions (these transitions will be fired one by one successively). Next, we give priority (priority 2) to transitions that are in conflict exclusively with firable transitions. When we fire such a transition, we also fire (from the current marking) the transitions that are in conflict with it in order to explore all possible nondeterministic cases (it corresponds to a branching in the trace automaton). All these possible cases lead to different nondeterministic branches (therefore, the *NDinfo* of the markings obtained after firing these transitions are distinct). Finally, priority 3 is given to firable transitions that are in conflict with at least one nonfirable transition (this is a situation of *confusion* [Rei85]).

When there remain only firable transitions of priority 3, we proceed as follows. Let t be one of these firable transitions. We know that t is in conflict with at least one transition x that is not firable from

---

1. Initialize: *Stack* is empty; *H* is empty;
        enter $(M_0, \emptyset, 0)$ in H;
        push $(M_0, \emptyset, 0)$ into *Stack*
2. Loop: while *Stack* $\neq \emptyset$ do
        begin
          pop $(M, Sleep, NDinfo)$ from *Stack*;
          $FT := \text{CHOOSE-AMONGST}(\{t_j \notin Sleep : \cdot t_j \subseteq M\}, Sleep, NDinfo)$
          for all $t_j \in FT$: add t_j to $succ(M)$ in H
          for all $t_j \in FT$ do
            begin
              $nextM := M - \cdot t_j + t_j \cdot$;
              $NDinfo := t_j\text{->}NDinfo$; $Sleep := t_j\text{->}Sleep$;
              if $nextM$ is already in H
                then $NDinfo$-old $:= NDinfo$ associated with $nextM$ in H;
                    if Same-ND-Branch($NDinfo$-old,$NDinfo$)
                      then $Sleep := Sleep \cup succ(nextM)$;
                          push $(nextM, Sleep, NDinfo)$ into *Stack*
                    else $Sleep$-old $:= Sleep$ associated with $nextM$ in H
                        if $Sleep$-old $\not\subseteq Sleep$
                          then enter $(nextM, Sleep \cap Sleep$-old,
                                      $NDinfo)$ in H;
                                push $(nextM, Sleep, NDinfo)$ into *Stack*
              else enter $(nextM, Sleep, NDinfo)$ in H;
                    push $(nextM, Sleep, NDinfo)$ into *Stack*
          end
        end

Figure 1: Algorithm

the current marking. But it is possible that x will become firable because of the evolution of some other tokens not in $\cdot t$. In that case, the firing of t could be replaced by the firing of x. In order not to miss this possibility, we consider several cases: we fire t from the current marking M, we fire the firable transitions $t_1, \ldots, t_n$ in conflict with t from M, if any, and then we "give the hand" to other tokens to see if they can make x firable (all these cases correspond to different nondeterministic branches). In the last case, since we have already considered the firing of t and $t_1, \ldots, t_n$ and since we are only interested by the "potential" firing of x, we do not consider $t, t_1, \ldots, t_n$ any longer and we put them in the *Sleep* set associated with the current configuration. If there still remain firable transitions of priority 3, we proceed again as described above with the new *Sleep* set, etc. In summary, when there remain only firable transitions of priority 3, all these transitions are fired from the current marking but with different *NDinfo* and *Sleep* sets. An example of such a situation is shown in Figure 2.a: from the current marking all the firable transitions have priority 3. The corresponding trace automaton is given below the net (the value of *Sleep* and *NDinfo* during the construction of the trace automaton is given between parentheses).

The function CHOOSE-AMONGST returns the next transition(s) to be fired according to the priority scheme. Moreover, this function associates two additional information to each chosen transition t_j to be fired: $t_j\text{->}Sleep$ and $t_j\text{->}NDinfo$ representing respectively the *Sleep* set and the *NDinfo* to be associated with the new marking obtained after firing t_j.

Another modification w.r.t. the classical depth-first search is that, if the current marking has already been reached, the search does not stop automatically. Indeed, suppose the current marking is M. The

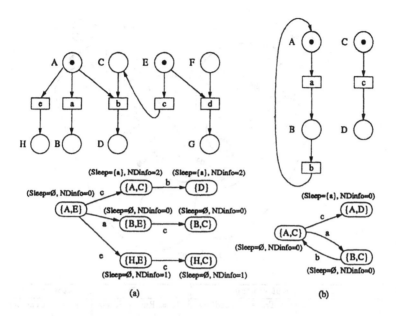

Figure 2: Nets and their corresponding trace automaton

search continues and after the firing of some transitions, the current marking becomes M again. In other words some of the tokens of M moved from their place and then returned back to their respective place. But it is possible that other tokens of M have not yet had the possibility of moving. In order to consider all possible behaviors, we have to "give the hand" to these tokens to see which other concurrent transitions could be fired (it is a kind of "fairness requirement" on the choices amongst independent firable transitions). This situation is illustrated in Figure 2.b: from the initial marking $\{A, C\}$, assume that we fire a and that next we fire b and go back to $\{A, C\}$; then we have to fire c although the marking $\{A, C\}$ has already been reached. This kind of situation can arise only with markings that can appear several times in the same nondeterministic branch of the trace automaton. Now, what happens if the next marking has already been reached in another nondeterministic branch ? The search from this next marking may stop if the Sleep set "Sleep-old" associated with this marking in the already reached marking table is included in the current Sleep set "Sleep". If this requirement is not satisfied, i.e. if there exists some transition t in Sleep-old (t is firable) such that t is not in Sleep, the search has to continue. Indeed, since t is not in Sleep, t (which is firable) has to be fired eventually and, since t is in Sleep-old, we know that t does not occur in the remainder of the search previously made starting from the next marking with Sleep-old as Sleep set.

Of course, there are many possible trace automata corresponding to a given contact-free one-safe P/T-net N. The algorithm described above constructs one of these possible automata. The order of the time complexity for our algorithm is given by the number of constructed transitions times the maximum number of simultaneous firable transitions. We do not claim that the trace automaton constructed by our algorithm is always the minimal one: it is often possible to further reduce the size of the trace automaton but at the cost of an increased time complexity.

Example 5.1 Let us consider the well-known dining philosophers problem. Figure 3 shows a net $dp2$ that represents two philosophers and their adjacent forks. The "classical" automaton, i.e. the one whose states correspond to all the reachable markings, and the trace automaton constructed by our algorithm that correspond to $dp2$ are presented in figure 4. The dotted part is not part of the trace automaton. ∎

The net $dp2$ is susceptible to deadlock: there is no firable transition from the marking $\{A_3, B_3\}$. Note

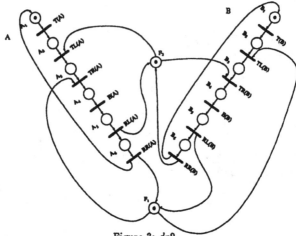

Figure 3: *dp2*

NET	Classical Automaton				Trace Automaton			
	Total Run Time (sec)	States	Trans.	Deadlock	Total Run Time (sec)	States	Trans.	Deadlock
dp2	0.76	21	34	13	0.7	13	14	13
dp3	3.64	99	240	25	1.1	25	28	20
dp4	43.52	465	1508	173	1.98	43	49	27
dp5	692.4	2163	8770	525	3.6	72	83	34

Table 1: Classical automaton versus trace automaton

that the state corresponding to the marking $\{A_3, B_3\}$ is a state of the trace automaton. Indeed, remember the main idea behind our algorithm is to choose some of the firable transitions to be fired whenever there are independent firable transitions. Thus, if there is *no* firable transition, we detect it (the deadlock-preserving property of partial order semantics was already pointed out in [Gai88], [Val88] among others).

Table 1 compares the performance of a depth-first search algorithm against the algorithm presented in this section for the nets *dp2* to *dp5* (five dining philosophers). The combinatorial explosion both in the number of states and transitions is clearly avoided by using trace automata. We also compare the run time needed to construct these automata (these results were obtained with a LISP prototype): our algorithm can be much faster than the classical one. Moreover, the deadlock is detected sooner: for *dp5*, the deadlock corresponds to the 34th reached state by our algorithm instead of the 525th one with a classical depth-first search.

Our method for modeling net behaviors has the potential of being *much more efficient, both in time and memory*, than the classical one. In the next sections, we show how to use trace automata to prove program correctness.

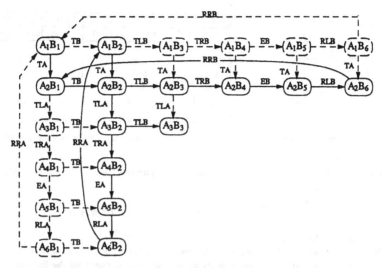

Figure 4: Classical automaton and trace automaton for $dp2$

6 Verification

Consider a set of processes P and a relation $\leq$ on this set such that $I \leq S$ iff I refines S (i.e., I is less nondeterministic than S). Let P be the set of all one-safe P/T-nets and $\leq$ the inclusion relation defined on the languages generated by these nets.

   To verify that the implementation N_I effectively meets the specification N_S, the classical method consists in comparing the automata I and S respectively generating the firing sequence language of the nets N_I and N_S. If these automata are deterministic, verifying the language inclusion reduces to "simulating" I by S, i.e. checking that all that I can do can also be done by S. The cost of checking the existence of this simulation increases with the size of the automaton I, i.e. with the total number of reachable markings of the net N_I. Our claim is that this verification can be done at a lower cost, by using a trace automaton for N_I.

   In the next section, we show how to express a criterion equivalent to language inclusion in terms of traces.

7 A Verification Criterion based on Traces

Let L_I (L_S) be the firing sequence language of a given one-safe P/T-net $N_I = (S_I, T_I, F_I, M_{0_I})$ (N_S). Let D_I (D_S) be the dependency in N_I (N_S) as defined previously and $\Sigma_I = (T_I, D_I)$ ($\Sigma_S = (T_S, D_S)$) the concurrent alphabet associated with N_I (N_S). We define $PO([w]_\Sigma)$ as the transitive closure of the relation $\{(a_i, a_j) : (a_i, a_j) \in D \text{ with } 1 \leq i < j \leq n\}$ if $w = a_1 a_2 \ldots a_n$ and D is the dependency of Σ. $PO([w]_\Sigma)$ represents the set of causality relations corresponding to $[w]_\Sigma$ ($PO([w]_\Sigma)$ is a partial order).

   Let I' denote a trace automaton corresponding to N_I and let $L_{I'}$ be the language generated by I'.

Theorem 7.1 $lin([w]_{\Sigma_I}) \subseteq lin([w]_{\Sigma_S}) \Leftrightarrow PO([w]_{\Sigma_I}) \supseteq PO([w]_{\Sigma_S})$.

   This theorem leads us directly to the following consideration: *to verify that the linearizations of a trace $[w']_{\Sigma_I}$ are included in the firing sequence language of a given one-safe P/T-net N_S, one needs to check if the word w' is a firing sequence of N_S and to verify that $PO([w']_{\Sigma_I}) \supseteq PO([w']_{\Sigma_S})$.*

Definition 7.1 *A one-safe P/T-net is called "restricted" if the following requirement is satisfied:*

$$\forall x, y \in T : (\cdot x \cap \cdot y) \neq \emptyset \Rightarrow ((\cdot x \cap \cdot y) \not\subseteq (y \cdot \cap x \cdot)). \tag{2}$$

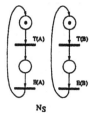

NET	Classical Simulation	Verification with P.O.
	Total Run Time (sec)	Total Run Time (sec)
dp2	0.96	0.74
dp3	5.08	1.18
dp4	60.1	2.36
dp5	940.4	4.58

Figure 5: Specification for $dp2$ Classical verification versus verification with P.O.

Theorem 7.2 *Let N_I be a one-safe P/T-net and N_S a "restricted" one-safe P/T-net. We have:*

$$L_I \subseteq L_S \Leftrightarrow \begin{cases} L_{I'} \subseteq L_S \\ \forall w' \in L_{I'} : PO([w']_{\Sigma_I}) \supseteq PO([w']_{\Sigma_S}) \end{cases}$$

Proof: Presented in the full paper. ∎

This theorem is of great interest because in most cases verifying the two inclusions of the second member can be done more efficiently than verifying the inclusion between all the linearizations. Indeed, the size of I' can be much smaller than the size of I, if N_I is a concurrent system.

8 The Verification Algorithm

We check that $L_{I'} \subseteq L_S$ as usual (see Section 6) by exploring the reachable states of $I' \times S$. A transition executed by I' must be simulated by S except if this transition is hidden. i.e. if this transition is not in the set T_S of transitions of the net N_S. Verifying the inclusion relation between the causality relations is done during the checking of the simulation of I' by S.

Algorithm: Presented in the full paper.

Example 8.1 Figure 5 shows a specification N_S for the two dining philosophers $dp2$ of figure 3. ∎

The table of figure 5 compares the run time needed to verify that $dp2$ is an implementation of N_S w.r.t. our criterion (theses results were obtained with a LISP prototype). In a similar way, this problem has been extended up to five philosophers and the corresponding results are presented in table 5: our algorithm can be much faster than the classical one. Moreover, from $dp3$ and beyond, the memory required to store the states and the transitions due to the combinatorial explosion in the classical case is larger than the additional memory required to check inclusion between causality relations in our method.

9 Conclusions

Most of the work on partial orders deals with semantic problems or with algebraic properties and remains in a theoretical framework. There are only a few papers related to concurrent system verification using partial order models [Gai88][KP86][KP88][Pen90][PL89][PP90][PW84][Val89]. Our method has the advantages of being simple, fully algorithmic, and of extending directly conventional verification techniques. Moreover, this method has the potential of being much more efficient, both in time and memory, than the classical one. The dining philosophers example emphasizes clearly the power of our method: we avoid the part of the combinatorial explosion due to the modeling of concurrency by interleavings. The verification framework presented in this paper is restricted to nonlabeled P/T-nets and safety properties. Interesting future work would be to broaden the scope of our method to cope with labeled P/T-nets and liveness properties.

Acknowledgements

I would like to thank Professor Pierre Wolper for his enthousiastic supervision, for fruitful discussions and thoughtful hints. I am grateful to Jean-Yves Pirnay and Philippe Simar for helpful comments and contributing to the typesetting of this paper. My thanks are also addressed to Marianne Baudinet, Froduald Kabanza, François Schumacker and Dr. Wojciech Penczek for reading drafts of this paper.

References

[Gai88]   H. Gaifman. Modeling concurrency by partial orders and nonlinear transition systems. In *Linear Time, Branching Time and Partial Order in Logics and Models for Concurrency*, LNCS 354, pages 467–488, 1988.

[God89]   P. Godefroid. Les modèles ordre partiel du parallélisme (partial order models for concurrency). Undergraduate thesis, Service d'Informatique, Université de Liège, June 1989.

[JK90]   R. Janicki and M. Koutny. On some implementation of optimal simulations. To appear in *Proc. Computer-Aided Verification Workshop*, Rutgers, 1990.

[KP86]   Y. Kornatzky and S. S. Pinter. A model checker for partial order temporal logic. EE PUB 597, Department of Electrical Enginering, Technion-Israel Institute of Technology, 1986.

[KP88]   S. Katz and D. Peled. An efficient verification method for parallel and distributed programs. In *Linear Time, Branching Time and Partial Order in Logics and Models for Concurrency*, LNCS 354, pages 489–507, 1988.

[Lam78]   L. Lamport. Time, clocks, and the ordering of events in a distributed system. *Communications of the ACM*, 21(7):558–564, 1978.

[Maz86]   A. Mazurkiewicz. Trace theory. In *Petri Nets: Applications and Relationships to Other Models of Concurrency, Advances in Petri Nets 1986, Part II; Proceedings of an Advanced Course*, LNCS 255, pages 279–324, 1986.

[Maz88]   A. Mazurkiewicz. Basic notions of trace theory. In *Linear Time, Branching Time and Partial Order in Logics and Models for Concurrency*, LNCS 354, pages 285–363, 1988.

[Pen90]   W. Penczek. Proving partial order properties using CCTL. Submitted to *Proc. Concurrency and Compositionality Workshop*, San Miniato, Italy, 1990.

[PL89]   D. K. Probst and H. F. Li. Abstract specification, composition and proof of correctness of delay-insensitive circuits and systems. Department of Computer Science, Concordia University, Montreal, Quebec Canada, 1989.

[Pra86]   V. Pratt. Modelling concurrency with partial orders. *International Journal of Parallel Programming*, 15(1):33–71, 1986.

[PP90]   D. Peled and A. Pnueli. Proving Partial Order Liveness Properties. ICALP, 1990.

[PW84]   S. S. Pinter and P. Wolper. A temporal logic for reasoning about partially ordered computations. In *Proc. 3rd ACM Symposium on Principles of Distributed Computing*, pages 28–37. Vancouver, 1984.

[Rei85]   W. Reisig. Petri nets: an introduction. EATCS Monographs on Theoretical Computer Science, Springer-Verlag, 1985.

[Roz86]   G. Rozenberg. Behaviour of elementary net systems. In *Petri Nets: Applications and Relationships to Other Models of Concurrency, Advances in Petri Nets 1986, Part II; Proceedings of an Advanced Course*, LNCS 254, pages 60–94, 1986.

[Val88]   A. Valmari. Error detection by reduced reachability graph detection. In *Proc. 9th International Conference on Application and Theory of Petri Nets*, pages 95–112, Venice, 1988.

[Val89]   A. Valmari. Stubborn sets for reduced state space generation. In *Proc. 10th International Conference on Application and Theory of Petri Nets*, vol. 2, pages 1–22, Bonn, 1989.

[Win86]   G. Winskel. Event structures. In *Petri Nets: Applications and Relationships to Other Models of Concurrency, Advances in Petri Nets 1986, Part II; Proceedings of an Advanced Course*, LNCS 255, pages 325–392, 1986.

[Wol89]   P. Wolper. On the relation of programs and computations to models of temporal logic. In B. Banieqbal, H. Barringer, and A. Pnueli, editors, *Proc. Temporal Logic in Specification*, LNCS 398, pages 75–123, 1989.

[Zie80]   W. Zielonka. Proving assertions about parallel programs by means of traces. ICS PAS Report 424, Institute of Computer Science, Polish Academy of Sciences, 1980.

Compositional Minimization of Finite State Systems*

Susanne Graf ♮   Bernhard Steffen ‡

Abstract

In this paper we describe a method for the obtention of the minimal transition system, representing a communicating system given by a set of parallel processes, avoiding the complexity of the non minimal transition system. We consider minimization with respect to observational equivalence, but the method may be adapted to any other equivalence.

An interesting method to achieve this goal is to proceed by stepwise composition and minimization of the components of the system. However, if no precautions are taken, the intermediate state graphs generated by this method may contain a lot of transitions which are impossible in the whole context. We give here a variant of this method which allows to avoid these impossible transitions by taking into account at each composition step a guess of the interface behaviour of the context. This "interface specification" must be provided by the user. The method is based on a *reduction operator* for the composition of a subsystem with its interface specification, which is similar to the parallel operator but introduces *undefinedness* predicates whereever the interface "cuts off" a transition. The parallel operator is defined in a way that these undefinedness predicates disappear again in the full context if and only if the corresponding transition is in fact impossible in the whole system.

The *efficiency* of the method depends in fact on the accuracy with which the designer is able to approximate the possible sequences of the context, but its *correctness* does not. The proof of the correctness of the method is based on a preorder relation similar to the one defined by Walker.

1 Motivation

Many tools for the automatic analysis or verification of finite state concurrent systems are based on the construction of the global state graph of the system under consideration (cf. [CES83,FSS83, CPS89,CPSb89]). Thus they often fail because of the *state explosion problem*: the state space of a system potentially increases geometrically in the number of its parallel components. To overcome this problem techniques have been developed in order to avoid the construction of the complete state graph (cf. [BFH90,CLM89,ClSte90,Fer88,Jos87,KuMcM89,Kr89,LaTh88,LaXi90,Pnu85,ShGr90, StGr89,Wa88,Win90,WoLo89]). In this paper we present a method for the compositional minimization of finite state concurrent systems, which is practically motivated by the following observation:

> For the verification of a system it is usually sufficient to consider an abstraction of its global state graph, because numerous computations are irrelevant from the observers point of view. Such abstractions often allow to reduce the state graph drastically by collapsing semantically equivalent states to a single state, which does not affect the observable behaviour. For example, the so obtained "minimization" of a complex communication protocol may be a simple buffer.

---

*This work has been partially supported by ESPRIT Basic Research Action 'Spec'
♮IMAG-LGI, BP 53X, F-38041 Grenoble
‡Department of Computer Science, University of Aarhus, DK-8000 Aarhus C

Let us refer to the size of the original state space of a system S as its *apparent* complexity, and to the size of the minimized state space as its *real* complexity. The intention of our method is to compositionally construct the minimal system representation and therefore to avoid the apparent complexity. Unfortunately, the straightforward idea to just successively combine and minimize the components of the system is not satisfactory, because "local" minimization does not take context constraints into account and therefore may even lead to subsystems with a higher real complexity than the apparent complexity of the overall system. This is mainly due to the fact that parts need to be considered that can never be reached in the global context. *Partial* or *loose* specifications allow to "cut off" these unreachable parts (see section 4.3). As in [ClSte90,Kr89,LaTh88,ShGr90,Wa88]) we will exploit this feature to take advantage of context information. Furthermore, we will refer to the size of the maximal transition system that is encountered by our method as the *algorithmic* complexity.

Our method is tailored for establishing $\mathcal{P} \models \phi$, i.e. whether $\mathcal{P}$ satisfies the property ϕ, when $\mathcal{P}$ is a system of the form $(p_1 \|_{I_1} \cdots \|_{I_{n-1}} p_n) \langle L \rangle$. Here we assume the processes p_i already as to be given as transition systems, I_i to be *interface specifications*, namely supersets of the set of sequences that can be observed at the associated interfaces, $\|$ to represent parallel composition and $\langle L \rangle$ a hiding operator that abstracts from the activities considered as internal by transforming them into the unobservable action τ (see Section 3.2).

The point of our method is the successive construction of *partially defined* transition systems P_i with the following properties:

1. P_i is less specified than $p_1 \| \cdots \| p_i$, i.e. is smaller in the sense of $\preceq$ (see Definition 4.2).

2. P_n is semantically equivalent to the full system $(p_1 \| \cdots \| p_n) \langle L \rangle$, whenever the interface specifications are correct.

3. P_i has the least number of states and transitions in its semantic equivalence class.

Subsequently, the validation of $P_n \models \phi$ completes the proof. Of course, this requires that ϕ is preserved by the semantic equivalence under consideration. In this paper, we are dealing with a refinement of observational equivalence (see Definition 4.1)). However, the method easily adapts to other equivalences as well (see Section 5.3).

Important factor in this approach are the interface specifications, which should be provided by the program designer. However, the correctness of the method does not depend on the correctness of these specifications. They are only used to "guide" the proof. Thus wrong interface specifications will never lead to wrong proofs. They may only prevent a successful verification of a valid statement. It should be noted that the total definedness (see Definition 3.1) of P_n already implies $P_n \approx^d \mathcal{P}$, which is enough to guarantee successful verification of all ($\approx^d$- consistent) properties satisfied by $\mathcal{P}$.

The power of our method is demonstrated by means of a setup that handles the mutually exclusive access of processes to a common resource. In fact, in this example, the apparent complexity is exponential, whereas the algorithmic and the real complexity are linear (see Section 6). This is illustrated by means of numerical results that have been obtained using the Aldebaran verification tool [Fer88]. The method can easily and efficiently be implemented in systems like the Edinburgh Concurrency Workbench [CPS89] or the Aldebaran verification tool [Fer88].

2   Related work

A great effort has been made in order to avoid the construction of the complete state graph, and therefore to avoid the state explosion problem. Roughly, the proposed methods can be split into two categories, the *compositional verification* and the *compositional minimization*. Characteristic for the former category is that the global system need not be considered at all during the verification process, and for the latter that a minimal semantically equivalent representation of the global system is constructed. This minimal representation can subsequently be used for all kinds of verification.

A pure approach to compositional verification has been proposed by Winskel in [Win90], where rules are given to decompose assertions of the form $\mathcal{P} \models \phi$ depending on the syntax of the program $\mathcal{P}$ and the formula ϕ. Unfortunately, the decomposition rules for processes involving the parallel operator are very restricted. Larsen and Xinxin [LaXi90] follow a similar line, however, their decomposition rules are based on an operational semantics of contexts rather than the syntax.

In order to deal with the problems that arise from parallel compositions Pnueli [Pnu85] proposed a "conditional" inference system, where assertions of the form $\phi \mathcal{P} \psi$ can be derived, meaning that the program $\mathcal{P}$ satisfies the property ψ under the condition that its environment satisfies ϕ. This inference system has been used by Shurek and Grumberg in [ShGr90], where a semi-automatic modular verification method is presented, which, like ours, is based on "guesses" for context specifications. However, in contrast to our method it requires a separate proof of the correctness of these guesses.

Josko [Jos87] also presented a method, where the assumptions on the environment of a component are expressed by a formula, which must be proved in a separate step. The main disadvantage of his method is that the algorithm is exponential in the size of the assumptions about the environment.

A method of the second category was proposed by Halbwachs et al. in [BFH90]. It constructs directly a transition system minimized with respect to bisimulation by successive refinement of a single state. In this method, which has been tailored for Lustre [CHPP87], symbolic computation is needed in order to keep the expressions small which, in general, may grow exponentially.

Another approach of this category was presented by Clarke et al. [CLM89]. They exploit the knowledge about the alphabet of interest in order to abstract and minimize the systems components. Using $\langle L \rangle$ operations together with an elementary rule for distributing them over the parallel operator (see Proposition 3.3) our method covers this approach.

Krumm [Kr89] considers minimization on the fly and reduction with respect to an interface specification that must be provided by the system designer. Whereas the former is limited, because the complete state graph need to be traversed, the latter requires a separate proof of the correctness of the interface specification. This contrasts with our method, where such a proof is unnecessary.

Larsen and Thomsen [LaTh88], and Walker [Wa88] use partial specifications in order to take context constraints into account. Our method is an elaboration of theirs. It uses a more appropriate preorder allowing to define a strategy for (semi-)automatic proofs where the required user support is kept to a minimum.

The methods proposed in [WoLo89,StGr89,KuMcM89] are tailored to verify properties of classes of systems that are systematically built from large numbers of similar processes. These methods are somewhat orthogonal to ours. This suggests to consider a combination of both types of methods.

3 Representation of Processes

In this section, we establish our framework in which processes (systems) are labelled transition systems extended by an undefinedness predicate with parallel composition and hiding defined on them. The extension provides a notion of partial definedness, which naturally leads to a specification-implementation relation between processes (see section 4.3).

3.1 Extended Transition Systems

An *extended finite state transition system* T is a quadruple $(S, \mathcal{A} \cup \{\tau\}, \rightarrow, \uparrow)$ where

1. S is a finite set of *processes* or *states*;

2. $\mathcal{A}$ is a finite *alphabet* of *observable actions*, and τ represents an internal or unobservable action not in $\mathcal{A}$;

3. $\rightarrow$ is a mapping associating with each $a \in \mathcal{A} \cup \{\tau\}$, a *transition relation* $\xrightarrow{a} \subseteq S \times S$;

4. $\uparrow \subseteq S \times 2^{\mathcal{A} \cup \{\tau\}}$ is a predicate expressing *guarded undefinedness*. Given $(p, L) \in \uparrow$ we write $p\uparrow a$ for $a \in L$.

Typically S is a set of program states, and the relationship $p \xrightarrow{a} q$ indicates that p can evolve to q under the observation of a. Finally, $p\uparrow a$ expresses that an a-transition would allow p to enter an undefined state. We say that p is a-*undefined* in this case. Thus, transition systems involving the undefinedness predicate are only *partially defined* or *specified*. It is this notion of partial specification together with its induced partial order which provides the framework for proving our method correct.

Processes are rooted transition systems, i.e. pairs consisting of a transition system and a designated start state. Given a transition system $T = (S, \mathcal{A} \cup \{\tau\}, \rightarrow, \uparrow)$, we identify (as usual) a state $p \in S$ with the process $((S_p, \mathcal{A}_p \cup \{\tau\}, \rightarrow_p, \uparrow_p), p)$, where

- S_p is the set of states that are reachable from p in T,

- $\mathcal{A}_p = \mathcal{A}$ and

- $\rightarrow_p$ and $\uparrow_p$ are $\rightarrow$ and $\uparrow$ restricted to S_p, respectively.

In future, obvious indices will be dropped. The following property characterizes the subset of "standard" transition systems.

Definition 3.1 *A process is* totally *defined if its undefinedness predicate $\uparrow$ is empty.*

3.2 Critical Processes

We now introduce a binary parallel operator $\|$ and unary hiding operators $\langle L \rangle$, where L is a set of observable actions. Intuitively, $p\|q$ is the parallel composition of p and q with synchronization of the actions common to both of their alphabets and interleaving of the others, and $p\langle L \rangle$ is the process in which only the actions in L are observable. The transition relations for the resulting processes are defined by:

1. $\dfrac{p \xrightarrow{a} p'}{p\langle L \rangle \xrightarrow{a} p'\langle L \rangle} \; a \in L$ 2. $\dfrac{p \xrightarrow{a} p'}{p\langle L \rangle \xrightarrow{\tau} p'\langle L \rangle} \; a \notin L$

3. $\dfrac{p \xrightarrow{a} p'}{p\|q \xrightarrow{a} p'\|q} \; (a \notin \mathcal{A}_q)$ 4. $\dfrac{q \xrightarrow{a} q'}{p\|q \xrightarrow{a} p\|q'} \; (a \notin \mathcal{A}_p)$

5. $\dfrac{p \xrightarrow{a} p' \quad q \xrightarrow{a} q'}{p\|q \xrightarrow{a} p'\|q'} \; a \neq \tau$

their undefinedness predicates by:

6. $\dfrac{p\uparrow a}{p\langle L \rangle \uparrow a} \; a \in L$ 7. $\dfrac{p\uparrow a}{p\langle L \rangle \uparrow \tau} \; a \notin L$

8. $\dfrac{p\uparrow a}{(p\|q)\uparrow a} \; (a \notin \mathcal{A}_q \text{ or } q \xrightarrow{a} q')$ 9. $\dfrac{q\uparrow a}{(p\|q)\uparrow a} \; (a \notin \mathcal{A}_p \text{ or } p \xrightarrow{a} p')$

10. $\dfrac{p\uparrow a \quad q\uparrow a}{(p\|q)\uparrow a}$

and their alphabets by $A_{p\langle L\rangle} = A_p \setminus L$ and $A_{p\|q} = A_p \cup A_q$. Finally $S_{p\langle L\rangle} = \{p'\langle L\rangle | p' \in S_p\}$ and $S_{p\|q}$ is the set of pairs $p'\|q'$ that are reachable in $p\|q$.

Thus $p \!\uparrow\! a$ $(q \!\uparrow\! a)$ implies $(p\|q) \!\uparrow\! a$, whenever q (p) does not preempt the execution of a, i.e. whenever $a \notin A_q$ or $q \xrightarrow{a} q'$ $(a \notin A_p$ or $p \xrightarrow{a} p')$. Remember that $\tau \notin A_p$ for any p, thus $\xrightarrow{\tau}$ is defined by the clauses 2, 3 and 4, and $\uparrow\!\tau$ by clauses 7, 8 and 9. The exact meaning of this definition will become clear in Section 5.2, where we introduce the reduction operator. We have:

Proposition 3.2 $\|$ *is associative and commutative.*

Thus processes of the form $(p_1\| \cdots \|p_n)\langle L\rangle$ are well-defined. Our method will concentrate on this form[1].

Proposition 3.3 $\forall p, q \; \forall L.$ $(p\|q)\langle L\rangle = (p\langle L \cup A_q\rangle \| q\langle L \cup A_p\rangle)\langle L \cap A_p \cap A_q\rangle$

This proposition is particularly important, because it allows to localize global hiding informations. In fact, this localization is the essence of the construction of the 'interface processes' in [CLM89].

4 Equivalence and Partial Order

In this section, we define a semantics of extended labelled transition systems in terms of observational equivalence (cf. [Mi80]) and establish a specification-implementation relation in terms of a preorder, which is compatible with this semantics. This preorder plays a key role in the correctness proof of our method.

4.1 Guarded Undefinedness

The $\rightarrow$ relation does not distinguish between observable and unobservable actions. In order to reflect that τ is internal, and hence not visible, we define the *weak* transition relation $\Rightarrow$ and the *weak* undefinedness predicate $\Uparrow$ for arbitrary $p, q \in S$ and $a \in A$ as the least relation defined by:

1. $p \xrightarrow{\tau}^* \xrightarrow{a} \xrightarrow{\tau}^* q$ implies $p \xRightarrow{a} q$

2. $p \xrightarrow{\tau}^* q$ implies $p \xRightarrow{\varepsilon} q$

3. $q \!\uparrow\! a \;\wedge\; p \xRightarrow{\varepsilon} q$ implies $p \!\Uparrow\! a$.

4. $q \!\uparrow\! \tau \;\wedge\; p \xRightarrow{\varepsilon} q$ implies $p \!\Uparrow\! \varepsilon$.

5. $q \!\Uparrow\! \varepsilon \;\wedge\; p \xRightarrow{a} q$ implies $p \!\Uparrow\! a$.

6. $p \!\Uparrow\! \varepsilon$ implies $p \!\Uparrow\! a$.

As usual, the effect of weakening is to swallow the invisible τ-actions.

4.2 Semantic Equivalence

Our notion of semantics is defined by means of the following equivalence relation[2]:

Definition 4.1 $\approx^d$ *is the union of all relations* $R \subseteq S \times S$ *satisfying that* pRq *implies for all* $a \in A$:

1. $p \!\Uparrow\! a$ *if and only if* $q \!\Uparrow\! a$

2. $p \xRightarrow{a} p'$ *implies* $\exists q'. \; q \xRightarrow{a} q' \wedge p'Rq'$

3. $q \xRightarrow{a} q'$ *implies* $\exists p'. \; p \xRightarrow{a} p' \wedge p'Rq'$

Note that $\approx^d$ coincides with the well-known observational equivalence $\approx$ (cf. [Mi80,Mi89]) if the first of the three defining requirements is dropped.

---

[1] This form is called *standard concurrent form* in CCS ([Mi80,Mi89]).
[2] A similar definition has been given in [ClSte90].

4.3 The Specification - Implementation Relation

The following preorder between processes is the basis of the framework in which we establish the correctness of our method:

Definition 4.2 $\preceq$ *is the union of all relations* R *satisfying that* pRq *implies for all* $a \in \mathcal{A}$:

1. $p \overset{a}{\Rightarrow} p'$ *implies* $\exists q'.\ q \overset{a}{\Rightarrow} q' \wedge p'Rq'$

2. $\neg p \Uparrow a$ *implies* $(\neg q \Uparrow a$ *and* $q \overset{a}{\Rightarrow} q'$ *implies* $\exists p'.\ p \overset{a}{\Rightarrow} p' \wedge p'Rq')$

$\preceq$ is a variant of the divergence preorder $\sqsubseteq$ (cf. [Wa88]) in which a-divergence does not require the potential of an a-move. Our modification serves for a different intend. We do not want to cover divergence, i.e. the potential of an infinite internal computation, but (guarded) undefinedness. This establishes $\preceq$ as a specification-implementation relation: a partial specification p is met by an implementation q iff $p \preceq q$, i.e. in contrast to [ClSte90,Wa88] we do not require an implementation to be able to pass these guards. This modification enhences the practicality of preorders as specification-implementation relations. In fact, similar definitions of preorders already appeared in [Ste89,Sti87], but have not been investigated as specification-implementation relations.

Observational equivalence $\approx$, divergence preorder $\sqsubseteq$, and our preorder $\preceq$ induce slightly different semantics on processes. However, it turns out that $\approx^d$ is a refinement of all of them:

Proposition 4.3 *If* $p \approx^d q$ *then* $p \approx q$ *and* $p \sqsubseteq q$ *and* $p \preceq q$.

Furthermore, on totally defined processes $\approx^d$ and $\preceq$ and $\approx$ all coincide. Finally we have the following monotonicity properties:

Proposition 4.4 *For all processes* p, p' *and* q, *and all sets of actions* L, *we have:*.

1. $p \preceq p'$ *implies* $p \| q \preceq p' \| q$

2. $p \approx^d p'$ *implies* $p \| q \approx^d p' \| q$

3. $p \preceq p'$ *implies* $p \langle L \rangle \preceq p' \langle L \rangle$

4. $p \approx^d p'$ *implies* $p \langle L \rangle \approx^d p' \langle L \rangle$

5 The Reduction Method

5.1 Interface Specifications

In this section we introduce our notion of *interface specification* together with a notion of *correctness*, which guarantees the success of our method. These notions concentrate on the set of sequences that may pass the interface. Thus the *exact* specification of the interface between p and q is the *language* of $(p \| q) \langle \mathcal{A}_p \cap \mathcal{A}_q \rangle$, i.e. its set of observable sequences. Denoting the language of a process p by $\mathcal{L}(p)$ we have:

Proposition 5.1 $\forall p, q.\ \mathcal{L}((p \| q) \langle \mathcal{A}_p \cap \mathcal{A}_q \rangle) = \mathcal{L}(p \langle \mathcal{A}_q \rangle) \cap \mathcal{L}(q \langle \mathcal{A}_p \rangle)$

We are going to use interface specifications in order to express context constraints. Thus interface specifications are correct or safe if the corresponding exact interface specification is more constraint. This motivates the following definition:

Definition 5.2 *Given two processes* p *and* q *we define:*

1. *A totally defined process* I *is an* interface specification *for* p *iff* $\mathcal{A}_I \subseteq \mathcal{A}_p$, *and it is an* interface specification *for* p *and* q *if it is an interface specification for both* p *and* q.

2. *An interface specification I for p and q is called* correct *for p and q iff*
$$\mathcal{L}((p\|q)\langle \mathcal{A}_p \cap \mathcal{A}_q\rangle) \subseteq \mathcal{L}(I).$$

The set of all interface specifications for p is denoted by $\mathcal{I}(p)$, and the set of all correct interface specifications for p and q is denoted by $\mathcal{I}(p,q)$.

Proposition 5.4 will show that these language-based definitions are adequate for our purpose.

5.2 The Reduction Operator

For a process p and an interface specification $I \in \mathcal{I}(p)$ the reduction $\Pi_I(p)$ of p wrt I is essentially the projection of $p\|I$ onto its first component; we define $\Pi_I(p) = ((S, \rightarrow, \mathcal{A} \cup \{\tau\}, \uparrow), p)$, where:

- $S = \{q \in S_p \,|\, \exists i' \in S_I.\, q\|i' \in S_{p\|I}\}$

- $\mathcal{A}_p = \mathcal{A}$

- $\forall q, q' \in S \; \forall a \in \mathcal{A} \cup \{\tau\}.\, q \xrightarrow{a} q'$ iff $\exists i, i' \in S_I.\, q\|i \xrightarrow{a}_{p\|I} q'\|i'$

- $\forall q \in S \; q\uparrow\tau$ iff $q\uparrow\tau$ in the transition system of p

- $\forall q \in S \; \forall a \in \mathcal{A}.\, q\uparrow a$ iff one of the following conditions holds:

  - $q\uparrow a$ in the transition system of p
  - $\exists q' \in S_p.\, q \xrightarrow{a}_p q'$ and $\neg\exists q' \in S.\, q \xrightarrow{a} q'$

The only difference between $\Pi_I(p)$ and the projection of $p\|I$ onto p are the undefinedness predicates: $\Pi_I(p)$ inherits all undefinedness predicates from p, and new ones are introduced where transitions of p have been "cut" away by I. The point of the reduction operator is that for correct interface specifications this second kind of undefinedness disappears again in the full context. Remember, if an a-transition of p has been replaced by $\uparrow a$, this predicate disappears again in $\Pi_I(p)\|q$ exactly if q in the corresponding state preempts the execution of an a-transition. This means that $\uparrow a$ are used as 'error' predicates, indicating in the full context where an interface specification is badly defined.

It is possible to show that $\Pi_I(p)$ can be constructed in time proportional to the product of the number of transitions of p and I. We have:

Proposition 5.3 $\forall p \; \forall I \in \mathcal{I}(p).\; \Pi_I(p) \preceq p$

Now we establish the promised independency of the reduction operator of the specific representation of the language specifying the interface.

Proposition 5.4

1. $\forall p \; \forall I, I' \in \mathcal{I}(p).\; \mathcal{L}(I) = \mathcal{L}(I')$ implies $\Pi_I(p) = \Pi_{I'}(p)$
2. $\forall p \; \forall I, I' \in \mathcal{I}(p).\; \mathcal{L}(I) \subseteq \mathcal{L}(I')$ implies $\Pi_I(p) \preceq \Pi_{I'}(p)$
3. $\forall p \; \forall I \in \mathcal{I}(p).\; \mathcal{L}(p) \subseteq \mathcal{L}(I)$ implies $\Pi_I(p) = p$

The correctness of our method is based on the following theorem:

Theorem 5.5 *Let p and q be processes and I an interface specification for p and q. Then $I \in \mathcal{I}(p,q)$ implies that $p\|q$ is isomorphic to $\Pi_I(p)\|q$, i.e. there exists a bijection $\iota : S_{p\|q} \rightarrow S_{\Pi_I(p)\|q}$, such that:*

1. $\forall p', p'' \in S_{p\|q} \; \forall a \in \mathcal{A}_{p\|q} \cup \{\tau\}.\; p' \xrightarrow{a}_{p\|q} p''$ iff $\iota(p') \xrightarrow{a}_{\Pi_I(p)\|q} \iota(p'')$
2. $\forall p' \in S_{p\|q} \; \forall a \in \mathcal{A}_{p\|q} \cup \{\tau\}.\; p'\uparrow a$ iff $\iota(p') \uparrow_{\Pi_I(p)\|q} (a)$

This result illustrates the generality of the reduction operator. It can be used for the verification of all properties, which are preserved by isomorphism. In particular, we have the following corollary concerning the semantic equivalence we are focusing on here:

Corollary 5.6 $\forall q, p \; \forall I.\; I \in \mathcal{I}(p,q)$ implies $p\|q \approx^d \Pi_I(p)\|q$

5.3 The Method

In this section, we show how the reduction operator can be used for the compositional minimization of processes of the form $\mathcal{P} = (p_1\|\cdots\|p_n)\langle L\rangle$. This form is of particular interest, because it is responsible for the state explosion problem and therefore characterizes the processes that are critical during analysis and verification[3]. Our method works by successive construction of minimal extended transition systems for components of $\mathcal{P}$. The point of this construction is that it exploits interface specifications expressing context constraints for the component under investigation. Thus, the *effect* of the method depends on the information, which has been be provided by the designer of the system. However, the *correctness* of the method does not depend on the correctness of these interface specifications. They are only used to "guide" the proof; and more precise interface specifications may allow more reductions. Thus, wrong interface specifications will never lead to wrong proofs. They may only prevent a successful verification of a valid statement.

The method expects $\mathcal{P}$ as to be annotated with interface specifications that describe the interface between the right hand process and the left hand process of the parallel operator they are attached to:

$$\mathcal{P} = (p_1\|_{\mathbf{I}_1} p_2\|_{\mathbf{I}_2} \cdots \|_{\mathbf{I}_{n-1}} p_n)\langle L\rangle$$

where $\mathbf{I}_i$ is an interface specification for $p_1\|\cdots\|p_i$ and $p_{i+1}\|\cdots\|p_n$. We proceed by successive construction of reductions P_i for the prefix processes of $\mathcal{P}$:

$$\underbrace{\underbrace{\underbrace{(p_1\|_{\mathbf{I}_1} p_2\|_{\mathbf{I}_2} \cdots \|_{\mathbf{I}_{n-1}} p_n}_{P_1})}_{P_2})\langle L\rangle}_{P_n}$$

where the P_i are defined as follows:

- $P_1 = \mathcal{M}(\Pi_{\mathbf{I}_1}(\mathcal{M}(p_1\langle\mathcal{A}_{\mathbf{I}_1}\cup L\rangle)))$

- $P_i = \mathcal{M}(\Pi_{\mathbf{I}_i}(\mathcal{M}((P_{i-1}\|p_i)\langle\mathcal{A}_{\mathbf{I}_i}\cup L\rangle)))$ for $2 \le i \le n-1$

- $P_n = \mathcal{M}((P_{n-1}\|p_n)\langle L\rangle)$

and $\mathcal{M}$ is a function that minimizes extended transition systems up to $\approx^d$.

The goal of this method is to avoid unnecessarily large intermediate transition systems during the construction of the minimal transition system representing the semantic equivalence class of $\mathcal{P}$. Thus it is important to minimize all the intermediate constructions as it is done above. Our method does not depend on a particular semantic equivalence. Other equivalences can be dealt with just by changing the minimization function accordingly (cf. Theorem 5.5). Of course, in order to prove the correctness of the method, this also requires to adapt the preorder definition.

Independently of the correctness of the interface specification, we obtain just by means of Propositions 4.4 and 5.3:

Proposition 5.7 *Let* $1 \le i \le n$ *and* Q *denote the parallel composition of all processes with index greater than* i. *Then we have:* $(P_i\|Q)\langle L\rangle \preceq \mathcal{P}$,

This is already enough to guarantee the correctness of our method, i.e. that the success of a subsequent validation of a ($\approx^d$-consistent) property for P_n proves this property for $\mathcal{P}$. The correctness of the interface specifications comes into play in order to guarantee the success of the method. The following theorem is an immediate consequence of Corollary 5.6:

---

[3]Of course, the method can be applied in a structured way to each of the p_i. Thus it is possible to analyze complex structures by successive construction of increasingly large transition systems.

Theorem 5.8 *For $1 \leq j \leq n$ let Q_j denote the parallel composition of all processes with index greater than j. Then we have for $1 \leq i \leq n$:*

$$\forall j \leq i. \ \mathbf{I}_j \in \mathcal{I}(p_1 \| \cdots \| p_j, Q_j) \ \text{implies} \ (P_i \| Q_i) \langle L \rangle \approx^d \mathcal{P}$$

The situation for totally defined overall systems is particularly simple, as can be inferred from:

Corollary 5.9 *Whenever P_n is totally defined, we have: $P_n \approx^d \mathcal{P}$.*

6 An Application

Finally, we demonstrate our method by means of a setup that is intended to ensure round robin access of n processes P_i to a common resource R. The idea is to pass a "token" via the communication channels tk_i in round robin manner and to allow access to R only for the process that currently possesses the token. This process then sends its request via ps_i to the resource R, which responds by transmitting the object requested. The corresponding transmission line is modelled by a buffer B_i. This is motivated by thinking of large objects whose transmission cannot be modelled just by an atomic "handshake" communication.

Let us now assume that we want to prove that the access is modelled as intended. For this purpose we can hide everything but the actions corresponding to the transmission of the token, and subsequently prove that the resulting process is equivalent to the process $Spec(n)$ that just iteratively executes the sequence $tk_1, \cdots, tk_n$, i.e. it is enough to show for

$$System(n) \ =_{def} \ (R \| P_1 \| B_1 \| \cdots \| P_n \| B_n) \langle \{tk_1, .., tk_n\} \rangle$$

that $System(n) \approx^d Spec(n)$.

It is easy to see that the apparent complexity of $System(n)$ is exponential in n, whereas its real complexity is linear. In fact, it is also possible to obtain an algorithmic complexity that is linear in n. This can be achieved by processing the system according to the structure indicated below, where the $\mathbf{I}_i$ denote the exact interface specifications:

$$(\overbrace{R \| P_1 \| B_1} \|_{\mathbf{I}_1} \overbrace{P_2 \| B_2} \|_{\mathbf{I}_2} \cdots \|_{\mathbf{I}_{n-1}} \overbrace{P_n \| B_n}) \langle \{tk_1, .., tk_n\} \rangle$$

The table below summarizes a numerical investigation of the efficiency of our method by means of the Aldebaran verification tool [Fer88]. It displays the size of the global state graph (its apparent complexity), the size of the maximal transition system constructed during stepwise minimization when exploiting the exact interface specifications (the algorithmic complexity), and finally the size of the minimized global state graph (its real complexity).

n	apparent complexity		algorithmic complexity		real complexity	
	states	trans.	states	trans.	states	trans.
4	144	368	20	29	4	4
5	361	1101	24	35	5	5
6	865	3073	28	41	6	6
7	2017	8177	32	47	7	7

It is worth mentioning that the method which works by stepwise composition and minimization of components encounters transition systems that are even larger than the global state graph: e.g. when using this method in order to construct the minimal state graph of the above system for $n = 7$, the largest intermediate state graph that must be constructed has 2916 states and 9801 transitions.

This stresses the importance of interface specifications for automatic proof techniques. It is our opinion that a software designer should always provide these specifications as part of the implementation. We believe that besides enabling automatic verification, this requirement also leads to a transparent and well structured programming[4].

7 Conclusions and Future Work

We have presented a method for the compositional minimization of finite state systems, which is intended to avoid the state explosion problem. This method can be used to support the verification of any property that is consistent with $\approx^d$. Even better, the choice of $\approx^d$ only affects the minimization step. Thus our method can be adapted to other semantic equivalences simply by modifying the minimization operator $\mathcal{M}$.

In this paper, we proposed a left to right strategy for the minimization of a highly parallel process. Of course, the correctness of our method does not depend on this particular choice.

The *effect* of our method depends on interface specifications, which we assume as to be given by the program designer. However, the *correctness* of the method does not depend on the correctness of these interface specifications. Wrong interface specifications will never lead to wrong proofs. They may only prevent a successful verification of a valid property. This is very important, because it allows the designer to simply "guess" interface specifications, while maintaining the reliability of a successful verification.

Another way to obtain interface specifications is by exploiting the property we are going to verify. This is what Clarke et al. [CLM89] had in mind. However, their approach only exploits the alphabet of the property under consideration. A refined treatment of property constraints using our notion of interface specification is under investigation.

Finally, it should be mentioned that our method can easily be implemented. In fact, an implementation in the Edinburgh Concurrency Workbench [CPS89] and in the Aldebaran verification tool [Fer88] is planned.

Acknowledgements

We would like to thank Rance Cleaveland and Ernst–Rüdiger Olderog for helpful discussions.

References

[BFH90] A. Bouajjani, J.-C. Fernandez, N. Halbwachs. *Minimal Model Generation*, this volume

[CES83] E.M. Clarke, E.A. Emerson, E. Sistla. *Automatic Verification of Finite State Concurrent Systems using Temporal Logic Specification: A Practical Approach*, POPL 1983

[CLM89] E.M. Clarke, D.E. Long, K.L. McMillan. *Compositional Model Checker*, LICS, 1989

[CPS89] R. Cleaveland, J. G. Parrow and B. Steffen. *The Concurrency Workbench*, Proceeding of the Workshop on Automatic Verification Methods for Finite State Systems, Grenoble, France, 1989, LNCS 407

[CPSb89] R. Cleaveland, J. G. Parrow and B. Steffen. *A Semantics based Verification Tool for Finite State Systems*, in the proceedings of the Ninth International Symposium on Protocol Specification, Testing, and Verification; North Holland, 1989

[ClSte90] R. Cleaveland, and B. Steffen. *When is "Partial" Complete? A Logic-Based Proof Technique using Partial Specifications*, in Proceedings of LICS'90, 1990

---

[4]This reminds for the situation for *while*-programs, where automatic verification depends on *loop invariants* that also need to be provided by the programmer.

[Fer88] Fernandez, J.-C. *Aldébaran: Un Système de Vérification par Réduction de Processus Communicants*, Ph.D. Thesis, Université de Grenoble, 1988

[FSS83] J.C. Fernandez, J.Ph. Schwartz, J.Sifakis. *An Example of Specification and Verification in Cesar*, Proceedings of 'The Analysis of Concurrent Systems', 1983, LNCS 207

[CHPP87] Caspi P., Halbwachs N., Pilaud N., Plaice J. *LUSTRE, a declarative language for programming synchronous systems*, Proceedings of 14th POPL, Munich, 1987

[Jos87] B. Josko. *MCTL - An extension of CTL for modular verification of concurrent systems*, Workshop on Temporal Logic in Specification 1987, LNCS 398

[KuMcM89] R.P. Kurshan, K. McMillan. *A Structural Induction Theorem for Processes*, in ACM Symposium on Principles of Distributes Computing, 1989

[Kr89] H.Krumm. *Projections of the Reachability Graph and Environment Models, two approaches to facilitate the functional analysis of Systems of cooperating finite state machines*, Proceedings of the Workshop on Automatic Verification of Finite State Systems, Grenoble 89, LNCS 407.

[LaTh88] Larsen, K.G., and B. Thomsen. *Compositional Proofs by Partial Specification of Processes*, in Proceedings LICS'88, 1988

[LaXi90] K.G. Larsen, L. Xinxin. *Compositionality through an Operational Semantics of Contexts*, in Proceedings ICALP'90, LNCS, 1990

[Mi80] R. Milner. *A Calculus for Communicating Systems*, LNCS 92, 1980

[Mi89] R. Milner. *Communication and Concurrency*, Prentice Hall, 1989

[Old90] E.-R. Olderog. *Nets, Terms and Formulas: Three Views of Concurrent Processes*, Habilitationsschrift, Universität Kiel, to appear in Tracts in Theoretical Computer Science, Cambridge University Press

[Pnu85] A.Pnueli. *In Transition from Global to Modular Temporal Reasoning about Programs*, in Logics and Models for Concurrent Systems, Nato ASI Series F, Vol. 13, Springer Verlag

[ShGr90] G. Shurek, O. Grumberg. *The Modular Framework of Computer-aided Verification: Motivation, Solutions And Evaluation Criteria*, this volume

[StGr89] Z. Stadler, O. Grumberg. *Network Grammars, Communication Behaviours and Automatic Verification*, in Proceeding of the Workshop on Automatic Verification Methods for Finite State Systems, Grenoble, France, 1989, LNCS 407

[Ste89] B. Steffen. *Characteristic Formulae*, in Proceedings ICALP 1989

[Sti87] C. Stirling. *Modal Logics for Communicating Systems*, TCS 49, pp. 311-347, 1987

[StiWa89] C. Stirling and D. J. Walker. *Local Model Checking in the Modal Mu-Calculus*, in Proceedings CAAP 1989

[Wa88] D.J. Walker. *Bisimulation and Divergence in CCS*, in Proceedings LICS 1988

[Win90] G. Winskel. *Compositional Checking of Validity on Finite State Processes*, Workshop on Theories of Communication, CONCUR, 1990

[WoLo89] P. Wolper, V. Lovinfosse. *Verifying Properties of Large sets of Processes with Network Invariants*, in Proceeding of the Workshop on Automatic Verification Methods for Finite State Systems, Grenoble, France, 1989, LNCS 407

Minimal Model Generation *

A. Bouajjani , J-C. Fernandez , N. Halbwachs
IMAG/LGI (U.A. CNRS 398)
B.P. 53, 38041 - Grenoble, France

Abstract

This paper adresses the problem of generating a minimal state graph from a program, without building first the whole state graph. The minimality is considered here with respect to bisimulation. A generation algorithm is presented and illustrated.

1 Introduction

Model generation consists of building a state graph from a program, a formula or any comprehensive expression of a transition system. It is used in program verification ("model checking" [6,11]) and compiling (scanner and parser generation [1], control structure synthesis [2,5],...). A crucial problem with model generation is the size of the graph, which can be prohibitive. This size can be large not only because of the intrinsic complexity of the model, but also because the graph contains a lot of states which are in some sense equivalent. Some solutions have been given to this problem, by applying reduction algorithms [8,9,10]. However, these algorithms can only be applied once the graph has been entirely generated. It is often the case that a tremendous amount of time and memory is necessary to generate a graph, which afterward reduces to a very simple one. It even happens that an infinite model reduces to a finite one. So, it would be interesting to reduce the graph during the generation, on one hand to improve the performances of model generation, and on the other hand, to allow finite model generation from infinite systems. This paper presents and illustrates an algorithm performing this task, when the equivalence considered on states is a bisimulation.

After fixing some terminology and notations (section 2), the algorithm is presented (section 3) and illustrated on a simple example (section 4).

2 Definitions and notations

Let $S = (Q, \rightarrow, q_{init})$ be a transition system, where Q is a set of states, $\rightarrow \subseteq Q \times Q$ is a transition relation, and q_{init} is the initial state. Let $\sim$ be an equivalence relation on Q. Our problem is to explicitely build the quotient of the set of reachable states from q_{init}, by the coarsest bisimulation compatible with $\sim$. Of course, this is only possible if this quotient has finitely many elements. Moreover, the method presented here only works if the quotient of Q by the coarsest bisimulation is finite (notice that Q itself can be infinite). The basic idea is to progressively build a partition of Q, by distinguishing two parts of Q only when their respective elements clearly don't bisimulate each other. Henceforth, we shall consider partitions instead of equivalence relations.

Let ρ be a partition of the set of states Q. The following notations will be used:

---

*This work was partially supported by ESPRIT Basic Research Action "SPEC"

For any state $q \in Q$, $post_\rho(q)$ is the set of classes in ρ immediately reachable from q

$$post_\rho(q) = \{X \in \rho \mid \exists q' \in X \text{ such that } q \to q'\}$$

An equivalence relation, noted $\overset{\mathcal{L}}{}$, is associated with ρ as follows:

$$q_1 \overset{\mathcal{L}}{} q_2 \iff post_\rho(q_1) = post_\rho(q_2)$$

A subset X of Q is said to be *stable with respect to* ρ if and only if it is included in some equivalence class of $\overset{\mathcal{L}}{}$. The partition ρ is said to be stable if and only if all of its classes are stable with respect to itself. In other words, a partition is stable if and only if it is the set of classes of a bisimulation.

A *refinement* of a partition ρ is a partition ρ' such that: $\quad \forall X \in \rho', \exists Y \in \rho$ such that $X \subseteq Y$

The *reduction* of a transition system S with respect to a stable partition ρ is the transition system $(\rho, \leadsto, [q_{init}]_\rho)$, where

$\quad [q_{init}]_\rho$ is the class of the initial state in ρ

$\quad X \leadsto Y \iff \exists q \in X, q' \in Y$ such that $q \to q'$

With the above terminology, given an initial partition ρ of Q, we are looking for the reduction of S with respect to the least stable refinement of ρ.

3 Algorithm

The algorithm consists of progressively refining the partition ρ. At each step, two subsets of classes will be distinguished:

- The set R of *reachable classes*, i.e. the classes containing at least one element which has been found reachable from q_{init}.

- The set S of *stable classes*, i.e. the reachable classes which have been found to belong to $\overset{\mathcal{L}}{}$.

The algorithm is the following:

$$
\begin{aligned}
&R = \{[q_{init}]_\rho\}; S = \phi; &&(1)\\
&\text{while } R \neq S \text{ do} &&(2)\\
&\quad \text{choose } X \text{ in } R - S; &&(3)\\
&\quad \text{let } N = X/\overset{\mathcal{L}}{}; &&(4)\\
&\quad \text{if } N = \{X\} \text{ then} &&(5)\\
&\quad\quad S := S \cup \{X\}; R := R \cup \{post_\rho(q) \mid q \in N\}; &&(6)\\
&\quad \text{else} &&(7)\\
&\quad\quad R := R - \{X\}; &&(8)\\
&\quad\quad \text{if } \exists Y \in N \text{ such that } q_{init} \in Y \text{ then } R := R \cup \{Y\}; &&(9)\\
&\quad\quad S := S - \{Y \in S \mid X \in post_\rho(Y)\}; &&(10)\\
&\quad\quad \rho := (\rho - \{X\}) \cup N; &&(11)\\
&\quad \text{fi} &&(12)\\
&\text{od} &&(13)
\end{aligned}
$$

Proof : Let Rea be the set of reachable states, that is the least subset X of Q containing q_{init}, and such that

$$(q \in X \wedge q \rightarrow q') \Longrightarrow q' \in X$$

Then,

(*i*) $X \in S \Longrightarrow X \in \mathcal{R}$
since a subset X is only put into S if $X = X/\mathcal{R}$ (line 6), and as soon as a refinement of ρ can involve a refinement of $X/\mathcal{R}$, X is extracted from S (line 10).

(*ii*) $X \in R \Longrightarrow X \cap Rea \neq \phi$
since a subset X is only put into R if either it contains q_{init} (line 9), or it contains successor states of a stable subset belonging to R (line 6).

(*iii*) When $R = S$, all the reachable classes are in R: If $X \in S$, all the classes directly reachable from X have been put in R (line 6).

(*iv*) So, when $R = S$, R defines a stable partition of Rea.

(*v*) The finiteness of the set of classes insures that the algorithm terminates.

Splitting a class : Line 4 of the algorithm splits a reachable class X into a partition $N = X/\mathcal{R}$, whose elements are stable with respect to the current ρ. Let us detail the computation of this partition. Let pre denote the precondition function $\lambda Y.\{q \in Q \mid \exists q' \in Y \text{ such that } q \rightarrow q'\}$. Then,

$$N = \{X \cap \bigcap_{Y \in \rho} Z_Y \mid Z_Y \in \{pre(Y), Q - pre(Y)\}\}$$

Instead of considering such an exponential number of intersections, most of which are generally empty, we propose to compute N as follows:

```
N = {X};
for each Y ∈ ρ do
    M := φ;
    for each W in N do
        let W₁ = W ∩ pre(Y);
        if W₁ = W or W₁ = φ then M := M ∪ {W}
        else M := M ∪ {W₁, W − W₁};
    od;
    N := M;
od
```

4   Example

Let us consider the following program, which could be a boolean abstraction of a more realistic program:

```
x := true; y := false; read(a);
loop
    write(x or y);
    z := a; read(a);
    w := x; x := not y; y := w or z;
end;
```

We want to examine all the possible input/output behaviours of this program. So, we consider it as a transition system, whose states are the values of the variables when the output is written. Now, since we are only interested in the output, we may consider as equivalent all the states which produce the same output. So, we start with the initial partition:

$$\{(a, w, x, y, z) \mid x \vee y = true\}, \{(a, w, x, y, z) \mid x \vee y = false\}$$

In the following, classes are represented by their characteristic formulas. The initial partition will be noted:

$$\rho: \quad C_1 = \{x \vee y\} \quad C_2 = \{\neg x \wedge \neg y\}$$

Standard rules of weakest precondition provide the precondition of a class X, with respect to the body of the loop:

$$pre(X) = X[w \vee z/y][\neg y/x][x/w] \downarrow a[a/z]$$

where $X \downarrow a = \exists a_0 X[a_0/a] = X[false/a] \vee X[true/a]$

So, $pre(C_1) = x \vee \neg y \vee a$, $pre(C_2) = \neg x \wedge y \wedge \neg a$

The successive partitions built by the algorithm are illustrated on figure 1.

Step 1: The only reachable class is C_1, since x is initially true. For splitting it, we compute:

$$C_1 \wedge pre(C_1) = (x \vee y) \wedge (x \vee \neg y \vee a) = x \vee (y \wedge a)$$

$$C_1 \wedge \neg pre(C_1) = \neg x \wedge y \wedge \neg a$$

$$C_1 \wedge pre(C_1) \wedge pre(C_2) = false$$

$$C_1 \wedge \neg pre(C_1) \wedge pre(C_2) = C_1 \wedge \neg pre(C_1)$$

So, C_1 is split into:

$$C_{11} = \{x \vee (y \wedge a)\} \quad C_{12} = \{\neg x \wedge y \wedge \neg a\}$$

and only C_{11} is reachable. We have: $pre(C_{11}) = x \vee \neg y \vee a$, $pre(C_{12}) = y \wedge (x \vee a)$

Step 2: For splitting C_{11}, we compute:

$$C_{11} \wedge pre(C_{11}) = (x \vee (y \wedge a)) \wedge (\neg y \vee x \vee a) = C_{11}$$

$$C_{11} \wedge pre(C_{11}) \wedge pre(C_{12}) = (x \vee (y \wedge a)) \wedge (y \wedge (x \vee a)) = y \wedge (x \vee a)$$

$$C_{11} \wedge pre(C_{11}) \wedge \neg pre(C_{12}) = x \wedge \neg y$$

$$C_{11} \wedge pre(C_{11}) \wedge pre(C_{12}) \wedge pre(C_2) = false$$

$$C_{11} \wedge pre(C_{11}) \wedge \neg pre(C_{12}) \wedge pre(C_2) = false$$

So, C_{11} is split into:

$$C_{111} = y \wedge (x \vee a) \quad C_{112} = x \wedge \neg y$$

and only C_{112} is reachable. We have: $pre(C_{111}) = x \vee a$, $pre(C_{112}) = \neg x \wedge \neg y \wedge \neg a$

Step 3: When splitting C_{112}, we get only:

$$C_{112} \wedge pre(C_{111}) \wedge \neg pre(C_{112}) \wedge \neg pre(C_{12}) \wedge \neg pre(C_2) = x \wedge \neg y = C_{112}$$

So C_{112} is stable, and leads to C_{111}. So, C_{111} is reachable.

Step 4: C_{111} is also found stable since:

$$C_{111} = C_{111} \wedge pre(C_{111}) \wedge \neg pre(C_{112}) \wedge pre(C_{12}) \wedge \neg pre(C_2)$$

It leads to itself and to C_{12}, which is found reachable.

Step 5: C_{12} is stable, and leads to C_2, since:

$$C_{12} = C_{12} \wedge \neg pre(C_{111}) \wedge \neg pre(C_{112}) \wedge \neg pre(C_{12}) \wedge pre(C_2)$$

Step 6: C_2 is split into:

$$C_{21} = C_2 \wedge pre(C_{111}) = \neg x \wedge \neg y \wedge a$$
$$C_{22} = C_2 \wedge pre(C_{112}) = \neg x \wedge \neg y \wedge \neg a$$

So C_{12} is removed from stable classes. We have: $pre(C_{21}) = pre(C_{22}) = \neg x \wedge y \wedge \neg a$

Step 7: C_{12} is again found stable, since:

$$C_{12} \wedge pre(C_{21}) \wedge pre(C_{22}) = C_{12}$$

It leads to C_{21} and C_{22}.

Step 8 and 9: From

$$C_{21} = C_{21} \wedge pre(C_{111})$$
$$C_{22} = C_{22} \wedge pre(C_{112})$$

we get that C_{21} and C_{22} are stable, and respectively lead to C_{111} and C_{112}.

We get a graph with 5 vertices (Fig. 2), instead of 16, which would be produced by standard generation (Fig. 3).

5 Conclusion

We have presented an algorithm combining generation and reduction methods. In our opinion, this algorithm is interesting for program verification: a state graph with several thousands (or even infinitely many) states may be reduced to one with a few number of states by considering an equivalence relation.

Of course, one must be capable to compute the function *pre* and intersections of classes, and to decide the inclusion of classes. Such a symbolic computation is achievable in the boolean case, with reasonable average cost [3,7].

Applying our algorithm to program verification appears very close to formal proof (in the Floyd/Hoare sense) or to what is now called "symbolic model checking" [4]. Concerning other applications, the algorithm is being implemented in the new version of the LUSTRE compiler.

We have not presented complexity measures. A comparison with classical reduction methods is difficult, mainly because the complexity of these methods is evaluated as a function of the size of the initial graph, whereas the cost of our method obviously depends on the size of the reduced graph.

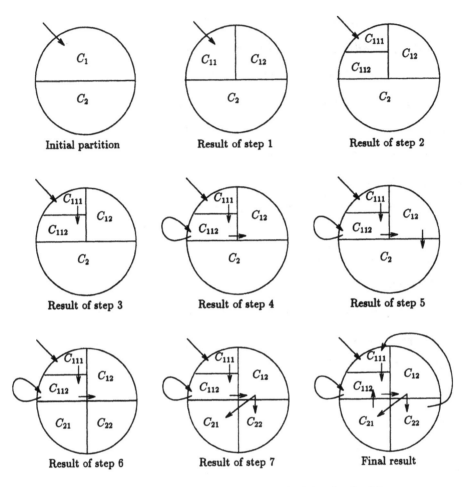

Figure 1: The successive partitions built by the algorithm

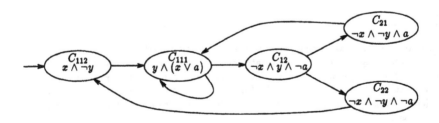

Figure 2: The reduced graph of the example

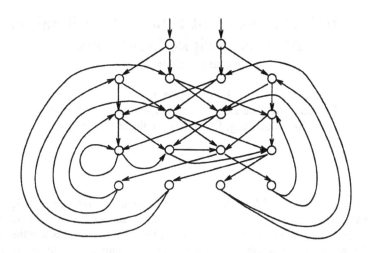

Figure 3: The complete graph of the example

References

[1] A. Aho, R. Sethi, J. Ullman. *Compilers: Principles, Techniques and Tools*. Addison-Wesley, 1986.

[2] G. Berry and G. Gonthier. *The synchronous programming language Esterel, design, semantics, implementation*. Tech. Report 327, INRIA, 1985. To appear in Science of Computer Programming.

[3] R. E. Bryant. Graph-based algorithms for boolean function manipulation. *IEEE Transactions on Computers*, C-35(8), 1986.

[4] J.R. Burch, E.M. Clarke, K.L. McMillan, D.L. Dill, J. Hwang. *Symbolic Model Checking:* 10^{20} *states and beyond*. Technical Report, Carnegie Mellon University, 1989.

[5] P. Caspi, D. Pilaud, N. Halbwachs, J. Plaice. Lustre: a declarative language for programming synchronous systems. In *14th POPL*, january 1987.

[6] E. Clarke, E.A. Emerson, A.P. Sistla. Automatic verification of finite state concurrent systems using temporal logic. In *10th. Annual Symp. on Principles of Programming Languages*, 1983.

[7] O. Coudert, C. Berthet, J. C. Madre. Verification of synchronous sequential machines based on symbolic execution. In *International Workshop on Automatic Verification Methods for Finite State Systems, LNCS 407*, Springer Verlag, 1989.

[8] J. C. Fernandez. An implementation of an efficient algorithm for bisimulation equivalence. *Science of Computer Programming*, 13(2-3), May 1990.

[9] P. Kanellakis and S. Smolka. CCS expressions, finite state processes and three problems of equivalence. In *Proceedings ACM Symp. on Principles of Distributed Computing*, 1983.

[10] R. Paige and R. Tarjan. Three partition refinement algorithms. *SIAM J. Comput.*, 16(6), 1987.

[11] J.L. Richier, C. Rodriguez, J. Sifakis, J. Voiron. Verification in Xesar of the sliding window protocol. In *17th International Workshop on Protocol Specification Testing and Verification*, 1987.

A Context Dependent Equivalence Relation Between Kripke Structures
(Extended abstract)

Bernhard Josko
Computer Science Department, University of Oldenburg
2900 Oldenburg, Federal Republic of Germany

Abstract
In [BCG87] Browne, Clarke and Grumberg define a bisimulation relation on Kripke structure and give a characterization of this equivalence relation in temporal logic. We will generalize their results to reactive systems, which are modelled by Kripke structures together with some constraints describing some requirements how the environment has to interact with the module. Our results subsume the result of [BCM87] by using the constraint **true**. Furthermore it answers the questions raised in that paper how the equivalence of Kripke structures with fairness constraints can be characterized.

Keywords: temporal logic, Kripke structures, bisimulation, modular specification, reactive systems, hierarchical design

1 Introduction

In a top down design step of a large system one component may be splitted in several subcomponents. Not only the tasks of the subcomponents have to be specified but also the interface between the subcomponents have to be defined. On the one hand the interface has to declare the interconnections of the subcomponents i.e. the inputs and outputs of the components, and on the other hand the protocols for the exchange of data have to be specified too. E.g. if a subsystem consists of two components which are coupled asynchronously together (Fig. 1), we can use a 4-cycle signalling protocol to send data from one component to the other (cf. Fig. 2). Module M_1 is responsible for the Req signal and it has to guarantee that this signal is set and reset according to the given protocol, and M_2 is responsible for the signal Ack. Proceeding in the design process we may define a more concrete representation of module M_2 by splitting this component in several subcomponents together with interface specifications or by defining an implementation. In such a design step we can use the fact that the environment (i.e. module M_1) acts according to the protocol and hence, it is only required that the module M_2 behaves correctly provided the environment guarantees the given interface constraint. Therefore a module specification is given by a pair (assm, spec) where assm describes some constraints on the environment and spec is the specification of the module, which has to be satisfied by the module provided the environment guarantees assm.

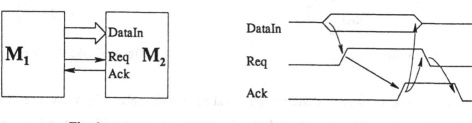

Fig. 1   Fig. 2: 4-cycle signalling protocol

In this paper we investigate system design in the framework of temporal logic. We will use the branching time temporal logic CTL* [EH86] as specification language. In the context of temporal logics an implementation of a system is modelled by a state/transition-graph, also called a Kripke structure. A temporal logic formula is then interpreted in the associated computation tree, which is obtained by unravelling the transition graph. As we are considering open systems (or reactive systems [Pn85]), a module will be modelled by a Kripke structure K together with an interface constraint assm. The constraint assm restricts the possible paths in the computation tree. Hence to check the validity of a specification spec only those paths are considered which satisfy the given assumption assm. Considering the example in Fig. 1 together with Fig. 2 the environment constraints used by module M_2 can be defined in temporal logic as follows:

$\Box$(Ack $\land$ ¬Req $\rightarrow$ [¬Req unless ¬Ack $\land$ ¬Req])

$\Box$(Req $\rightarrow$ [stable(DataIn) unless Ack])

$\Box$(Req $\rightarrow$ [Req unless Ack $\land$ Req])

$\Box$(Ack $\land$ Req $\rightarrow$ $\Diamond$¬Req)

Fig. 3 shows a state/transition graph for M_2 (only the protocol is implemented).

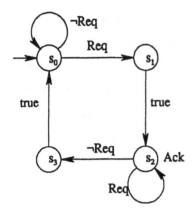

Fig. 3: Kripke structure K_1

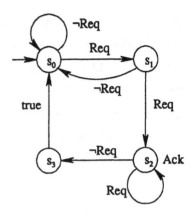

Fig. 4: Kripke structure K_2

During the design process a module implementation may be replaced by a new one. If a replaced module has already been verified to be correct w.r.t. its specification and one has derived properties of a composed module which contains that specific module, one is interested in the question whether the derived properties remain valid. Hence one is interested in an equivalence notion on Kripke structures, which guarantees that the replacement of a submodule by an equivalent one does not violate the correctness of derived properties of a composed (sub)system.

In Fig. 4 another implementation for module M_2 is given. In this implementation the request signal is checked after reading the input data. If the request signal is low - this is an violation of the protocol - the input data is ignored. In an environment which guarantees a correct behaviour according to the protocol, this step will never occurs. Thus both implementations, K_1 and K_2, are equivalent with respect to an environment satisfying the constraints.

Two structures will be called equivalent if they cannot be distinguished by any formula. In [BCG87] a bisimulation ([Mi83], [Pa81]) for closed systems, i.e. Kripke structures without constraints on the environment, is given. Two states s and s' are called bisimulation equivalent if they are labelled with the same atomic propositions and for every transition $s \rightarrow s_1$ in one structure there is a corresponding transition in the other structure leading to a state s'_1 which is equivalent to s_1. It is shown that this relation can be characterized by the fact that both structures satisfy the same temporal logic formulae. In this paper we will show how the results of [BCG87] can be generalized to open systems. We will relativize the correspondence of equivalent steps by requiring that only for those transitions which are on a path satisfying the environment constraints there are corresponding transitions in the other module. We will show that this bisimulation with respect to an assumption assm characterizes the structures satisfying the same set of CTL* formulae relatively to the given assumption assm. i.e. two structures K_1 and K_2 are bisimulation equivalent w.r.t. assm iff the modules $(K_1,assm)$ and $(K_2,assm)$ are not distinguishable by any CTL* formula.

For action based transition systems like CCS Larsen has given a notion of bisimulation relatively to some contexts [La86], [LM87], but there is no relation to temporal logic specifications.

2   Basic definitions

A *Kripke structure* is given by $K = (S, R, s^0, L)$, where S is a finite set of states, $s^0 \in S$ is the initial state and R is a transition relation $R \subseteq S \times S$, such that for every state s there is some state s' with $(s, s') \in R$, and L is a labelling function which associates with every state $s \in S$ a set of atomic propositions, which are valid in that state.

Dealing with reactive systems one should use labelled transitions, as a transition depends on the input signals. A *Kripke structure with inputs* is given by $K = (S, R, s^0, IN, OUT, L)$ where S is a set of states, s^0 is the initial state, IN and OUT are disjoint set of propositions, L is a labelling of states with subsets of OUT, and R is a transition relation with $R \subseteq S \times BExpr(IN) \times S$, such that for every satisfiable boolean expression b and every state s there is some state s' with $(s, b, s') \in R$. IN is a set of propositions on the input signals (usually the input signals themselves) and OUT are propositions on the internal and output signals. The transition labels are constraints on the input signals, restricting a transition to the instances where the actual input signals satisfy the given boolean expression. For every Kripke structure K with inputs there is a corresponding Kripke structure K' (without inputs) [Br86]. Therefore we will use the usual notion of Kripke structure in this paper. Fig. 5 resp. Fig. 6 shows the transformed Kripke structures of Fig. 3 resp. Fig. 4.

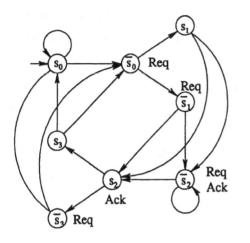

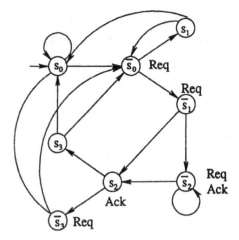

Fig. 5: Kripke structure K_1 transformed   **Fig. 6:** Kripke structure K_2 transformed

Given a Kripke structure $K = (S, R, s^0, L)$, $\pi = (s_i \mid i \in \mathbf{N})$ is a *path* in K if $(s_i, s_{i+1}) \in R$ for all $i \in \mathbf{N}$. We will refer to the i-th state s_i also by $\pi(i)$. The suffix of a path starting at state $\pi(i)$ will be denoted by π^i. The (infinite) *computation tree* of K is obtained by unravelling the Kripke structure (considered as a graph) starting with the initial state s^0.

PTL [Pn77] is the linear propositional temporal logic defined by

$$f ::= a \mid \neg f \mid f_1 \wedge f_2 \mid f_1 \vee f_2 \mid Xf \mid [f_1 \ U \ f_2] \mid \square f \mid \lozenge f,$$

where a denotes an atomic proposition. $[f_1$ unless $f_2]$ will be used as an abbreviation for $\square f_2 \vee [f_1 \ U \ f_2]$. The validity of a PTL formula f along a path π will be denoted by $\pi \models f$. We also call a path π *f-good* if π satisfies the formula f. A state s is called *f-good* iff there is some path π starting at s with is f-good.

CTL and CTL* are branching time temporal logics which are defined by the following rules:

(1)   sf $::=$ $a \mid \neg sf \mid sf_1 \wedge sf_2 \mid sf_1 \vee sf_2 \mid \forall pf \mid \exists pf$

(2)   pf $::=$ $sf \mid \neg pf \mid pf_1 \wedge pf_2 \mid pf_1 \vee pf_2 \mid Xpf \mid [pf_1 \ U \ pf_2]$

(3)   pf $::=$ $Xsf \mid [sf_1 \ U \ sf_2]$

The rules in (1) describe the building of state formulae and the rules in (2) and (3) describe the construction of path formulae. CTL is the set of all state formulae defined by (1) and (3) and CTL* is the set of all state formulae defined by (1) and (2). Furthermore we will use the path formula $\lozenge f$ as an abbreviation for $[\mathbf{true} \ U \ f]$, and $\square f$ as an abbreviation for $\neg \lozenge \neg f$. The semantics of CTL and CTL* for closed systems is as usual (cf. [EH86]).

A module is given by a pair (K,assm), where K is a Kripke structure and assm is a PTL formula. The interpretation of a CTL* formula spec in a module (K,assm) can be described as follows:

(1)   Construct the computation tree with root s^0.

(2)   Mark all paths starting at s^0 which satisfy the constraint assm.

(3)   Interpret the formula spec as in the usual interpretation of CTL* formulae but with the restriction that all path quantifiers are restricted to the marked paths.

To give an inductive definition of the semantics we need a description of the behaviour of the environment at a specific instance on a computation path. The expected behaviour of the environment at instance i on a path π is determined by the given interface constraint assm and the computation history along π. E.g. if $\Box$(Req $\rightarrow$ [Req unless Req $\wedge$ Ack]) is a conjunctive of the interface constraint and assume that the signal Req is set to high at instance j (i.e. the proposition Req is valid at instance j) then [Req unless Req $\wedge$ Ack] is also a part of the expected behaviour of the environment at the next time instance. Using the computation history along a path π we can define a formula $assm_i(\pi, assm)$ which describes the expected behaviour of the environment at instance i on the computation path π.

Given a PTL formula assm and a state s we can define a PTL formula next-assm(assm,s) with the following property:

$$\pi \models assm \quad \text{iff} \quad \pi^1 \models \text{next-assm(assm,s)} \qquad \text{for any path } \pi \text{ starting at s.}$$

This defintion can be extended to all time instances along a path. Given a path π and a formula assm, we can determine for every $i \in N$ a formula $assm_i$ which is valid on π^i iff π satisfies assm. We define

$$assm_0(f,\pi) := f \qquad assm_{k+1}(f,\pi) := \text{next-assm}(assm_k(f,\pi), \pi(k))$$

Now we can give an inductive definition of the validity of a CTL* formula in a module (K,assm) using the notion of next-assm(assm,s) and $assm_i$(assm,π). Given a Kripke structure K and a context assm, the validity of a formula f at a state s, denoted by $(K, assm, s) \models f$, is defined as follows:

$(K, assm, s) \models \forall pf$ iff   for all assm-good paths π starting at s:

$\qquad\qquad\qquad\qquad\qquad\qquad (K, assm, \pi) \models pf$

$(K, assm, s) \models \exists pf$ iff   there is some assm-good path π starting at s with

$$(K, assm, \pi) \models pf$$

$(K, assm, \pi) \models sf$	iff	$(K, assm, \pi(0)) \models sf$
$(K, assm, \pi) \models Xpf$	iff	$(K, assm_1(assm,\pi), \pi^1) \models pf$
$(K, assm, \pi) \models [pf_1 \ U \ pf_2]$	iff	there is some k with $(K, assm_k(assm,\pi), \pi^k) \models pf_2$
		and for all j, $0 \le j < k$: $(K, assm_j(assm,\pi), \pi^j) \models pf_1$

The other cases are straightforward. Furthermore, we say that a module $(K, assm)$ satisfies a specification f, denoted by $(K, assm) \models f$, iff $(K, assm, s^0) \models f$.

3   Relativized Equivalence of Kripke Structure

Usually two Kripke structures K and K' are equivalent if both satisfy the same set of formulae. Dealing with modules embedded in an environment which has to guarantee a behaviour according to an interface specification assm, we use a relativized equivalence relation, which demands only that both modules should have the same behaviour in such an environment, i.e. (K,assm) and (K',assm) cannot be distinguished by any formula.

Two Kripke structures K and K' are *equivalent relatively to assm* $\in$ PTL, denoted by $K \equiv_{assm} K'$, iff for all CTL* formulae f: $(K, assm) \models f$ iff $(K', assm) \models f$.

This (semantical) equivalence relation can be defined by a syntactical relation on the structures (bisimulation) as we will show in the sequel. Furthermore we will characterize the equivalence class of a given Kripke structure by a CTL formula. We will define the relativized bisimulation by a chain of approximations considering paths up to depth i.

Given two Kripke structures $K = (S,s^0,R,L)$ and $K' = (S',s'^0,R',L')$ with the same set ATOM of atomic propositions and given a PTL formula assm we define the relativized equivalence relations $BISIM_i(assm) \subseteq S \times S'$ as follows:

(1) $(s,s') \in BISIM_0(assm)$ iff $L(s) = L'(s')$

(2) $(s,s') \in BISIM_{i+1}(assm)$ iff

    (a) $L(s) = L'(s')$ and

    (b)   for every next-assm(assm,s) good successor s_1 of s there is a successor s'_1 of s' with $(s_1,s'_1) \in BISIM_i(next\text{-}assm(assm,s))$

    (c)   for every next-assm(assm,s') good successor s'_1 of s' there is a successor s_1 of s with $(s_1,s'_1) \in BISIM_i(next\text{-}assm(assm,s))$

(3) $(s,s') \in BISIM(assm)$ iff $(s,s') \in BISIM_i(assm)$ for all $i \in N$.

(4)   K BISIM(assm) K'   iff $(s^0, s'^0) \in$ BISIM(assm).

Example

Consider the Kripke structures of Fig. 5 and Fig. 6. Without consideration of the environment constraints the two structures are different, as K_1 can proceed from state s_1 to $\bar{s}_2$ or s_2, whereas in K_2 the next states of s_1 are s_0 or $\bar{s}_0$ as the input signal Req is low in state s_1. But as the environment has to guarantee that the signal Req remains high unless the response Ack occurs - this is expressed by the assumption assm = $\Box$(Req $\rightarrow$ [Req unless Req $\wedge$ Ack]) -, both structures are equivalent, as under the constraint assm the state s_t in K_1 and the state s_t in K_2 are not reachable.

The bisimulation BISIM(assm) and the equivalence relation $\equiv_{assm}$ coincide. To prove this fact we first show that two Kripke structures are equivalent relatively to a given constraint assm if they are bisimulation equivalent.

Theorem 1

If $(s,s') \in$ BISIM(assm) then for all $f \in$ CTL*: $((K, assm, s) \models f$ iff $(K', assm, s') \models f)$.

For the reverse direction we give a characterization of the bisimulation class of a Kripke structure by a CTL formula. Basicly this formula describes the computation tree of the structure. Two states which are bisimulation equivalent have corresponding computation trees w.r.t. the given constraint on the environment. For a state s, let $CT_n(s)$ denote the computation tree of depth n rooted at s. We can describe the computation tree $CT_n(s)$ w.r.t. an assumption assm by a CTL formula. The formula $F_{CT,n}(s,assm)$ is defined by:

$\qquad F_{CT,0}(s,assm) = a_1 \wedge ... a_n \wedge \neg b_1 \wedge ... \wedge \neg b_m$,

$\qquad\qquad$ where $L(s) = \{a_1, ..., a_n\}$ and ATOM $\setminus L(s) = \{b_1, ..., b_m\}$

$\qquad F_{CT,k+1}(s,assm) = F_{CT,0}(s,assm) \wedge$

$\qquad \bigwedge \{ \exists X\ F_{CT,k}(s',\text{next-assm}(assm,s))\ |\ s'$ is a next-assm(assm,s)-good successor of s $\} \wedge$

$\qquad \forall X (\bigvee \{ F_{CT,k}(s',\text{next-assm}(assm,s))\ |\ s'$ is a next-assm(assm,s)-good successor of s$\})$

As there is a finite depth m such that $CT_m(s^0)$ w.r.t. assm characterizes the (infinite) computation tree we obtain:

Theorem 2

Given a Kripke structure K with initial state s^0 and an assumption assm, then there is a CTL formula $F_{BISIM}(K,assm)$ that characterizes that structure up to BISIM(assm)-equivalence.

Combining the results of Theorem 1 and Theorem 2 we obtain the following characterizations.

Corollary 1

(1)   Given two structures K and K' then it holds:

$(K,K') \in BISIM(assm)$ iff $(\forall f \in CTL^* : (K,assm) \models f$ iff $(K',assm) \models f)$

(2)   Given two structures K and K' then it holds:

$(K,K') \in BISIM(assm)$ iff $(\forall f \in CTL : (K,assm) \models f$ iff $(K',assm) \models f)$

(3)   Given two structures K and K' then it holds:

If there is some CTL* formula f with $(K,assm) \models f$ but $(K',assm) \models \neg f$, then there is already a CTL formula f' which distinguishes both structures.

4   Conclusion

In this paper we have defined a relativized bisimulation between Kripke structures and we have characterized this relation in temporal logic. Our equivalence relation is a generalization of the relation given in [BCG87], the results of [BCG87] are subsumed by our results by using the assumption **true**. Our bisimulation is a strong bisimulation, but we can also define a weak bisimulation by weakening the condition of a corresponding step: for every step $s \rightarrow s_1$ on an assm-good path there should be a corresponding finite path in the other structure. To give a temporal characterization of the weak bisimulation we have to omit the next operator. This leads to a generalization of the corresponding result of [BCG87]. As fairness constraints may be specified as an assumption, we can characterize the equivalence of Kripke structures with fairness constraints, too. This solves a problem raised in [BCG87].

If a temporal specification of a system composed of several modules is derived from the specifications of the modules, and the modules are proved correct w.r.t. their specifications, one can replace every module by an equivalent one without loosing the correctness of the derived specification. This can be done as two equivalent modules can not be distinguished by any temporal formula.

Furthermore the equivalence relation is decidable, as the formula $F_{BISIM}(K,assm)$ is constructable and the validity of a formula spec in a module (K',assm) is decidable.

If the given constraints on the environment are safety properties, the definition of an assm-good state can be weakened to the requirement that the state has to satisfies only the disjunction of all sets literal(C), where C is a closure of assm. Hence this condition can be checked locally.

In the context of the computer architecture design language AADL we use a temporal logic MCTL for module specification [DD89], [DDGJ89], [Jo89]. MCTL consists of pairs (assm, spec) where assm is a restricted PTL formula and spec is a CTL formula. By this paper we have an appropriate notion of equivalence for that logic, which can be used in the design process within AADL.

5   References

[BCG87]   M.C. Browne, E.M. Clarke, O. Grumberg: Charactering Kripke Structures in Temporal Logic. Tech. Report CMU-CS-87-104, Carnegie Mellon University, Pittsburgh (1987)

[Br86]   M. Browne: An improved algorithm for the automatic verification of finite state systems using temporal logic. Symp. Logics in Computer Science, pp. 260 - 266, 1986

[DD89]   W. Damm, G. Döhmen: AADL: A net based specification method for computer architecture design. in: de Bakker (Ed.): Languages for Parallel Architectures: Design, Semantics, and Implementation Models. Wiley & Sons, 1989

[DDGJ90]   W. Damm, G. Döhmen, V. Gerstner, B. Josko: Modular verification of Petri nets: The temporal logic approach. REX Workshop on Stepwise Refinement of Distributed Systems: Models, Formalisms, Correctness, LNCS 430, pp. 180 - 207, 1990

[EH86]   E.A. Emerson, J.Y. Halpern: "Sometimes" and "not never" revisited: On branching versus linear time temporal logic. Journal of the ACM 33, pp. 151-178, 1986

[Jo90]   B. Josko: Verifying the correctness of AADL modules using model checking. REX Workshop on Stepwise Refinement of Distributed Systems: Models, Formalisms, Correctness, LNCS 430, pp. 386 - 400, 1990

[Kr87]   F. Kröger: Temporal Logic of Programs. EATCS-Monographs, Springer, 1987

[La86]   K.G. Larsen: Context-dependent bisimulation between processes. Ph.D. Thesis, Edingburgh, 1986

[LM87]   K.G. Larsen, R. Milner: Verifying a protocol using relativized bisimulation. ICALP 87, LNCS 267, pp. 126-135, 1987

[LP85]   O. Lichtenstein, A. Pnueli: Checking that finite state concurrent programs satisfy their linear specification. 12th ACM Symp. on Principles of Programming Languages, pp. 97-107, 1985

[Mi83]   R. Milner: Calculi for synchrony and asynchrony. TCS 25, 1983

[Pa81]   D. Park: Concurrency and automata on infinite sequences. LNCS 104, pp167-183, 1981

[Pn77]   A. Pnueli: The temporal logic of programs. 18th Annual Symposium on Foundations of Computer Science, 1977

[Pn85]   A. Pnueli: Linear and branching structures in the semantics and logics of reactive systems. ICALP 85, LNCS 194, 1985

THE MODULAR FRAMEWORK OF COMPUTER-AIDED VERIFICATION

Gil Shurek and Orna Grumberg
Computer Science Department, The Technion
Haifa 32000, Israel
E-address: orna@techsel (BITNET), orna@sel.technion.ac.il (CSNET)

1. INTRODUCTION

Temporal logic model checking procedures have been proven to be a feasible approach to automatic verification of relatively small finite-state systems. As the expressive power of temporal logic makes it adequate for specifying properties of reactive systems and as interesting examples are typically those of concurrent and distributed systems, it seems natural to consider the applicability area of automatic model checking as that of reactive finite-state concurrent systems. This context, however, exposes the major disadvantage of model checkers, which is their global nature. By this we mean that only a complete system can be verified to meet its temporal specifications. This globality, on the one hand, induces the "state explosion problem" and, on the other hand, prevents the realization of the "pre-verified component methodology", presented below.

The *state explosion problem* arises in systems composed of many loosely coupled components. The size of such a system may grow as the product of the sizes of its components. Thus, space requirement might become too large for a model checking procedure to be applicable. Obviously, state explosion affects the time-complexity as well.

The *pre-verified component methodology* refers to the verification of systems containing an already verified component, that avoids redoing the component's proof. When verifying a specific system, it should be possible to exploit the pre-verification of the component. As a result, the average verification time-complexity over the set of systems containing this pre-verified component should be reduced.

It seems natural to consider the use of *compositional* reasoning as a solution to the state explosion problem. Exploiting the modular structure of a system, a compositional verification method applies model checking procedures to individual components, to check properties from which the specification of the global system is deduced. The construction of the global system is avoided, and thus space requirements are reduced. Hopefully time-complexity is reduced as well. The desirability of using pre-verified components raises the need for our method to be *modular* too. By this we mean that it should be possible to apply a verification procedure to an individual component when no specific environment is available. The difference between compositionality and modularity lies in the amount of information that the verification procedure has about one component when verifying the other. In a compositional approach complete knowledge about the environment may be used, while in the modular approach no information about the actual environment is available. Note that our terminology adapts the terms "compositionality" and "modularity", yielding notions that differ slightly from the traditional ones.

This paper presents a modular, semi-automatic method for the verification of finite-state programs with temporal logic specifications. The *purpose* of this work, however, goes beyond the search for a technical solution to modular verification. The additional goal is to gain a wider understanding of the modular framework. In a related work [SG90] we discuss motivation, solutions, solutions' properties, and evaluation criteria in that context, e.g.:

• An obvious quality of a global model checker that should be preserved in a modular procedure is the ability to verify arbitrary properties of a distributed system, not being restricted to properties of single components within the compound system.

• Modular procedures have no knowledge about the environment. Hence, applying the checking procedure to one component, an *assumption* on the behavior of the environment is often necessary in order to deduce a global property. Assumptions and their effect on the verification process are thoroughly discussed in [SG90].

Following the previous discussion, this work is a search for a partially automated modular verification procedure supported by a model checking procedure, for reactive, finite-state distributed systems. The systems are modeled by finite structures described as state-transition graphs, where asynchronous actions have an interleaving semantics. We assume that specifications are given by propositional temporal logic formulas that are interpreted over the computation tree of the system's transition graph.

Several works on compositional techniques for reasoning about reactive concurrent systems established the background for our research: [P85],[J87], and [CLM89]. As we regard the "*Assume – Guarantee*" paradigm [P85] to be the preferable modular scheme, we begin with an exploration of its underlying mechanism. This is done by presenting a deductive proof system scheme (AG) which is an abstraction of that paradigm. Considering a further abstraction (the AGS) we claim that, such paradigms essentially manipulate infinite sets of models. The main difficulty in implementing an AG-like scheme as a semi-automatic procedure, is identified to be the need to represent these sets so that they can be handled automatically.

We present an AG-like proof system scheme, denoted AGM, in which a set of models is represented by a single model. In this scheme assumptions are expressed by finite state-transition structures, while the specifications are given in a temporal language. The scheme defines complete proof systems, in which soundness depends on conditions that should be satisfied in an actual framework (i.e., an actual model of computation and a temporal language). Implementations of this scheme as semi-automatic procedures are based on an automatic model checker that is not necessarily modular.

We show how a proof system can be derived from the AGM scheme. As the model of computation we choose a simplified version of CCS [M80] which describes asynchronous communicating systems. The specification language is a sublanguage of CTL^* (CTL^*_{-E}) which is more expressive than LTL. In CTL^*_{-E} negation is restricted to the atomic propositions and the E ("there exists") path quantifier is eliminated. A generalized version of AGM (AGM^*) can be realized in this framework. The AGM^* enables simultaneous use of partial information on both components of the distributed system. In AGM^*, verification can be viewed as state-reduction of the global model without construction of the model. The reduction is relative to the property to be verified (as opposed to minimization which guarantees semantic equivalence of models). Proof systems derived from the AGM and the AGM^* schemes can be implemented as semi-automatic procedures for modular verification. Properly used, these procedures enable the use of pre-verified components, and guarantee reduced memory requirement and even improved time complexity, compared to a global model checker.

In Section 2 we describe the AG and AGS proof system schemes. In Section 3 the AGM and its generalized version AGM^* are presented. Section 4 presents the actual proof system derived from these schemes. In [SG90] this section includes an application of that proof system to a program.

2. THE AG AND AGS SCHEMES

2.1 The AG Proof System Scheme

The AG deductive proof system scheme presented in this section should be viewed as an abstraction of Pnueli's assume-guarantee paradigm [P85].

Let L be a logic used as a reasoning and specification language for distributed systems. Let M be a set of models, containing component models and composed models, for which ' $\|$ ' denotes the composition operation. Each composed model m is accompanied with a fixed decomposition into two component models m_A, m_B, such that $m = m_A \| m_B$. For a component model $m \in M$, $C_A[m]$ denotes the set of all models in M of the form $m_A \| m_B$, for which m_A is m (i.e., $C_A[m] = \{m_A \| m_B \mid m_A = m\}$). $C_B[m]$ is defined similarly.

For $\varphi_1, \varphi_2 \in L$, the notation $\varphi_1 \underset{C_i[m]}{\models\!=\!} \varphi_2$, $i \in \{A,B\}$ means that every model in $C_i[m]$ which satisfies φ_1, satisfies φ_2 as well. Thus, φ_2 is the consequence of φ_1 in $C_i[m]$. $\underset{C_i[m]}{\vdash\!\!\!\!-}$ is a proof system that validates L formulas in $C_i[m]$. The soundness of this system is defined as: $\forall \varphi_1, \varphi_2 \in L \ [\varphi_1 \underset{C_i[m]}{\vdash\!\!\!\!-} \varphi_2 \Rightarrow \varphi_1 \underset{C_i[m]}{\models\!=\!} \varphi_2]$. When $\varphi_1 \equiv$ true, the definition results in $\underset{C_i[m]}{\vdash\!\!\!\!-} \varphi_2 \Rightarrow \underset{C_i[m]}{\models\!=\!} \varphi_2$ which means that if φ_2 is provable in $\underset{C_i[m]}{\vdash\!\!\!\!-}$ then φ_2 is true in the composition of m with *any* environment.

Given the proof systems $\underset{C_A[m_1]}{\vdash\!\!\!\!-}$ and $\underset{C_B[m_2]}{\vdash\!\!\!\!-}$ we present below a set of AG proof systems (one for each model $m_1 \| m_2$). The proof system $\underset{AG-m_1 \| m_2}{\vdash\!\!\!\!-}$ for the verification of L-formulas in the composed model $m_1 \| m_2$ is defined as follows:

Axioms:

$$\{\varphi \mid \underset{C_A[m_1]}{\vdash\!\!\!\!-} \varphi\} \cup \{\varphi \mid \underset{C_B[m_2]}{\vdash\!\!\!\!-} \varphi\}$$

Inference rules:

$$\frac{\underset{AG-m_1\|m_2}{\vdash\!\!\!\!-} \varphi \qquad \varphi \underset{C_A[m_1]}{\vdash\!\!\!\!-} \psi}{\underset{AG-m_1\|m_2}{\vdash\!\!\!\!-} \psi} \qquad\qquad \frac{\underset{AG-m_1\|m_2}{\vdash\!\!\!\!-} \varphi \qquad \varphi \underset{C_B[m_2]}{\vdash\!\!\!\!-} \psi}{\underset{AG-m_1\|m_2}{\vdash\!\!\!\!-} \psi}$$

The soundness of $AG-m_1 \| m_2$ is defined as: $\forall \psi \in L \ [\underset{AG-m_1\|m_2}{\vdash\!\!\!\!-} \psi \Rightarrow m_1 \| m_2 \models \psi]$.

Theorem 1: Given sound proof systems $\dfrac{\quad\quad}{C_A[m_1]}$ and $\dfrac{\quad\quad}{C_B[m_2]}$, $\dfrac{\quad\quad\quad\quad}{AG-m_1 \| m_2}$ is sound.

The $AG-m_1 \| m_2$ is a scheme for a proof system, capable of verifying properties of the distributed system $m_1 \| m_2$, without being restricted to the behavior of a single component in the context of the global system. Moreover, it can be considered modular according to our terminology. Examining a typical AG proof, its modularity can be observed:

$$\frac{\quad\quad}{C_A[m_1]}\ \varphi_n\ ,\ \ \varphi_n \frac{\quad\quad}{C_B[m_2]}\ \varphi_{n-1}\ \cdots\ \varphi_1 \frac{\quad\quad}{C_A[m_1]}\ \varphi$$

We are interested in a semi-automatic implementation of the AG scheme, where the user is required to supply ("guess") the sequence of assumptions φ_n to φ_1 and the validation in $C_A[m_1]$ and in $C_B[m_2]$ is done automatically. The m_1 proof-block is the set of proof stages in which $\dfrac{\quad\quad}{C_A[m_1]}$ is applied. When assumptions are given, the m_1 proof block can be carried out without having to consider any specific "m_2". Given a component model m_i , if the proof-block consisting of all $\dfrac{\quad\quad}{C_B[\ ]}$ stages can be carried out successfully with respect to $\dfrac{\quad\quad}{C_B[m_i]}$, then m_1 proof-block can be completed to a full AG proof, deducing $m_1 \| m_i \models \varphi$.

2.2 The AG-sets (AGS) Proof System Scheme

In this section we present another proof system scheme, AG-sets (AGS), which is an abstraction of AG, meant to explore the underlying mechanisms of AG. In AGS, sets of models are used directly as the reasoning language (similar to the *Semantic Model* of [AL89]). However, we still assume temporal specification. Therefore, an inference rule that derives L formulas (from a previously "derived" model set) is necessary. Defined below is an AGS proof system for $m_1 \| m_2$.

$gs[m_1 \| m_2]$ is a set of model sets, defined inductively by:
Atoms ("axioms"):

$$\{C \mid\ C \supseteq C_A[m_1]\} \cup \{C \mid\ C \supseteq C_B[m_2]\}$$

Closure operations ("inference rules"):

$$\frac{\begin{array}{c} C \in gs[m_1 \| m_2] \\[4pt] C' \supseteq C \cap C_A[m_1] \end{array}}{C' \in gs[m_1 \| m_2]} \qquad\qquad \frac{\begin{array}{c} C \in gs[m_1 \| m_2] \\[4pt] C' \supseteq C \cap C_B[m_2] \end{array}}{C' \in gs[m_1 \| m_2]}$$

The inference rule:

$$\frac{\begin{array}{c} \varphi \in L \\[6pt] C \in gs[m_1 \| m_2] \\[6pt] \dfrac{\quad\quad}{C}\ \varphi \end{array}}{\dfrac{\quad\quad\quad\quad}{AGS-m_1 \| m_2}\ \varphi}$$

Theorem 2: $AGS - m_1 \parallel m_2$ is sound and complete.

Noticing that both *AG* and *AGS* basically manipulate potentially infinite sets of models, leads us to identify the major obstacle to the implementation of *AG*. In *AG*, sets of models are represented either by themselves ($C_A[m_1]$ and $C_B[m_2]$) or by L-formulas (i.e., φ defines and represents the set of all models satisfying φ). The lack of uniform representation for these sets forces *AG* implementations (e.g., [J87]) to provide conversion procedures (which are, in general, cumbersome and time-consuming) to enable constructive manipulation of model sets. In *AGS* we abstract away from the *representation problem*. The *AGS* proof system is theoretic and has no explicit representation for sets. This system is not intended to be implemented directly, but to inspire the definition of other proof system schemes.

In the next section we present an AGS-like proof system, in which sets are represented in a way they can be handled constructively.

3. THE *AG*-MODELS (*AGM*) SCHEMES

3.1 The *AGM* Proof System Scheme

The *AGM* is an *AGS*-like proof system scheme, in which the basic elements handled by the deductive procedure are models (i.e., finite state-transition structures). These basic elements are used to represent the potentially infinite sets of models, discussed in the previous section. The *AGM* proof system is a modular model checking system for finite state distributed system, based on their decomposition into two components.

Let *M* be a model domain consisting of component models m_i and composed models m, which may differ syntactically. Let Σ be an alphabet and let Σ_m denote the alphabet associated with the model m. Σ_m consists of state labels and/or transition labels. Let $\parallel$ denote a composition operator that applied to two component models m_i, m_j results in a composed model m, i.e., $m = m_i \parallel m_j$. Also, let $|_\Sigma$ denote a restriction transformer, that applied to a composed model results in a component model with alphabet set restricted to Σ.

Let **L** be a temporal language interpreted over the domain of models *M*. Let $\mathbf{L}(\Sigma)$ denote the set of formulas in **L** defined over the set of atomic propositions in Σ. We assume (for the semantics of **L**) that if $\varphi \notin L(\Sigma_m)$ then $m \not\models \varphi$.

Let $\sqsupseteq$ denote a *Preorder* (reflexive and transitive relation) defined over *M*, which has the following semantic property: $m_A \sqsupseteq m_B$ implies that all **L** formulas that are satisfied in m_A, are satisfied in m_B (i.e., if $m_A \sqsupseteq m_B$ than $\forall \varphi \in \mathbf{L}$, $m_A \models \varphi \Rightarrow m_B \models \varphi$).

Let $\vdash_{MC}$ denote a proof system for checking the satisfaction of **L** formulas in *M* models (i.e., a *model checker*).

The *AGM* definition inherits the structure of the *AGS* definition, where inference rules which do not deduce temporal formulas, are referred to as closure operations. Defined below is an *AGM* proof system for $m_1 \parallel m_2$.

gm[m₁ ∥ m₂] is a set of models, defined inductively by:

Atoms : $\qquad \{m' \parallel m_2 \mid m' \sqsupseteq m_1\} \cup \{m_1 \parallel m'' \mid m'' \sqsupseteq m_2\}$

closure operations :

$$\frac{m_1 \parallel m'' \in gm[m_1 \parallel m_2] \qquad m' \sqsupseteq (m_1 \parallel m'')|_{\Sigma_{m_1}}}{m' \parallel m_2 \in gm[m_1 \parallel m_2]}$$

$$\frac{m' \parallel m_2 \in gm[m_1 \parallel m_2] \qquad m'' \sqsupseteq (m' \parallel m_2)|_{\Sigma_{m_2}}}{m_1 \parallel m'' \in gm[m_1 \parallel m_2]}$$

The inference rule :

$$m \in gm[m_1 \parallel m_2]$$

$$\varphi \in L(\Sigma_m)$$

$$\frac{\vdash_{MC} m \vDash \varphi}{\vdash_{AGM-m_1 \parallel m_2} \varphi}$$

The soundness of the *AGM* scheme depends on the following conditions (regarding the language, the models domain, the composition operator, the restriction transformer and the semantic preorder) :

Soundness Conditions :
1. For all m', m_A, m_B component models in M, if $m_A \sqsupseteq m_B$ then $m_A \parallel m' \sqsupseteq m_B \parallel m'$.
2. For all m', m_A, m_B component models in M, if $m_A \parallel m' \sqsupseteq m_B \parallel m'$ then $(m_A \parallel m')|_{\Sigma_{m'}} \sqsupseteq (m_B \parallel m')|_{\Sigma_{m'}}$.
3. For all m_A, m_B component models in M , $((m_A \parallel m_B)|_{\Sigma_{m_A}}) \parallel m_B \sqsupseteq m_A \parallel m_B$.

When the composition operator is *not commutative* the dual conditions are required as well.

Theorem 3: Given that the soundness conditions are fulfilled then:
3.1 Using a sound proof system $\vdash_{MC}$, $\vdash_{AGM-m_1 \parallel m_2}$ is sound.
3.2 Using a complete proof system $\vdash_{MC}$, $\vdash_{AGM-m_1 \parallel m_2}$ is complete.

Comment: Systems which are composed of more than two components, can be handled by *AGM*-like schemes.

3.1.1 Evaluation

The *AGM* scheme is capable of verifying any property of the global distributed system, which is expressible in **L**. Examining a typical $AGM-m_1 \parallel m_2$ proof, verifying that $m_1 \parallel m_2 \vDash \varphi$, its modularity can be observed. Regarding the state explosion problem, this proof system is feasible as long as the size of the structures representing the assumptions is significantly less than the size of the components' models. Applying the pre-verified components methodology, feasibility depends on the assumptions strength. As "greater" ($\sqsupseteq$) structures make weaker assumptions, the *AGM* enables the user to weaken the supplied assumptions provided that the knowledge needed for the verification of the temporal specification is preserved.

3.1.2 Resemblance to the *AGS* Scheme

In the *AGM* scheme, a single model of a composed system represents an *AGS* basic element: a set of models. The model represents the set of *composed* models in M which are *smaller* with respect to the semantic relation $\sqsupseteq$ (i.e., m represents a set of models in which every formula satisfied by m is valid). The "greater" model can be viewed as holding partial information about the "small" one. Thus, the preorder between two basic elements of the *AGM* is analogous to the subset relation between *AGS* sets. The task of combining partial knowledge about the checked system with full information about one of the components, which is carried out by the intersection operator in the *AGS* scheme, is achieved here by applying the composition operator itself (aided by the restriction transformer).

3.1.3 Implementation

A computer-aided implementation of the *AGM* proof system is expected to carry out automatically the following actions: **a.** Models composition, **b.** Applying the restriction transformer, **c.** Verifying the satisfaction of L formulas in M models, **d.** Verifying the preorder between models.
It is not difficult to imagine implementable definitions of a composition operator and a restriction transformer which make tasks **a.** and **b.** algorithmic. To automate task **c.** , a global model checker can be applied. The only task which is not clearly decidable even for the finite-state case is the verification of the semantic preorder (task **d.**). Any useful implementation of the *AGM* system should suggest a semantic preorder which is syntactically identified by an efficient algorithm.

In section **4.** we suggest an implementation of *AGM* system in which all four tasks are algorithmically handled. As can be seen there, the need for a decidable preorder affects the choice of the temporal specification language.

3.2 The AGM^* Proof System Scheme

Setting stricter soundness conditions, the AGM^* which is a generalized version of *AGM*, can be defined. The AGM^* enables simultaneous use of partial information on both components of the distributed system. Using this scheme it is possible to conduct a proof, along which only the knowledge that is essential for the verification of the final specification is preserved.

$gm^*[m_1 \parallel m_2]$:

Atoms : $\{ m_A \parallel m_B \mid m_A \sqsupseteq m_1 , m_B \sqsupseteq m_2 \}$

Closure operation :

$$\bar{m}, \bar{\bar{m}} \in gm^*[m_1 \parallel m_2]$$

$$m' \sqsupseteq \bar{m}|_{\Sigma_{m_1}}$$

$$m'' \sqsupseteq \bar{\bar{m}}|_{\Sigma_{m_2}}$$

$$\overline{\qquad\qquad\qquad\qquad}$$

$$m' \parallel m'' \in gm^*[m_1 \parallel m_2]$$

The inference rule :

$$m \in gm^*[m_1 \| m_2]$$

$$\varphi \in L(\Sigma_m)$$

$$\frac{\qquad}{MC} \, m \models \varphi$$

$$\frac{\qquad}{AGM^*-m_1 \| m_2} \, \varphi$$

In the case where "smaller" ($\sqsupseteq$) models can often be represented by smaller structures, the AGM^* enables further reduction in the size of the structures handled by the proof system (in comparison with AGM). Thus, further reduction in time and space complexity is achieved.

4. ASYNCHRONOUS COMMUNICATING SYSTEMS

4.1 The Model of Computation

We use here a simple computation model for distributed systems. The components of these systems are asynchronous processes, that communicate through synchronized actions. This model is similar to the CCS model [M80]. Our version is restricted to one application of the composition operation (i.e., every model of a compound system is composed of two process models). In this version, compound models are syntactically guaranteed to be deadlock free, i.e., when a composed system is verified using this type of models, all the deadlocked sub-structures of the computation tree of the system are ignored.

4.2 The Specification Language

Defined below is a propositional temporal logic which is a sublanguage of CTL^* [CE81]. Let Σ be the disjoint union of AP_Σ and AC_Σ. The language $CTL^*_{-E}(\Sigma)$ is defined as the smallest set of state formulas such that:
(1) If $A \in AP_\Sigma$, then A and $\neg A$ are state formulas.
(2) If φ and ψ are state formulas, then $\varphi \vee \psi$ and $\varphi \wedge \psi$ are state formulas.
(3) If φ is a path formula, then $A\varphi$ is a state formula.
(4) If φ is a state formula, then φ is a path formula.
(5) If φ and ψ are path formulas, then $\varphi \vee \psi$, $\varphi \wedge \psi$, $\varphi \, U \, \psi$, $X\varphi$ and $G\varphi$ are path formulas.

Note that in CTL^*_{-E} negations are applied only to atomic formulas. Thus, the path quantifier E ("for some computation path") is eliminated *semantically* from CTL^*_{-E}. The semantics of $CTL^*_{-E}(\Sigma_m)$ is defined in the usual way, with respect to the Kripke structure K_m [HC77], generated from the (process or system) model m by ignoring the model's transition labels.
The satisfaction of a state formula $\varphi \in CTL^*_{-E}(\Sigma)$ in a model m (with s_0 the initial state), denoted $m \models \varphi$, is defined by :
(1)   If $\varphi \in CTL^*_{-E}(\Sigma_m)$ then $m \models \varphi$ iff $m, s_0 \models \varphi$.
(2)   If $\varphi \notin CTL^*_{-E}(\Sigma_m)$ then $m \not\models \varphi$.

We say that one language is *more expressive* then the other if there exists a formula in the one that does not have an equivalent formula in the other. Let $\triangleright$ denote "more expressive" and let CTL_{-E} be $CTL \cap CTL^*_{-E}$. Comparing CTL^*_{-E} with CTL^*, CTL and LTL

[CG87] we get the following lemma:

Lemma 3: (a) $CTL^* \vdash CTL^*_{-E}$;   (b) $CTL^*_{-E} \vdash CTL_{-E}$;   (c) $CTL^*_{-E} \vdash LTL$.

We can now choose a model checking procedure $\vdash\!\!\!\underset{MC}{\qquad}$, needed in our modular proof systems (AGM and AGM^*). As CTL^*_{-E} is a sub language of CTL^* there is a model checking algorithm which is exponential in the size of the specification and linear in the size of the Kripke structure (i.e., $O(2^{|\varphi|}(|S|+|R|))$) [EL85]. One may prefer to restrict the specifications to CTL_{-E}, gaining a linear model checking algorithm (i.e., $O(|\varphi|(|S|+|R|))$) [CES86].

4.3 The Preorder

The choice of CTL^*_{-E} as a specification language is motivated by a unique semantic quality: The validity of CTL^*_{-E} formulas is preserved over "*pruned*" computation trees. A description of this property is presented below. In this context, we view the semantics of CTL^*_{-E} as defined over computation trees. The originating transition graphs (if exist) of these trees are omitted. Let T be a computation tree, and $\varphi \in CTL^*_{-E}$ be a temporal formula. $T,v \vDash \varphi$ denotes that φ is satisfied at node v in T. A *branching node* in T is a node that has at list two immediate descendants. Eliminating a subtree of T rooted at v, such that v is an immediate descendant of a branching node, is defined as *pruning T at v*.

Theorem 5: If T' is generated by pruning T at nodes which are descendants of v, then
$$\forall \varphi \in CTL^*_{-E} \ [\ T,v \vDash \varphi \ \Rightarrow \ T',v \vDash \varphi \ ].$$

Long [L89] defined a syntactic preorder over Kripke structures, which partially identifies the *pruning relation* of their computation trees. Basically, we adopt his relation as our choice of the required semantic preorder $\sqsupseteq$. Assimilated in our system, the definition is enhanced to encompass more cases for which the semantic property holds, including more cases for which the pruning relation holds. In addition, it is modified to consider transition labels and to verify the containment of the atomic proposition sets.

Let $m = \langle AC,AP,S,R,s_0,L \rangle$ and $m' = \langle AC',AP',S',R',s'_0,L' \rangle$ be either both process models, or both system models, such that $AP \subseteq AP'$. A sequence of relations F_i, such that $\forall i \geq 0 \ \ F_i \subseteq S \times S'$, is defined as follows:

For all $s \in S, s' \in S'$
1. $s F_0 s'$ iff $L(s) = L'(s') \cap AP$.
2. $s F_{n+1} s'$ iff -
    2.1 $s F_n s'$.

    2.2 $\forall s'_1 \in S'$, for all action $\alpha \ [\ s' \overset{\alpha}{\longrightarrow} s'_1 \ \Rightarrow \ \exists s_1 \in S \ [\ s \overset{\alpha}{\longrightarrow} s_1 \ \wedge \ s_1 F_n s'_1 \ ]].$

$F \subseteq S \times S'$ is defined by: For all $s \in S, s' \in S'$, $s F s'$ iff $\forall i \geq 0 \ s F_i s'$.

Let $m = \langle AC,AP,S,R,s_0,L \rangle$ and $m' = \langle AC',AP',S',R',s'_0,L' \rangle$. The relation $\geq$ over the model domain M is defined as follows:
$m \geq m'$ iff -

i. m and m' are either both process models, or both system models.

ii. $AP \subseteq AP'$.

iii. $s_0 F s'_0$.

Theorem 6: If $m_1 \geq m_2$ then for every Σ, and for every formula $\varphi \in CTL^*_{-E}(\Sigma)$, $m_1 \vDash \varphi \Rightarrow m_2 \vDash \varphi$.

There is an algorithm to verify that two models are preordered, which is polynomial-time in their size. Using $\geq$ as the required syntactically-identified semantic preorder, note that "greater" ($\geq$) models are, in general, expressed by smaller structures. The implications of this property were previously discussed.

Theorem 7: The AGM (and AGM^*) soundness conditions are satisfied by the model of computation, the language CTL^*_{-E}, and the preorder $\geq$.

Establishing that, the presentation of a semi automatic model checking system derived from these schemes, is completed. In [SG90] we apply these proof rules to a program.

REFERENCES

[AL89]   M. Abadi and L. Lamport, "Composing Specifications", REX Workshop on Step-wise Refinement of Distributed Systems: Models, Formalisms, Correctness, Mook, May 1989.

[CE81]   E.M. Clarke, and E.A. Emerson, "Synthesis of Synchronization Skeletons for Branching Time Temporal Logic", Proc. of Workshop on Logic of Programs, Yorktown-Heights, 1981.

[CES86] E.M. Clarke, E.A. Emerson, and A.P. Sistla, "Automatic Verification of Finite-State Concurrent Systems using Temporal Logic Specifications", ACM Transactions on Programming Languages and Systems 8, 2, pp. 244-263, 1986.

[CG87]   E.M. Clarke, O. Grumberg, "Research on automatic verification of finite-state concurrent systems", Annual Reviews of Computer Science, Vol. 2, 269-290, 1987 (J.F. Traub, editor).

[CLM89] E.M. Clarke, D.E. Long, and K.L. McMillan, "Compositional Model Checker". Proc. of the 4th IEEE Symp. on Logic in Computer Science, Asilomar, June 1989.

[EL85]   E.A. Emerson, and C. Lei, "Modalities for Model Checking: Branching Time Strikes Back", 12th Symposium on Principles of Programming Languages, New Orleans, La., January 1985.

[HC77]   G.E. Hughes, and M.J. Creswell, "An introduction to Modal Logic", London: Methuen, 1977.

[J87]   B. Josko, "MCTL - An Extension of CTL for Modular Verification of Concurrent Systems". Workshop on Temporal Logic, (H. Barringer, ed.) University of Manchester, April 1987, LNCS 398, pp. 165-187.

[L89]   D.E. Long, Private communication.

[M80]   R. Milner, A Calculus of Communicating Systems, Springer Lecture Notes on Computer Science, Vol. 92, 1980.

[P85]   A. Pnueli, "In Transition from Global to Modular Temporal Reasoning about Programs". Logics and Models of Concurrent Systems, (K. Apt, ed.), Vol. 13 of NATO ASI Series F: Computer and System Sciences, Springer-Verlag, 123-144, 1985.

[SG90]   G. Shurek and O. Grumberg, "THE MODULAR FRAMEWORK OF COMPUTER-AIDED VERIFICATION Motivation, Solutions, and Evaluation Criteria", Workshop on Computer-Aided Verification, Rutgers, NJ., June 1990, to appear in ACM/AMS DIMACS series.

Verifying Liveness Properties
By
Verifying Safety Properties
(Extended Abstract)

Jerry R. Burch
School of Computer Science
Carnegie Mellon University
Pittsburgh, PA 15213

Abstract

Conventional techniques for automatically verifying liveness properties of circuits involve explicitly modeling infinite behaviors with either infinite paths through a Kripke structure or with strings in an ω-regular language. This paper describes how *timed trace structures* [2, 3] can be used to convert liveness properties (including unbounded liveness properties such as strong fairness) to safety properties. Such properties can then be modeled and verified using only finite traces. No new algorithms are needed. All that is required is a new interpretation of what behaviors are represented by the finite traces. A mapping is defined between timed trace structures and *complete trace structures* [5], which contain infinite traces, to show that this new interpretation makes sense. The method is demonstrated on a fair mutual exclusion circuit.

1 Introduction

The primary goal of research in automated circuit verification is to devise algorithms that check whether circuits enjoy certain desirable properties. It is common to classify these properties into *safety* and *liveness* properties. Informally, safety properties state that nothing bad happens, while liveness properties state that something good happens.

In formalisms that explicitly model time (such as timed trace structures, described below), it is also possible to express *bounded liveness* properties. Bounded liveness properties state that something good happens within a bounded amount of time. This is in contrast to *unbounded liveness* properties, which state that something good happens eventually without specifying a time bound. A bounded liveness property can be viewed as a form of safety property; conventional methods used to automatically verify safety properties are also adequate for bounded liveness properties.

There are many systems described in the literature that can automatically verify unbounded liveness properties. All of these explicitly model behaviors as being infinite: most often as infinite paths through a Kripke structure or as strings in an ω-regular language. This paper describes a method for converting unbounded liveness properties to bounded liveness properties, and thereby to safety properties. This makes it possible to verify unbounded liveness properties (such as strong fairness) without explicitly modeling infinite behaviors. Instead, methods based on prefix-closed trace structures of finite traces are shown to be adequate. This result adds new insight concerning Black's proof [1] that delay-insensitive fair mutual exclusion cannot be represented using conventional finite traces.

The method merely requires a new interpretation of the behaviors represented by timed trace structures. It was not necessary to develop new algorithms or modify the automatic verifier in any way. The method is demonstrated using a simple buffer circuit as an example. In addition, a mutual exclusion circuit is used to show how strong fairness properties can be modeled and verified.

Using this method a mapping can be defined from timed trace structures on finite traces to complete traces structures [5] on infinite traces. Conformation between trace structures is preserved under this mapping. Thus, the mapping can be used to check whether a specification properly captures the desired liveness properties.

2 Timed Trace Theory

We can only give a short overview of timed trace theory in the space of an extended abstract; the interested reader may refer to [2] and [3]. The formalism is based on Dill's trace theory [5, 6]. We begin by describing trace theory, and then show how it can be extended to include timing information.

In trace theory based verification digital circuits and their specifications are modeled by *trace structures*, which are ordered 4-tuples of the form $T = (I, O, S, F)$. The set I is the set of input wire names of the circuit; O is the set of output wire names. The set $A = I \cup O$ is called the *alphabet* of T. The sets S (the *success set*) and F (the *failure set*) are regular sets of finite strings, called *traces*, over the alphabet A. Each trace models a possible behavior of the circuit by having each symbol in the trace represent a transition on the corresponding wire.

The S set and the F set are used to give partial specifications. A partial specification of a device describes requirements for the proper use of the device, and specifies the behavior of the device given that those requirements are satisfied. In a trace structure, the set S describes the behaviors of a device when it is used properly. The set F describes behaviors resulting from improper use. For example, in asynchronous circuits a gate is typically modeled so that the F set contains all behaviors that cause a hazard and the S set contains all behaviors that do not cause a hazard.

The set $P = S \cup F$ is the set of all *possible traces*. The S and P sets must be prefix-closed, and P must be non-empty. For some non-deterministic devices (such as the vending machine example in [8], which is also discussed in [5]) a given trace can be both a success and a failure, so the S and F sets need not be disjoint. Also, no circuit can control its inputs, which is modeled by requiring that $PI \subseteq P$ (if A and B are sets of strings or symbols, then $AB = \{ab : a \in A \text{ and } b \in B\}$). This is called the *receptiveness* requirement.

As an example, consider how a buffer might be modeled by a trace structure $T = (I, O, S, F)$. Let b be the name of the input wire of the buffer, and let x be the name of the output wire. Then $I = \{b\}$ and $O = \{x\}$. Assume that both the input and the output of the buffer are initially low. We define the set of successful behaviors of the buffer to be those behaviors in which the environment does not cause a hazard. Thus, the set S of successful behaviors is equal to $(bx)^*(b + \epsilon)$. If the environment does cause a hazard, then we do not restrict the ensuing behavior of the circuit, so $F = (bx)^*bb(b + x)^*$.

Consider two variations of the above buffer. The first buffer always eventually fires when it is put in a firable state. The second buffer sometimes might never fire. Both of these buffers would be modeled by exactly the same trace structure. This shows that liveness properties cannot be modeled with this form of trace theory.

A composition operation (denoted by "$\circ$") can be defined on trace structures. Let $T_0 = (I_0, O_0, S_0, F_0)$ and $T_1 = (I_1, O_1, S_1, F_1)$ be trace structures. The composition of T_0 and T_1 is defined when O_0 and O_1 are disjoint. The set of outputs of the composition is $O_0 \cup O_1$, the set of inputs is $(I_0 \cup I_1) - (O_0 \cup O_1)$. Let $A_0 = I_0 \cup O_0$ and $A_1 = I_1 \cup O_1$. If x is a trace in $(A_0 \cup A_1)^*$, define the projection $x |_{A_0}$ to be the trace formed from x by removing all symbols not in A_0. The S and F sets of the composition are given by

$$
\begin{aligned}
S &= \{x \in (A_0 \cup A_1)^* : x |_{A_0} \in S_0 \wedge x |_{A_1} \in S_1\}, \\
F &= \{x \in (A_0 \cup A_1)^* : (x |_{A_0} \in F_0 \wedge x |_{A_1} \in F_1) \vee \\
&\quad (x |_{A_0} \in F_0 \wedge x |_{A_1} \in S_1) \vee \\
&\quad (x |_{A_0} \in S_0 \wedge x |_{A_1} \in F_1)\}.
\end{aligned}
$$

The operation **hide** is also defined on trace structures. The trace structure denoted by $\text{hide}(D)[(I, O, S, F)]$ (where D is a subset of O) is equal to $(I, O - D, S |_{A'}, F |_{A'})$, where $A' = I \cup (O - D)$ and the projection operation is extended to sets of traces in the normal manner.

A trace structure is said to be *failure free* when its F is empty. If the composition of several trace structures is failure free, then the components have been composed in such a way that each of their environmental requirements has been satisfied. The trace theory verifier can efficiently check whether the composition of a set of trace structures is failure free. The S and F sets of trace structures are represented in the verifier with deterministic finite automata. Checking if the composition of two trace structures is failure free is done in time linear in the product of the sizes of these automata. If a composition is not failure free, then usually only a small fraction of the states of the composition need to be explored before a short error trace can be given.

Let T_0 and T_1 be trace structures. Then, T_0 *conforms to* T_1 (written $T_0 \preceq T_1$) if $I_0 = I_1$ and $O_0 = O_1$ and for any trace structure T, the composition of T and T_1 being failure free implies that the composition of T and T_0 is failure free. The idea captured here is that if a circuit works correctly with T_1 as a component, then it works correctly with T_1 replaced by T_0 (up to safety properties). If $T_0 \preceq T_1$ and $T_1 \preceq T_0$, then T_0 and T_1 are *conformation equivalent*. This means that T_0 and T_1 are functionally interchangeable. There exists a canonical form for trace structures such that two canonical trace structures are conformational equivalent if and only if they are equal.

A trace structure T_1 can also be used to represent a specification. A circuit T_0 satisfies the specification T_1 if T_0 conforms to T_1. Checking that a circuit satisfies a specification would be very expensive if it required checking the failure freedom of compositions with all possible trace structures. However, there exists an operation on trace structures, called *mirroring*, that makes checking conformation practical. If $T = (I, O, S, F)$ is a canonical trace structure, then $T^M = (O, I, S, A^* - (S \cup F))$ is its mirror. If T' has the

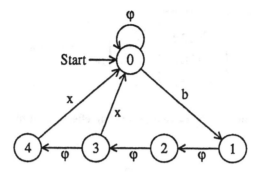

Figure 1: Automata that accepts the S set of a 3/2 rule buffer.

same inputs and outputs as T, then the composition of T' and T^M is failure free if and only if T' conforms T. Thus, checking if a circuit satisfies a specification only requires checking if the composition of the circuit with the mirror of the specification is failure free. This is done in time linear in the product of the sizes of the automata representing the circuit and the specification. If there is an error in the circuit, then usually on a small fraction of the state space needs to be explored before a failure is found. When a failure is found, a short trace of the transitions that led to the failure is output to the user.

2.1 Adding Timing Information

In standard trace theory, only speed-independent circuits can be represented. However, trace theory can be extended to allow the representation of a large class of timing models for different circuit components. We call the extended theory *timed trace theory*. The extension allows traces to contain additional symbols not corresponding to physical wires. The presence of such a symbol in a trace is interpreted as representing the passage of some fixed amount of time. (Later in the paper, we describe an alternative interpretation that allows for the modeling of unbounded liveness properties.) More formally, we allow trace structures of the form (I, O, V, S, F), where V is a set of symbols (disjoint from I and O), and where S and F are sets of traces over the alphabet $A = I \cup O \cup V$. The elements of V do not correspond to any physical wires, so they are called *virtual wire names*. In order to give an intuitive explanation of the intended meaning of such structures, we will consider the case in which V contains a single element, call it φ.

The presence of a φ in a trace indicates the passage of a unit of time, call it τ. The trace $\varphi\varphi x\varphi$ represents a single behavior in which a transition occurs on wire x at time $T_0 + 2\tau$, where T_0 is the time at which the behavior described by the trace began.

Figure 1 is an automata describing the S set of a buffer with input b, output x, and virtual wire φ. Notice that since S is prefix-closed, all of the states in this automata are accepting states. The F set of this buffer is equal to $(SI - S)A^*$. Interpreted in discrete time, this buffer clearly has a minimum delay of 2τ and a maximum delay of 3τ.

Consider a circuit formed using the output of a pulse generator as the input to a buffer as in Figure 2. The pulse generator has a period of 8τ and a 50% duty cycle. Its S set is described by the automata in Figure 3 and its F set is empty. The buffer is the same as

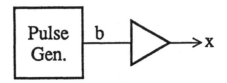

Figure 2: Example circuit for demonstrating effects of timing assumptions.

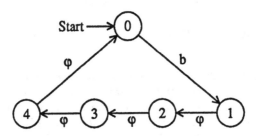

Figure 3: Automata that accepts the S set of a pulse generator.

that represented in Figure 1. The trace structure representing the resulting circuit has an empty F set and its S set is given by the automata in Figure 4. Since the maximum delay of the buffer is 3τ and the pulse generator waits 4τ between outputting transitions, there are never two consecutive b transitions without an x transition in between. This fact could not be represented with the speed-independent timing model.

It is clear how to modify the buffer in Figure 1 to have a minimum delay of $m\tau$ and a maximum delay of $n\tau$ for any non-negative integers m and n such that $m \le n$. The buffer can also be made to have an unbounded maximum delay by removing state 4 and adding a transition on φ from state 3 back to state 3.

The trace theory verifier has been extended to include virtual wires in this way. Checking that a circuit satisfies a specification has the same complexity as before. Examples of using the verifier to check the correctness of circuits can be found in [2] and [3].

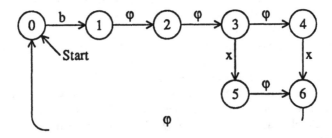

Figure 4: Automata representing the behavior of the example circuit in Figure 2.

3   Converting Liveness to Safety

Let trace structure T model a buffer with input b, output x, and virtual wire φ_s that has a minimum delay of zero and a maximum delay of one. Normally, the presence of a φ_s in a trace in T indicates the passage of a unit of time. For the purpose of converting liveness properties to safety properties, we propose a different interpretation. A transition on b can be seen as placing an obligation on the buffer to eventually toggle x in response. We interpret the nth φ_s in a trace to mean that the buffer has met all such obligations that the environment placed on it before the $(n-1)$th φ_s in the trace. In this interpretation, there need not be a constant, or even bounded, amount of time between each φ_s.

As an example of how this interpretation can be used to verify liveness properties, consider verifying that two buffers in series conform to a single buffer. Let trace structure T_1 model a buffer with input b and output c, and let T_2 model a buffer with input c and output x. Both trace structures have a single virtual wire φ_i, with a minimum delay of zero and a maximum delay of one. In order to show that $T_1 \circ T_2$ conforms to the specification, we compose $T_1 \circ T_2$ with a constraining trace structure C. The trace structure C has no inputs or outputs, but has virtual wires φ_i and φ_s. The sole purpose of C is to constrain the transitions on the virtual wires so that there must be at least three transitions of φ_i between any two transitions of φ_s. Three is the minimum necessary to assure that the two buffers in series satisfy their obligation to transition x before a transition of φ_s. The automatic verifier can be used to show that

$$\text{hide}(\{c, \varphi_i\})[T_1 \circ T_2 \circ C] \preceq T.$$

If C is chosen badly, then the verifier may incorrectly report that the circuit does not conform to the specification. The user is required to find a C that allows the verification to go through. Regardless of what C is used, however, the verifier will never report that a circuit conforms to a specification when in fact it does not. In this sense, the verification method is *conservative*.

4   Mapping to Complete Traces

In the previous section, we described a specification for a live buffer in terms of a timed trace structure. But it may not be intuitively clear that the timed trace structure captures the properties of a live buffer. This section describes a way of addressing that problem.

In [5], Dill describes a version of trace theory, called complete trace theory, where liveness properties are modeled using infinite traces. Dill describes a complete trace structure that is a specification for a live buffer, as a special case of a live gate. It is easy to give an intuitive argument that this complete trace structure accurately captures the properties of a live buffer. If it could be shown that the timed trace specification for a live buffer was in some way equivalent to the complete trace specification, that would be strong evidence that the timed trace specification is correct.

One way to do this is to define a mapping Φ from timed trace structures to complete trace structures such that

$$T_1 \prec T_2 \Rightarrow \Phi(T_1) \prec \Phi(T_2).$$

and such that the timed trace specification of the live buffer is mapped by Φ to the complete trace specification. As an example of why the implication is in only one direction, let T_1 be a buffer with a maximum delay of one and let T_2 be a buffer with a maximum delay of two. Both T_1 and T_2 have a minimum delay of zero, and have one virtual wire φ. In this case $\Phi(T_1) = \Phi(T_2)$, so $\Phi(T_2) \preceq \Phi(T_1)$. However T_2 clearly does not conform to T_1.

The definition of such a mapping Φ, and proofs of its properties, are included in the full version of the paper [4].

5 Fair Mutual Exclusion Example

The verification example above for buffers only deals with a simple kind of liveness property. In this section to use the verifier to check a strong fairness property, which is a more difficult kind of liveness property.

The specification for a fair mutual exclusion element has inputs ai and bi and outputs ao and bo. It requires that any transition on ai be followed by a transition on ao with a maximum delay of one, as measured against the virtual wire φ_s. A similar requirement is placed on bi and bo. Again, this does not represent a bounded response time. The response time is unbounded since there is not assumed to be a bound on the amount of time between transitions of φ_s. However, this specification can not be satisfied by any implementation unless it is assumed that the environment will always eventually release the mutual exclusion element. This assumption is represented by introducing another virtual wire φ_r. After a request is granted by the mutual exclusion element, the environment must release it with a maximum delay of one, as measured against φ_r. Otherwise, it is a failure. This does not actually bound the time for which a channel can hold the mutual exclusion element. It simply means that the environment must constrain φ_r so that the mutual exclusion element is released before there are to many φ_r transitions. Let T be the trace structure representing this specification. Also, let T_1 represent a buffer with input ai and output ao and a maximum delay of one, as measured against φ_s. Finally, let T_2 represent an inverter with input bo and output bi and a maximum delay of one, as measured against φ_r. The automatic verifier can be used to show that

$$\mathbf{hide}(\{bi, bo, \varphi_r\})[T \circ T_2] \preceq T_1.$$

This is evidence that the formal specification captures the desired informal specification for a fair mutual exclusion element.

The verifier was used to verify the speed-independent fair mutual exclusion circuit described in [2]. This involves modeling each component of the circuit as having a maximum delay of one relative to a virtual wire φ_i, and then constraining φ_s relative φ_i and φ_r.

It is straightforward to define a complete trace structure that is a specification for fair mutual exclusion. The mapping Φ described above maps the timed trace theory specification for fair mutual exclusion to such a complete trace specification. This shows that the notion of fair mutual exclusion is accurately modeled by the timed trace theory specification.

In [1] Black shows that conventional finite traces cannot model delay-insensitive fair mutual exclusion. His arguments can be used to show that conventional finite traces cannot model speed-independent fair mutual exclusion if there is no bound on the number requests granted to one channel before a request is granted to the other channel, as in the specification above. Black proposed that infinite traces be used to model fair mutual exclusion. We have shown how trace theory can be modified to model fair mutual exclusion without introducing infinite traces.

6   Conclusions

We have shown how liveness properties, such as unbounded fairness, can be converted to safety properties. This makes possible the automatic verification of liveness properties using finite traces. No new algorithms are needed, just a reinterpretation of the meaning of timed traces. We showed this new interpretation makes sense by defining a mapping Φ from timed trace structure to complete trace structures that have infinite traces. This mapping preserves conformation, and maps specifications in timed trace theory to specifications in complete trace theory that are more obviously correct.

Unfortunately, tests indicate that the method presented here may not be efficient in practice. The same fair mutual exclusion circuit that was verified in this paper was verified more quickly using CTL in [2]. Verifying the circuit with timed traces required that a much larger state space be explored than was necessary using CTL. Thus, the results in this paper may be of more theoretical than practical interest. However, there may be areas of practical application. For example, in machine aided verification techniques like those in [9], a circuit is proved correct by proving it satisfies an invariant. Previously, only safety properties could be verified in this way. But the method described here might be adapted to allow verification techniques based on invariants to also verify unbounded liveness properties.

References

[1] David L. Black. On the existence of fair delay-insensitive arbiters: Trace theory and its limitations. *Distributed Computing*, 1(4):205–225, 1986.

[2] Jerry R. Burch. Combining CTL, trace theory and timing models. In Joseph Sifakis, editor, *Automatic Verification Methods for Finite State Systems, International Workshop, Grenoble, France*, volume 407 of *Lecture Notes in Computer Science*. Springer-Verlag, June 1989.

[3] Jerry R. Burch. Modeling timing assumptions with trace theory. In *IEEE International Conference on Computer Design*, October 1989.

[4] Jerry R. Burch. Verifying liveness properties by verifying safety properties. In Robert Kurshan and Edmund M. Clarke, editors, *Computer-Aided Verification, Proceedings of the 1990 Workshop*, volume 3 of *DIMACS Series in Discrete Mathematics and Theoretical Computer Science*. American Mathematical Society, 1991. To appear.

[5] David L. Dill. *Trace Theory for Automatic Hierarchical Verification of Speed-Independent Circuits*. PhD thesis, Carnegie Mellon University, Pittsburgh, PA 15213, 1988. Also appeared as [7].

[6] David L. Dill. Trace theory for automatic hierarchical verification of speed-independent circuits. In Jonathan Allen and F. Thomson Leighton, editor, *Advanced Research in VLSI: Proceedings of the Fifth MIT Conference*. MIT Press, 1988.

[7] David L. Dill. *Trace Theory for Automatic Hierarchical Verification of Speed-Independent Circuits*. ACM Distinguished Dissertations. MIT Press, 1989.

[8] Robin Milner. *A Calculus of Communicating Systems*, volume 92 of *Lecture Notes in Computer Science*. Springer-Verlag, 1980.

[9] Jorgen Staunstrup, Stephen J. Garland, and John V. Guttag. Localized verification of circuit descriptions. In Joseph Sifakis, editor, *Automatic Verification Methods for Finite State Systems, International Workshop, Grenoble, France*, volume 407 of *Lecture Notes in Computer Science*. Springer-Verlag, June 1989.

Memory Efficient Algorithms for the Verification of Temporal Properties

C. Courcoubetis*
Inst. of Comp. Sci. of Crete[†]

M. Vardi
IBM Almaden[‡]

P. Wolper[§]
Un. de Liège[¶]

M. Yannakakis
AT&T Bell Labs[‖]

Abstract

This paper addresses the problem of designing memory efficient algorithms for the verification of temporal properties of finite-state programs. Both the programs and their desired temporal properties are modeled as automata on infinite words (Büchi automata). Verification is then reduced to checking the emptiness of the automaton resulting from the product of the program and the property. This problem is usually solved by computing the strongly connected components of the graph representing the product automaton. Here, we present algorithms which solve the emptiness problem without explicitly constructing the strongly connected components of the product graph. By allowing the algorithms to err with a small probability, we can implement them with a randomly accessed memory of size $O(n)$ bits, where n is the number of states of the graph, instead of $O(n \log n)$ bits which the presently known algorithms require.

1 Introduction

Reachability analysis is one of the most successful strategies for analyzing and validating computer protocols. It was first proposed by West [Wes78], and further studied by many researchers (cf. [Liu89, Rud87]. Reachability analysis is applied to a protocol by systematically exercising all the protocol transitions. Such analysis can detect syntactical errors such as *static deadlock*, *unspecified reception*, or *unexercised code*. The simplicity of the strategy lends itself to easy implementation. Indeed, automated reachability analyses detected errors in published standards such as the X.21 (cf. [WZ78]). The approach is less successful when it comes to *protocol verification*, i.e., verifying that the given protocol achieves its functional specification. This limitation is due to the fact that a functional specification cannot be directly checked by reachability analysis. To apply reachability analysis to such a task, one first has to manually translate the functional specification to a property of the protocol state graph. While this can be done for some specific specifications (cf. [RW82]), it is not a general approach.

---

*The work of this author is partially supported by ESPRIT-BRA project SPEC (3096)

[†]Address: 36 Dedalou Street, P.O. Box 1385, 71110 Iraklio, Crete, Greece. Email: courcou@ariadne.uucp

[‡]Address: Department K55/802, 650 Harry Road San Jose, California 95120-6099, U.S.A. Email: vardi@ibm.com

[§]The work of this author is partially supported by ESPRIT-BRA project SPEC (3096)

[¶]Address: Institut Montefiore, B28, B-4000 Liège Sart-Tilman, Belgium. Email: pw@montefiore.ulg.ac.be

[‖]Address: 600 Mountain Avenue, Murray Hill, New Jersey 07974, U.S.A. Email: mihalis@research.att.com

A general approach to protocol verification is to use a *theorem-prover* for an appropriate logic. Early systems used to focus on input/output behavior of protocols rather than on ongoing behavior (cf. [Sun83]), but systems that are based on temporal logic overcame this shortcoming (cf. [Hai85]). Unfortunately, theorem-proving systems are semi-automated at best, and their success at dealing with real-life protocols is not as impressive as that of reachability analysis (cf. [GJL84]).

A new approach that emerged in the 1980's is the so-called *model-checking* approach [CES86, CG87, LP85, QS81]. Model checking is based on the idea that verifying a propositional temporal logic property of a finite-state program amounts to evaluating that formula on the program viewed as a temporal interpretation. The algorithms for doing this are quite efficient, since their time complexity is a linear function of the size of the program. As was shown later in the automata-theoretic approach of [Var89, VW86, Wol89], model checking can be viewed as an augmented reachability analysis; the model-checking algorithm uses the temporal logic specification to guide the search of the protocol state space in order to verify that the protocol satisfies its functional specification. Model checking thus seems to solve one of the limits of reachability analysis: the inability to automatically verify functional specifications.

Model checking suffers, however, from the same fundamental problem plaguing the reachability-analysis approach: the ability to explore only limited-size state spaces. This problem, called the *state-explosion* problem, is the most basic limitation of both approaches. It has been the subject of extensive research both in the context of reachability analysis (cf. [Liu89, Rud87]) and in the context of model checking (cf. [CG87]). A recent development [Hol88] has substantially pushed back the state-explosion limit for reachability analysis. The main idea behind this development is that, at the price of possibly missing part of the state space, the amount of randomly accessed memory necessary for exploring a state space of a given size could be substantially reduced (essentially from $O(n \log(n))$ to $O(n)$ for a graph with n states). The essence of the method is the use of hashing without collision detection.

In this paper, we show that model checking can also benefit from a similar reduction in the required random memory. This result is obtained by a combination of techniques. We approach model checking from the automata-theoretic perspective of [Var89, VW86, Wol89]. This has the advantage of essentially reducing model checking to reachability analysis, though on a state space that is the cross product of the original state space with the state space of an automaton describing the functional specification. It is then possible to adapt techniques inspired by those of [Hol88] to solve this problem. However, while Holtzmann's technique is suitable for searching for "bad" states in the state space, model checking involves searching for "bad" cycles. We thus had to develop some special purpose algorithms that are presented here.

This paper is organized as follows. We first review some background on model checking using the automata-theoretic approach and define the corresponding graph-theoretic problem. Then, we discuss the requirements that algorithms for solving this problem have to satisfy. Next we present our solutions. Finally, we present some extensions and some final remarks.

2 Temporal Logic Verification using Büchi Automata

The model-checking problem we consider is the following. Given a program P described as a product of finite-state transition systems P_i and a temporal logic formula f, check that all infinite computations of P satisfy f. To solve this problem, we use the following steps:

1. Build the finite-automaton on infinite words for the *negation* of the formula f (one uses the negation of the formula as this yields a more efficient algorithm). The resulting automaton is $A_{\neg f}$.

2. Take the product of the program $P = \prod P_i$ and the automaton $A_{\neg f}$.

3. Check if the product automaton is non-empty.

The approach we have just outlined has one major advantage over other model-checking approaches: it does not build the entire state graph for the program (the product of the P_i) before checking that it satisfies the temporal property f. Indeed, the product $\prod P_i \times A_{\neg f}$ can be computed in one pass. This can lead to more efficiency for various reasons. In the first place, the product of P and $A_{\neg f}$ only accepts sequences that do not satisfy the requirement. One expects few of these (none if the program is correct). It is thus possible that the product of P and $A_{\neg f}$ will have fewer reachable states than P. Furthermore, when building $P \times A_{\neg f}$, it is not necessary to store the whole state-graph. It is sufficient to keep just enough information for checking that condition 3 above is satisfied. This is exactly what the algorithms we present in Section 3 will do. The advantages of reducing model checking to a reachability problem are also investigated in [JJ89], but only for pure safety properties. In that case, it is sufficient to check that some states are simply reachable and the algorithms we develop in this paper are not needed.

To be able to describe our algorithms, we need more details about Büchi automata and how to check their emptiness. A *Büchi automaton* is a tuple $A = (\Sigma, S, \rho, S_0, F)$, where

- Σ is an alphabet,

- S is a set of states,

- $\rho : S \times \Sigma \rightarrow 2^S$ is a nondeterministic transition function,

- $S_0 \subseteq S$ is a set of starting states, and

- $F \subseteq S$ is a set of designated states.

A *run* of A over an infinite word $w = a_1 a_2 \ldots$, is an infinite sequence $s_0, s_1, \ldots$, where $s_0 \in S_0$ and $s_i \in \rho(s_{i-1}, a_i)$, for all $i \geq 1$. A run $s_0, s_1, \ldots$ is *accepting* if there is some designated state that repeats infinitely often, i.e., for some $s \in F$ there are infinitely many i's such that $s_i = s$. The infinite word w is *accepted* by A if there is an accepting run of A over w. The set of denumerable words accepted by A is denoted $L(A)$.

From the definition of Büchi automata, it is relatively easy to see that a Büchi automaton is nonempty iff it has some state $f \in F$ that is reachable from the initial state and reachable from itself (in one or more steps) [VW88]. In graph theoretic terms, this means that the graph representing the automaton has a reachable cycle that contains at least one state in F. In what follows, we will give a memory-efficient algorithm to solve this problem.

To formalize our verification approach, we define a program P as being a finite-state transition system consisting of

- a state space V,

- a nondeterministic transition function $\sigma : V \times \Sigma \to 2^V$ (Σ is the alphabet common to the program and the automaton for the property $A_{\neg f}$) and

- a set of starting states $V_0 \subseteq V$.

The accepting runs of P are defined by viewing P as a restricted type of Büchi automaton in which the set of designated states is the whole set of states V.

According to the definitions above, if $A_{\neg f} = (\Sigma, S, \rho, S_0, F)$, the product $P \times A_{\neg f}$ is a Büchi automaton with

- state set $V \times S$,

- transition function $\tau : V \times S \to 2^{V \times S}$ defined by $(v_2, s_2) \in \tau((v_1, s_1), a)$ iff $v_2 \in \sigma(v_1, a)$ and $s_2 \in \rho(s_1, a)$,

- and set of designated states $V \times F$.

This product automaton accepts all runs which are possible behaviors of P (accepted by the automaton P) and violate the formula f (are accepted by the automaton $A_{\neg f}$). Hence we have reduced the problem of proving that the program P satisfies the formula f to the problem of checking the emptiness of the Büchi automaton $P \times A_{\neg f}$.

It is interesting to note that the product automaton $P \times A_{\neg f}$ has the Büchi type of acceptance condition because the acceptance condition for P is the trivial one. In the case in which the program P is modeled as an arbitrary Büchi automaton, the problem of checking the emptiness of $P \times A_{\neg f}$ is different and will be examined in Section 4.

3 Verification Algorithms

3.1 Requirements on the Algorithms

We characterize the memory requirements of any verification algorithm as follows. We consider the data structures used by the algorithm. The total amount of space used by these data structures corresponds to the total space requirements of the algorithm. The above space can be divided into memory that is *randomly accessed* and into memory that is *sequentially accessed*. For example, for implementing a hash table we need randomly accessed memory, while a stack can be implemented with sequentially accessed memory.

As correctly pointed out in [Hol88], the bottleneck in the performance of most verification algorithms is directly related to the amount of the randomly accessed memory these algorithms require, and is due to the significant amount of paging involved during the execution of the algorithm. Holzmann observed that there is a tremendous speed-up for an algorithm implemented so that its randomly accessed memory requirements do not exceed the main memory available in the system (since sequentially accessed memory can be implemented in secondary storage).

The basic problem that Holzmann considered is how to perform reachability analysis by using the least amount of randomly accessed memory. For a graph with n states, his scheme involves a depth-first search in the graph, where the information about the states visited is stored in a bit-array of size m as follows. When a new state is generated, its name is hashed into an address in the array; if the bit of the corresponding location is on, then the algorithm considers that the above state has already been visited; if the bit is off then it sets the bit and adds the state on the stack used by the depth-first search. Since there is no collision detection it follows that the above search is partial; there is always a possibility that a state will be missed.

The key assumption behind this method, see [Hol88], is that in general one can choose the value of m large enough and construct a hash function so that the number of collisions becomes arbitrarily small. Furthermore, since the limiting factor in reachability analysis is usually the space required by the computation rather than the time required to do the computation, one could significantly reduce the probability of error by running the algorithm a few times with different hash functions. Indeed, Holzmann claims that, for most practical applications, choosing a hash table of size $m = O(n)$ together with appropriate hash functions is sufficient for the effect of collisions to become insignificant. Is this really so?

To answer this question, let us consider the memory requirements of the general reachability problem defined as follows. We assume that the states of the graph G have names from a name space U. In many applications (for example protocols), $|U|$ is many orders of magnitude larger than the number n of reachable states of G. In this case, complete reachability analysis (no missed states whatever the input graph) appears to require $O(n \log |U|)$ bits of randomly accessed memory, and probably can not be done with less memory (unless the names of the reachable states of G are not randomly selected from U). Indeed, representing each state with less than $\log |U|$ bits amounts to mapping the state space U to a smaller state space. Now, for any such mapping there will always be subsets of U on which it is not one-to-one and hence on which complete reachability will not be guaranteed.

The situation is different is one analyses the problem from a probabilistic point of view. Consider all possible mappings from the set $S = \{1, \ldots, n\}$ into the set $\{1, \ldots, m\}$. There are m^n such mappings of which $m!/(m-n)!$ are one-to-one. Thus, if one assumes that the mapping implemented by a hash function is randomly selected, the probability that it is one-to-one (no collisions) is $m!/((m-n)!m^n)$ which for $n << m$ can be approximated by $e^{-n^2/m}$. This implies that in the case of a name space U and a graph with n reachable states, we can do partial reachability (with arbitrarily small probability of missing reachable states) by using $O(n \log n)$ bits of randomly accessed memory (instead of $O(n \log |U|)$ bits for complete reachability) as follows. First hash the n reachable states into a set $1, \ldots, m$ with an arbitrarily small probability of collision. As we have just seen, this is possible if we take $m = O(n^2)$. Then, do complete reachability using the set $1, \ldots, m$ as the name space for the states.

Holtzmann's technique goes one step further and only uses one bit of randomly accessed memory per reachable state. This is equivalent to assuming that there exists a hash function mapping U into $1, \ldots, m$, $m = O(n)$, with a small probability of collisions. As the analysis above shows, this is not possible if we just assume that the hash function is random. It can however be possible if the state space U is only of size $O(n)$ or if the set of reachable states has a particular structure that can be used by the hash function. In these cases, the gain in randomly accessed memory, size $O(n)$ instead of size $O(n \log |U|)$ is quite significant for large state spaces.

However, this gain in memory use is only obtained for straightforward reachability analysis. To verify general temporal properties we have to check nonemptiness of the product automaton. One way to accomplish this is to construct the strongly connected component of the product automaton state graph and then to check whether one of the strongly connected component contains an accepting state. Unfortunately, we cannot apply Holtzmann's method to the standard algorithm for constructing the strongly connected components of the graph [AHU74]. Indeed, although in that algorithm the states of the components are stored in a stack, it requires access to information (depth-first and low-link number) about states randomly placed in the stack, which implies the need of at least $O(n \log n)$ bits of randomly accessed memory. Hence, given a fixed amount of memory, the size of the problems we could efficiently analyze with the above algorithm is substantially smaller than the size of the problem that can be analyzed with the technique of [Hol88].

From the previous discussion the following problem emerges. Assuming that reachability analysis in graphs of size n can be efficiently done with randomly accessed memory of size $O(n)$, can we solve the emptiness problem for Büchi automata using only randomly accessed memory of size $O(n)$? The answer to this problem is positive and the corresponding algorithms are described in the following section.

3.2 The Algorithms

In this section we provide algorithms for the following problem.

Problem 1 (nonemptiness of Büchi automata) *Given directed graph G, start node s_0, distinguished set of accepting nodes F, determine whether there is a member of F which is reachable from s_0 and belongs to a cycle, or equivalently, to a nontrivial strong component.*

We make the following representation assumptions. The graph G is given by a *successor* function: a function that takes a node as argument and returns an ordered list of its successors. The set F is specified by a membership routine. We assume that we have a function h mapping one-to-one every node to an integer in the range $1, \ldots, m$.

Algorithm A:

The algorithm consists of two depth-first-searches (DFS's). The two searches can be performed one after the other, or can be done together in an interleaved fashion. It is simpler to describe first the noninterleaved execution. The purpose of the first DFS is to (1) determine the members of F that are reachable from s_0, and (2) order them according to last visit (i.e., in postorder) as $f_1, \ldots, f_k$.[1] The second DFS explores the graph using this ordering; it does not perform k searches but only one. In more detail, the main data structures are as follows: a stack S (to hold the path of DFS from root to current node), a (FIFO) queue Q to hold the reachable members of F in postorder and a bit-array M indexed by the hash values $1, \ldots, m$ for the "marked" bit (whether the node has been visited). The two passes share the same structures S and M.

The first DFS is as follows:

1. Initialize: $S := [s_0]$, $M := 0$, $Q := \emptyset$.
2. Loop: while $S \neq \emptyset$ do
       begin
           $v := \text{top}(S)$;
           if $M[h(w)] = 1$ for all $w \in \text{succ}(v)$
             then begin
               pop v from S;
               if $v \in F$ insert v into Q

---

[1] f_1 is the first postorder reachable accepting state and f_k is the last

```
                        end
                else begin
                        let w be the first member of succ(v) with M[h(w)] = 0;
                        M[h(w)] := 1;
                        push w into S
                end
        end
end
```

The second DFS is as follows:

1. Initialize: $S := \emptyset$, $M := 0$.
2. Loop: while $Q \neq \emptyset$ do
```
        begin
                f := head(Q);
                remove f from Q;
                push f into S;
                while S ≠ ∅ do
                        begin
                                v := top(S);
                                if f ∈ succ(v) then halt and return "YES";
                                if M[h(w)] = 1 for all w ∈ succ(v)
                                        then pop v from S
                                        else begin
                                                let w be the first member of succ(v) with M[h(w)] = 0;
                                                M[h(w)] := 1;
                                                push w into S
                                        end
                        end
        end
```

The correctness of the algorithm is based on the following claims.

Lemma 1 *Let $f_1, \ldots, f_k$ be the members of Q after the first DFS, i.e., the members of F that are reachable from s_0 in postorder (f_1 is the first member of F to be reached in postorder, f_k the last). If for some pair f_i, f_j with $i < j$ there is a path from f_i to f_j, then node f_i belongs to a nontrivial strong component.*

Proof: Suppose that there is a path from f_i to f_j. If no node on this path was marked before f_i, then the DFS would have reached f_j from f_i, so f_j would have come before f_i in the postorder. Thus, some node p on the path was marked before f_i. If p comes before f_i in the postorder, then f_j also should come before f_i in the postorder. Since p was marked before f_i, but comes after f_i in the postorder, it must be an ancestor of f_i. Thus, f_i can reach an ancestor and therefore belongs to a nontrivial strong component. □

Theorem 1 *If the second DFS halts and returns "YES", then some reachable node of F belongs to a nontrivial strongly connected component. Conversely, suppose that some reachable node of F belongs to a nontrivial strongly connected component. Then the second DFS will return "YES".*

Proof: The first part is clear: suppose the second DFS returns "YES" while processing node f_j of Q. Then, it is building a tree with root f_j and discovers a back edge to the root f_j, and therefore f_j is obviously in a cycle. For the converse, let f_j be a reachable member of F that belongs to a nontrivial strongly connected component and has the smallest index j among all such members. Consider a path p from f_j to itself. We claim that no node of p is reachable from a f_i with a smaller i. For, if some node was reachable, then f_i would also reach f_j, which by Lemma 1 contradicts the choice of f_j. Therefore, no node of the path p is marked when we push f_j into S in the second DFS, and thus we will find a back edge to the root f_j. $\square$

Note that the creation of both S and Q and access to them in both searches are sequential. Hence, both can be stored in secondary memory as needed.

So far we analyzed the algorithm under the assumption that the hash function f is perfect. One of the main features of our algorithm is its behavior in the presence of hash collisions. In that case, although the algorithm might erroneously conclude (due to collisions) that the Büchi automaton does not accept any word, it will never mistakenly conclude that the automaton accepts some word. In terms of the underlying verification problem, this means that our algorithm might miss some errors, but will never falsely claim that the protocol is incorrect. Thus, the algorithm should be viewed more as a systematic debugging tools rather than as a verification tool.

An alternative is to do away with the queue Q and instead immediately start the second depth-first search each time a final state is encountered. Once the second search from a state is finished, the first search is resumed. To do this, one needs a second stack S_2 and a second bit array M_2 and hence one uses twice as much space as that required by the first algorithm. The advantage is that if the automaton is found to be nonempty, an accepted word can be extracted from the stacks S_1 and S_2. In verification terms, this means that, if the protocol is found to be incorrect by the algorithm, a sample incorrect path can be produced. This is essential for debugging to be possible.

4   Extensions and Concluding Remarks

An extension of the verification problem described in Section 2 is the verification of programs with liveness conditions, see [ACW90]. In this case the program is given in terms of components, each having it own liveness conditions. Each such component is modeled as a Büchi automaton. Hence, the product $P \times A_{\neg f}$ corresponds to an automaton whose transition table G is the product of the corresponding transition tables and its acceptance condition is given in terms of a set of sets of designated states $\{F_1, \ldots, F_k\}$. A run is accepting if it repeats some state from each of these sets infinitely often. Clearly, checking the emptiness of $P \times A_{\neg f}$ is equivalent with checking for the existence of a strongly connected component in the product transition table which is reachable from the initial state and intersects all these sets. Let S be the state space of the product transition table. We can construct a Büchi automaton B with $k|S|$ states, such that the emptiness of B is equivalent with the emptiness of $P \times A_{\neg f}$ (see for instance [VW86]).

- The graph of B consists of k copies of G with the transitions modified as follows. Consider the k copies $G_1, \ldots, G_k$ of G. For $i = 1, \ldots, k$, replace the transitions from every state $f \in F_i$ of G_i by similar transitions to the states in $G_{(i \bmod k)+1}$.

- The initial states of B are those of one copy of G, say G_1.

- The accepting states of B are the states F_i of the copy G_i of G, for some arbitrary i. For instance we can take $F_1 \subset G_1$.

Hence, if we apply the algorithms of the previous section to B, we can do verification with $O(k|S|)$ bits of randomly accessed memory.

Another remark is the following. In many applications it is reasonable to assume that the predecessor function of the graph is given as well. In this case one can use the algorithm in Section 6.7 in [AHU82] for constructing the strongly connected components of the graph G by using randomly accessed memory of size $O(n)$. Let G_r be the directed graph corresponding to G by reversing its edges. This algorithm performs first a DFS on G and numbers the states in order of completion of the recursive calls (in postorder). This can be implemented by pushing the states in a stack according to their postorder visit by the DFS; this stack can use sequentially accessed memory. Then the algorithm performs a DFS on G_r (by using the predecessor function of G) starting with the state with the highest postorder sequence number (top of stack). This DFS on G_r must be restricted to the states reached during the first DFS, and uses a hashing mechanism for marking the states already visited. If the search does not reach all states, the algorithm starts the next DFS on G_r from the highest-numbered state which has not been already visited by the previous DFS. This can be easily done by poping the postorder stack until a state which has not been visited (the corresponding bit in the hash table is zero) is found. Since each tree in the resulting spanning forest is a strongly connected component, one can easily check for the properties of each such component while it is being generated.

References

[ACW90] S. Aggarwal, C. Courcoubetis, and P. Wolper. Adding liveness properties to coupled finite-state machines. *ACM Transactions on Programming Languages and Systems*, 12(2):303–339, 1990.

[AHU74] Alfred V. Aho, John E. Hopcroft, and Jeffrey D. Ullman. *The Design and Analysis of Computer Algorithms*. Addison Wesley, Reading, 1974.

[AHU82] Alfred V. Aho, John E. Hopcroft, and Jeffrey D. Ullman. *Data Structures and Algorithms*. Addison Wesley, Reading, 1982.

[CES86] E.M. Clarke, E.A. Emerson, and A.P. Sistla. Automatic verification of finite-state concurrent systems using temporal logic specifications. *ACM Transactions on Programming Languages and Systems*, 8(2):244–263, January 1986.

[CG87] E. M. Clarke and O. Grümberg. Avoiding the state explosion problem in temporal logic model-checking algorithms. In *Proc. 6th ACM Symposium on Principles of Distributed Computing*, pages 294–303, Vancouver, British Columbia, August 1987.

[GJL84] R. Grotz, C. Jard, and C. Lassudrie. Attacking a complex distributed systems from different sides: an experience with complementary validation tools. In *Proc. 4th Work. Protocol Specification, Testing, and Verification*, pages 3–17. North-Holland, 1984.

[Hai85] B.T. Hailpern. Tools for verifying network protocols. In K. Apt, editor, *Logic and Models of Concurrent Systems, NATO ISI Series*, pages 57–76. Springer-Verlag, 1985.

[Hol88]   G. Holzmann. An improved protocol reachability analysis technique. *Software Practice and Experience*, pages 137–161, February 1988.

[JJ89]   C. Jard and T. Jeron. On-line model-checking for finite linear temporal logic specifications. In *Automatic Verification Methods for Finite State Systems, Proc. Int. Workshop, Grenoble*, volume 407, pages 189–196, Grenoble, June 1989. Lecture Notes in Computer Science, Springer-Verlag.

[Liu89]   M.T. Liu. Protocol engineering. *Advances in Computing*, 29:79–195, 1989.

[LP85]   O. Lichtenstein and A. Pnueli. Checking that finite state concurrent programs satisfy their linear specification. In *Proceedings of the Twelfth ACM Symposium on Principles of Programming Languages*, pages 97–107, New Orleans, January 1985.

[QS81]   J.P. Quielle and J. Sifakis. Specification and verification of concurrent systems in cesar. In *Proc. 5th Int'l Symp. on Programming*, volume 137, pages 337–351. Springer-Verlag, Lecture Notes in Computer Science, 1981.

[Rud87]   H. Rudin. Network protocols and tools to help produce them. *Annual Review of Computer Science*, 2:291–316, 1987.

[RW82]   H. Rudin and C.H. West. A validation technique for tightly-coupled protocols. *IEEE Transactions on Computers*, C-312:630–636, 1982.

[Sun83]   C.A. Sunshine. Experience with automated protocol verification. In *Proceedings of the International Conference on Communication*, pages 1306–1310, 1983.

[Var89]   M. Vardi. Unified verification theory. In B. Banieqbal, H. Barringer, and A. Pnueli, editors, *Proc. Temporal Logic in Specification*, volume 398, pages 202–212. Lecture Notes in Computer Science, Springer-Verlag, 1989.

[VW86]   M.Y. Vardi and P. Wolper. An automata-theoretic approach to automatic program verification. In *Proc. Symp. on Logic in Computer Science*, pages 322–331, Cambridge, june 1986.

[VW88]   M.Y. Vardi and P. Wolper. Reasoning about infinite computation paths. IBM Research Report RJ6209, 1988.

[Wes78]   C.H. West. Generalized technique for communication protocol validation. *IBM J. of Res. and Devel.*, 22:393–404, 1978.

[Wol89]   P. Wolper. On the relation of programs and computations to models of temporal logic. In B. Banieqbal, H. Barringer, and A. Pnueli, editors, *Proc. Temporal Logic in Specification*, volume 398, pages 75–123. Lecture Notes in Computer Science, Springer-Verlag, 1989.

[WZ78]   C.H. West and P. Zafiropulo. Automated validation of a communication protocol: the ccitt x.21 recommendation. *IBM Journal of Research and Development*, 22:60–71, 1978.

A Unified Approach to the Deadlock Detection Problem in Networks of Communicating Finite State Machines

WUXU PENG

Dept of Computer Science

Southwest Texas State University

San Marcos, TX 78666

S. PURUSHOTHAMAN

Department of Computer Science

The Pennsylvania State University

University Park, PA 16802

Abstract

We consider the deadlock detection problem (DDP) of networks of communicating finite state machines (NCFSMs). The DDP problem is known to be undecidable for NCFSMs. In this paper, we provide a characterization of those subclasses of networks for which the deadlock problem is decidable. We also provide a proof technique based on our characterization and illustrate our technique on an example.

1 Introduction

Communicating finite state machines are a very useful abstract model for specifying, verifying and synthesizing communication protocols [2, 3]. In this model a system of communicating finite state machines can communicate typed messages asynchronously with each other over uni-directional, unbounded FIFO channels.

A central issue in this model is whether a network of communicating finite state machines (NCFSMs) is free of progress errors. Several widely addressed progress properties are: freedom from deadlocks, freedom from unspecified receptions, and freedom from unbounded communication. The problem of checking for non-progress in NCFSMs is known to be undecidable [1, 2, 3]. Because of this negative conclusion a natural question is: "For what classes of NCFSMs is the progress problem decidable?" A large amount of literatures (e.g. [8, 6, 7]) has been devoted to identifying classes of NCFSMs for which some of the progress problems are decidable. Specific classes of NCFSMs are usually obtained by placing restrictions on the structure of the systems. For instance, the number of machines in the system, the number of message types allowed, and channel capacity (maximum number of pending messages allowed in channels) etc. However the underlying question

For what classes of NCFSMs are the progress problems decidable?

has not been answered before.

In this paper, we investigate the deadlock detection problem from the language-theoretic point of view. Techniques developed here are applicable to other progress problems. Specifically, we give a necessary and sufficient condition on execution sequences of a class of network such that the deadlock detection problem is decidable for that class of NCFSMs. We believe that our work provides insight into the underlying nature of decidability of deadlock detection problem. We provide an example to illustrate our technique.

This paper is organized as follows: In Section 2 we introduce necessary notations and definitions, we present our main result in Section 3, in Section 4 we give an example to illustrate the main result and conclude in Section 5.

2 Preliminaries

A communicating finite state machine (CFSM) is a labeled directed graph with a distinguished initial state, where each edge is labeled by an event. The events of a CFSM are **send** and **receive** commands over a finite set of message types M. The communication between CFSMs is assumed to be asynchronous (i.e., non-blocking sends and blocking receives). Consequently, we assume the availability of an infinite buffer between each pair of machines to store pending messages.

Let $I = \{1, \cdots, n\}$, where $n \geq 2$ is some constant. Formally, we have

Definition 2.1 (*CFSM*) *A CFSM P_i is a four-tuple $(S_i, \langle M_{i,j}\rangle_{j\in I} \cup \langle M_{j,i}\rangle_{j\in I}, \delta_i, p_{0i})$, where*

(1) *S_i is the set of local states,*

(2) *$M_{i,j}$ is the set of message types that P_i can send to machine P_j, and $M_{j,i}$ is the set of message types that P_i can receive from machine P_j. It is assumed that $M_{i,i} = \emptyset$, since P_i can not directly send messages to or receive messages from itself.*

(3) *Let $-M_{i,j} = \{-m | m \in M_{i,j}\}$ and $+M_{j,i} = \{+m | m \in M_{j,i}\}$.*
δ_i is a partial mapping, $\delta_i: S_i \times (\langle -M_{i,j}\rangle_{j\in I} \cup \langle +M_{j,i}\rangle_{j\in I}) \times I \longrightarrow 2^{S_i}$. $\delta_i(p, -m, j)$ is the set of new states that machine P_i can possibly enter after sending message of type m to machine P_j, and $\delta_i(p, +m, j)$ is the set of new states that machine P_i can possibly enter after receiving message of type m from machine P_j.

(4) *p_{0i} is the initial local state.*

A state p in P_i is said to be a send (receive, resp.) state iff all of its outgoing edges are send (receive, resp.) edges. p is said to be a mixed state iff it has both outgoing send and receive edges. Let $RMsg(p)$ be the set of message types that can be received in state p, i.e. $RMsg(p) = \{m | \exists p' \; \exists j \; p' \in \delta_i(p, +m, j)\}$. Define

$$M_i = \cup_{j\in I}(M_{i,j} \cup M_{j,i}),$$
$$M = \cup_{i\in I} M_i, \text{ and}$$
$$-M_i = \cup_{j\in I} - M_{i,j}, \qquad +M_i = \cup_{j\in I} + M_{j,i},$$
$$\pm M_i = -M_i \cup +M_i,$$
$$-M = \cup_{i\in I} - M_i, \qquad +M = \cup_{i\in I} + M_i,$$
$$\pm M = +M \cup -M.$$

Without loss of generality, we assume that $M_{i,j} \cap M_{k,l} = \emptyset$ if $(i,j) \neq (k,l)$. Due to this assumption, for any $a \in \pm M_i$ we can simplify the notation $\delta_i(p, a, j)$ to $\delta_i(p, a)$.

Definition 2.2 (*Network of communicating finite state machines*) *A network of communicating finite state machines (NCFSM) is a tuple* $N = \langle P_1, \cdots, P_n \rangle$, *where each* P_i ($i \in I$) *is a CFSM.*

A global state of N *is a tuple* $[\langle p_i \rangle_{i \in I}, \langle c_{i,j} \rangle_{i,j \in I}]$, *where* p_i *is a local state of machine* P_i, $c_{i,j}$ *is the sequence of messages in the channel from machine* P_i *to* P_j. *Let* V *be the cartesian-product of the sets* $S_1, \cdots, S_n$, *i.e.* $V = S_1 \times \cdots \times S_n$, *and let* C *be the cartesian-product of the sets* $M_{1,2}^*, \cdots, M_{1,n}^*, M_{2,1}^*, \cdots, M_{n,n-1}^*$.

Initially, N *is in its initial state* $[\langle p_{0i} \rangle_{i \in I}, \langle c_{i,j} \rangle_{i,j \in I}]$, *where* $c_{i,j} = \varepsilon$ ($i \neq j$). *Let* $[\langle p_i \rangle_{i \in I}, \langle c_{i,j} \rangle_{i,j \in I}]$ *be a global state. The global state transition function* $\delta_N : (V \times C) \times \pm M \longrightarrow 2^{V \times C}$ *is a partial function defined as:*

(1) if $\exists i, j \in I$ ($i \neq j$) *such that* $p_i' \in \delta_i(p_i, -m, j)$ *then*
$$[\langle p_i' \rangle_{i \in I}', \langle c_{i,j}' \rangle_{i \in I}'] \in \delta_N([\langle p_i \rangle_{i \in I}, \langle c_{i,j} \rangle_{i,j \in I}], -m), \text{ where}$$
$p_k = p_k'$ ($k \neq i$), $c_{k,l} = c_{k,l}'$ ($k \neq i$ *or* $l \neq j$), *and* $c_{i,j}' = c_{i,j}.m$.

(2) if $\exists i, j \in I$ ($i \neq j$) *such that* $p_i' \in \delta_i(p_i, +m, j)$ *then*
$$[\langle p_i' \rangle_{i \in I}', \langle c_{i,j}' \rangle_{i \in I}'] \in \delta_N([\langle p_i \rangle_{i \in I}, \langle c_{i,j} \rangle_{i,j \in I}], +m), \text{ where}$$
$c_{k,l} = c_{k,l}'$ ($k \neq j$ *or* $l \neq i$), *and* $m.c_{j,i}' = c_{j,i}$.

We use $P_i \rightarrow P_j$ to denote the channel from P_i to P_j. In essence, the first case in Definition 2.2 denotes the event that P_i sends a message m to P_j, which causes the message m to be appended to the end of channel $P_i \rightarrow P_j$. The second case represents the event that P_i receives a message of type m sent by P_j, which has the effect of removing the first message (which must be of type m, or error (unspecified reception) would occur) in the channel $P_j \rightarrow P_i$. In both cases, after the successful completion of the event, P_i enters local state p_i' while all other machines remains in the same local states and the contents of all other channels are unchanged.

To simplify the expressions, we will use the notation $[v, c]$ to denote a global state whenever necessary, where by convention $v = \langle p_i \rangle_{i \in I} \in V$, and $c = \langle c_{i,j} \rangle_{i,j \in I} \in C$. $[v_0, c_0]$ will be used to denote the initial state.

Definition 2.3 (*Reachability function*) *Let* $N = \langle P_1, \cdots, P_N \rangle$ *be an NCFSM. The global state transition function* δ_N *can be easily extended to the following reachability function* $\delta_N^* :$ $(V \times C) \times \pm M^* \longrightarrow 2^{V \times C}$,

(1) $\delta_N^*([v, c], \varepsilon) = \{[v, c]\}$.

(2) $\delta_N^*([v, c], e.a) = \{[v', c'] \mid \exists [v'', c''] \in \delta_N^*([v, c], e), [v', c'] \in \delta_N([v'', c''], a)\}$.

We often write $\delta_N^*([v_0, c_0], e)$ *as* $\delta_N^*(e)$. *Furthermore, we will drop the subscripts in* δ_N *and* δ_N^* *if no confusion arises.*

Given a reachability function δ^*, we define the set of all states reached as a result of some execution to be the *reachability set*. More formally,

Definition 2.4 (*Reachability sets*) *Let* $N = \langle P_1, \cdots, P_n \rangle$ *be an NCFSM. The reachability set* $RS(N)$ *is the set of all reachable global states,* $RS(N) = \{[v, c] \mid [v, c] \in \delta^*(e), e \in \pm M^*\}$.

In the rest of this paper sequences of events will form the back bone of our discussions. Hence, we will use the following abbreviations:

Event Sequence A string $e \in \pm M^*$ is an *event sequence*.

Executable An event sequence e is *executable*, notated as $\delta^*(e) \neq \emptyset$, if $\delta^*(e)$ is defined.

Feasible An event sequence $e \in \pm \Sigma^*$ is *feasible*, if [1]

1. $\forall e' \in \text{pref}(e)\ \forall g \in \Sigma\ |e'|_{+g} \leq |e'|_{-g}$; and

2. $\forall i, j \in I$, if $+g_{i,j}$ is the kth receive event in $f_{i,j}(e)$, then $-g_{i,j}$ is the kth send event in $f_{i,j}(e)$.

Feasibility is a very strict requirement. We can easily show that feasible sequences are context-sensitive. We will use $F(N)$ to denote the set of all feasible event sequences of a network N.

Stable An event sequence e is *stable*, if it is feasible and in addition it contains the same number of send and receive events of any message type, i.e., $\forall a \in \Sigma\ |e_{+a}| = |e_{-a}|$. We will use $SE(N)$ to denote the set of all stable event sequences of a network N.

It should be clear to the reader that a definition of deadlock (or any progress error) can be stated in terms of execution sequences. Of course, execution sequences capture the semantics (and causality) of processes and buffers present in a network. The causality constraints among possible actions in a network can be split into those imposed by the behavior of FIFO buffers and those imposed by the sequencing constraints of the processes. The notion of feasible and stable (event) sequences defined above capture the constraints of the FIFO buffers in the network. The sequencing constraints of the network can be easily captured as a shuffle of the sequencing constraints of the individual processes. Formally, we define:

Definition 2.5 (*Shuffle-product of NCFSMs*) *Let* $N = \langle P_1, \cdots, P_n \rangle$ *be an NCFSM. The shuffle-product of* N, *notated as* $SP(N)$, *is a four-tuple* (V, M, Δ, v_0), *where*

(1) $V = S_1 \times S_2 \times \cdots \times S_n$.

(2) $v_0 = [p_{01}, p_{02}, \cdots, p_{0n}] \in V$.

(3) The transition function $\Delta: V \times \pm M \longrightarrow 2^V$ *is defined as*

---

[1]For a string w, $|w|$ is the length of w, $|w|_a$ is the number of occurrence of letter a in w and $pre(w)$ is the set of all prefixes of w.

$v' \in \Delta(v, a)$, where $a \in \pm M_i \subseteq \pm M$, iff $v'_j = v_j$ ($j \in I$ & $j \neq i$) and $v'_i \in \delta_i(v_i, a)$.

The shuffle-product $SP(N)$ can be viewed as a (nondeterministic, in general) finite automaton by identifying some subset of V as final state set. We use $SP(N)(F)$ to notate the finite automaton obtained from the shuffle-product with $F \subseteq V$ as the final state set.

A tuple $v \in V$ is a *receive* node if for each $i \in I$ v_i is a receive state in P_i. $REV(N)$ denotes the set of all receive nodes in V.

A number of progress properties have received wide attention and one of the well known progress properties is the deadlock detection problem (DDP).

Definition 2.6 (*Deadlock*) *Let* $N = \langle P_1, \cdots, P_n \rangle$ *be an NCFSM and* $[v, c] \in RS(N)$ *be a global state.*

$[v, c]$ *is a* deadlock state *if the predicate*

$$d([v, c]): \quad v \in REV(N) \ \& \ c = c_0$$

holds.

The network N *is* free of deadlocks, *if the predicate*

$$\forall [v, c] \in RS(N) \ (\text{not } d([v, c]))$$

holds.

It is well known that in general it is undecidable whether an NCFSM is free of deadlocks ([2, 3]). We state this fact in following theorem.

Theorem 2.1 *DDP is undecidable.*

3 A Unified Approach to DDP

Theorem 2.1 states that the problem of detecting deadlocks in a network is undecidable in general. To cope with this negative result, many special classes of NCFSMs have been identified for which the DDP is decidable. This usually involves finding sufficient conditions under which the problem becomes decidable.

In this section we take a different approach to this problem. Instead of trying to find special classes of NCFSMs with decidable DDP, we give a necessary and sufficient condition under which the DDP of a given class of NCFSM is decidable.

First let us formalize the concept of *classes* of NCFSMs.

A *class* $\mathcal{N}$ of NCFSMs is a tuple (Q, Σ, T), where

1. Σ is a finite (or countably infinite) set of message types,

2. Q is a collection of NCFSMs each of which draws message types from Σ,

3. T is a predicate which characterizes the properties of NCFSMs in the Q. For instance, T may be the predicate: "each $N \in Q$ has only two CFSMs", which is the class of NCFSMs with two CFSMs.

Let $\mathcal{N} = (Q, \Sigma, T)$ be a class of NCFSMs, and let $N = \langle P_1, \cdots, P_n \rangle \in \mathcal{N}$. We say that the DDP is *decidable* for a network class $\mathcal{N}$, if the predicate

$$D(\mathcal{N}) : \forall N \in \mathcal{N} \; \exists [v, c] \in RS(N) \; (v \in RV(N) \& c = c_0).$$

is decidable. In the following we would like to relate the decidability of deadlock to conditions on execution sequences. It is easy to see that if an execution sequence e leads to a deadlock state, then it should have the property that for every send event in e there should be a corresponding receive event. In fact, e should be a stable event sequence.

In general the set $SE(N)$ is context-sensitive, since we can easily construct a linear bounded automaton (LBA) that accepts $SE(N)$. Even for some trivial classes of NCFSMs $SE(N)$ remains context-sensitive. However in order to check if DDP is decidable for N, it is not necessary to know every member in $SE(N)$. This observation is based on the fact that there a number of event sequences that are really interleavings of the same set of actions of the processes, and hence lead to the same global state. We should therefore consider these interleavings as being equivalent. Formally we have, two event sequences e_1 and e_2 are *equivalent*, notated as $e_1 \simeq e_2$, iff

1. e_1 and e_2 are permutations of each other,

2. $\delta^*(e_1) = \delta^*(e_2)$.

Since we are really interested in stable event sequences, we will say two event sequences e_1 and e_2 are *stable equivalent*, notated as $e_1 \simeq_{st} e_2$, iff $e_1 \simeq e_2$ and both e_1 and e_2 are stable.

Define $class(e) = \{e' \mid e' \simeq_{st} e\}$. It is easy to see that $\simeq_{st}$ is an equivalence relation on $SE(N)$ and $class(e)$ is an equivalence class.

As mentioned earlier, we only need a *representative* from each equivalence class of $SE(N)$ to check for existence of deadlock in N. To that end, we define a language $C_N \subseteq SE(N)$ to be a *stable cover set* for N if

$$\forall e \in SE(N)(equiv(e) \cap C \neq \emptyset). \tag{1}$$

With the concepts of stable cover sets and shuffle-product automata, we have the following theorem regarding the decidability of DDP for classes of NCFSMs.

Theorem 3.1 *Let $\mathcal{N} = (Q, \Sigma, T)$ be a class of NCFSMs. Let $\mathcal{R} = \{L(M_v) \mid M_v = SPA(N, \{v\}) : v \in V_N\}$, i.e. $\mathcal{R}$ is the collection of shuffle-product automata, each of which has some node $v \in V_N$ as the single final state. The DDP is decidable for $\mathcal{N}$ if and only if for every network $N \in \mathcal{N}$, there exists a stable cover set $C_N \subseteq SE(N)$ such that for every $M_v \in \mathcal{R}$, the predicate $C_N \cap L(M_v) = \emptyset$ is decidable.*

Proof: Let $\mathcal{N} = (Q, \Sigma, T)$ be a class of NCFSMs.

If. Let N be a member in $\mathcal{N}$, and assume that we have a stable cover set $C_N \subseteq SE(N)$ which satisfies the condition listed in the theorem.

Claim: $[v, c_0]$ is a deadlock state if and only if $C_N \cap L(M_v) \neq \emptyset$.

Proof of the claim:

Assume that $C_N \cap L(M_v) \neq \emptyset$. Since C_N is a stable cover, each $e \in C_N$ is stable. Moreover C_N contains at least one element from each equivalence class $equiv(e)$. As $L(M_v)$ contains all executable event sequences which can lead to a reachable global state of the form $[v, c]$, there must be at least one executable event sequence $e \in C_N \cap L(M_v)$. such that $[v, c_0] \in \delta(e)$.

Conversely assume that $[v, c_0]$ is a reachable deadlock state. Therefore there exists an executable sequence e where $[v, c_0] \in \delta(e)$. As C_N is a stable cover, there must exist $e' \simeq_{st} e$ and $e' \in C$. Obviously $e' \in L(M_v)$, hence $C_N \cap L(M_v) \neq \emptyset$.

By definition DDP is decidable for N. Since DDP is decidable for any $N \in \mathcal{N}$ we conclude that DDP is decidable for the class $\mathcal{N}$.

Only If. Assume that DDP is decidable for $\mathcal{N}$. We must show that for every network $N \in \mathcal{N}$ there exists a stable cover set C_N for N such that the emptiness problem of $C_N \cap L(M_v)$ is decidable for all $v \in V_N$. In the following argument, the symbol $\Longrightarrow$ is used to express logical implication. "$A \Longrightarrow B$" means that the decidability of A implies the decidability of B. We have following logical reasoning:

$D(\mathcal{N})$
$\Longrightarrow \forall\, N \in \mathcal{N}\ (\exists\, [v, c] \in RS(N)\ (v \in RV(N)\ \&\ c = c_0))$
$\Longrightarrow \forall\, N \in \mathcal{N}\ (\exists\, e \in NL(N)\ \exists\, [v, c] \in \delta(e)\ (v \in RV(N)\ \&\ c = c_0))$
$\Longrightarrow \forall\, N \in \mathcal{N}\ (\exists\, e \in SE(N)\ \exists\, [v, c_0] \in \delta(e))$
$\Longrightarrow \forall\, N \in \mathcal{N}\ (\exists\, e \in SE(N)\ \exists\, v \in RV(N)\ (e \in L(M_v))$
$\Longrightarrow \forall\, N \in \mathcal{N}\ (\exists\, v \in RV(N)\ (e \in SE(N) \cap L(M_v)))$
$\Longrightarrow \forall\, N \in \mathcal{N}\ (\exists\, v \in RV(N)\ (SE(N) \cap L(M_v) \neq \emptyset)).$

However, $SE(N)$ itself is a stable cover set.

$\square$

It is obvious that the problem of finding a cover set with the stated properties is undecidable. However Theorem 3.1 presents a unified view of DDP for NCFSMs. It sheds new light on the decidability of DDP for NCFSMs in following sense: Given a specific class $\mathcal{N}$ of NCFSMs, to test if DDP is decidable for $\mathcal{N}$ we try to find a cover set satisfying the theorem. If we can find such a cover, we can conclude that DDP is decidable for $\mathcal{N}$. We shall illustrate this idea in next section.

In the methodology engendered by this characterization, we expect to be able to define the cover set C_N for a network N independent of the transitions (or semantics) of a particular network N. Furthermore, we expect to be able to check that such a cover set has the necessary properties. Based on the fact that the language of a shuffle-product automaton is regular, we have the following:

Corollary 3.1 *For a class of networks $\mathcal{N}$, let $\mathcal{L}$ be a family of languages such that the cover set C_N for every network $N \in \mathcal{N}$ belongs to $\mathcal{L}$. If*

1. *$\mathcal{L}$ is closed under intersection with regular languages, and*

2. *The emptiness problem is decidable for $\mathcal{L}$*

then the deadlock detection problem is decidable for the class of networks $\mathcal{N}$.

The Corollary given above provides a very tight sufficient condition to show that a class of networks has decidable DDP.

4 Applications

An NCFSM $N = \langle P_1, \cdots, P_n \rangle$ is *cyclic* if the topology graph of the network N is a simple cycle. More formally, N is a cyclic network if there exists a permutation $\{i_1, i_2, \cdots, i_n\}$ of the set I such that P_{i_j} can only send message to $P_{i_{j+1}}$ and receiving message from $P_{i_{j-1}}$ ($1 \leq j \leq n$, module $n + 1$).

All two-machine networks are cyclic. As it is known that DDP is even undecidable for the class of two-machine networks, so DDP is undecidable for general cyclic networks.

However DDP is decidable for cyclic networks where only one channel is unbounded. Let $\mathcal{N}^1_{1-u-cyc}$ be the class of cyclic networks of which only one channel is unbounded (referred to as 1-U cyclic NCFSMs, for short).

Let $N = \langle P_1, \cdots, P_n \rangle$ be a 1-U cyclic NCFSM. To simplify the discussion we assume without loss of generality that P_i can only send to P_{i+1} and receive from P_{i-1} ($1 \leq i \leq n$, module $n + 1$) and only the channel $P_1 \rightarrow P_2$ is unbounded. Let $L = (A^*B)^*H$, where

$A = (\cup_{2 \leq i \leq n-1} - \Sigma_{i,i+1}) \cup (-\Sigma_{n,1}) \cup (\cup_{2 \leq i \leq n-1} + \Sigma_{i,i+1}) \cup (+\Sigma_{n,1})$,

$B = (-\Sigma_{1,2})(+\Sigma_{1,2})$,

$H = (A \cup (-\Sigma_{1,2}))^*$.

The following lemma is a generalization of a lemma from [4].

Lemma 4.1 *Let $N = \langle P_1, \cdots, P_n \rangle$ be a 1-U cyclic NCFSM. For each executable event sequence e there exists another event sequence $e' \simeq e$ and $e' \in L$.*

Proof: Let $N = \langle P_1, \cdots, P_n \rangle$ be a 1-U cyclic NCFSM. The proof is by induction on the number k of receive events from $+\Sigma_{1,2}$.

Basis: $k = 0$. Since e does not contain events from $+\Sigma_{1,2}$, $e \in H$. Clearly the conclusion is true.

Induction: Assume that the conclusion holds for some $k > 0$. Let e be an executable event sequence which contains $k + 1$ receive events from $+\Sigma_{1,2}$. Let $s_{1,2}$ be the first send event from $-\Sigma_{1,2}$ and $r_{1,2}$ be the first receive event from $+\Sigma_{1,2}$ in e. We can rewrite e as $e = w_1.s_{1,2}.w_2.r_{1,2}.w_3$, where w_1 and w_2 do not contain any events from $(+\Sigma_{1,2})$.

As the network is cyclic (P_2 can only send to P_3 and receive from P_1), we can move all the send events from $(-\Sigma_{2,3})$ in w_2 before $s_{1,2}$. Let $w = w_1'.s_{1,2}.w_2'.r_{1,2}.w_3$ be the event sequence after such a reordering of e, where w_2' does not contain any events from $(-\Sigma_{2,3}) \cup (+\Sigma_{1,2})$. It is easy to see that $w \simeq e$. Since w_2' does not contain events initiated by P_2, we can move the event $r_{1,2}$ as the immediate successor of $s_{1,2}$. Let $w' = w_1'.s_{1,2}.r_{1,2}.w_2'.w_3$ be the event sequence after such reordering of w. Still $w' \simeq e$ holds.

Since w_3 contains k receive events from $+\Sigma_{1,2}$, by induction hypothesis there exists another event sequence $w_3' \simeq w_3$ and $w_3' \in L$. Hence $e \simeq e' \in L$, where $e' = w_1'.w_2.s_{1,2}.r_{1,2}.w_3'$.

□

Lemma 4.1 says that for any executable sequence e there is another sequence $e' \simeq e$ such that when e' is executed, there is at most one pending message in the only unbounded channel $P_1 \to P_2$ as long as P_2 can still receive. Therefore to check for deadlocks, we need only concentrate on the behavior of other bounded channels.

Theorem 4.1 *There exists a regular stable cover set for each network* $N \in \mathcal{N}_{1-u-cyc}$.

Proof: Let $N = \langle P_1, \cdots, P_n \rangle$ be a 1-U cyclic NCFSM.

To find a stable cover set for N, we need to construct a language which contains at least one element from each equivalent class $SE(e)$ where e is a stable event sequence.

Let $L' = (A^*B)^*A^*$, where A, B are defined as above. Notice that each $e \in L'$ contains the same number of events from $-\Sigma_{1,2}$ and $+\Sigma_{1,2}$. We can conclude from Lemma 4.1 that for each stable event sequence e, there exists $e' \simeq e$ and $e' \in L'$.

We can construct a finite state automaton F to accept the set of all stable events in L' as follows. Let each state in F record the number of pending event types, message types and message positions it has seen so far. The only exception is that when F sees an event $-g \in -\Sigma_{1,2}$, it will expect another event $+g \in +\Sigma_{1,2}$ in its next step. Notice that since all the channels except $P_1 \to P_2$ are bounded and the number of message types each machine can send is finite, the number of states in F is finite. The final state set in F includes only those states in which F has seen a stable event sequence (only the start state need be in the final state set). Therefore $L(F)$, the language accepted by F is a cover set for N.

□

By Corollary 3.1, the DDP is decidable for $\mathcal{N}_{1-u-cyc}$.

5 Conclusions

Deadlock detection problem for networks of communicating finite state machines has been known to be undecidable. The undecidability stems from the fact that even a two-machine NCFSM has the same computing power as a Turing machine. With special restrictions, specific classes of NCFSMs have been found for which the deadlock detection problem is decidable. However the underlying question "For what classes of NCFSMs is the DDP decidable," has not been answered before.

In this paper we considered the deadlock detection problem in NCFSMs from the formal language point of view. We have given a necessary and sufficient condition to show decidability of detecting deadlocks in classes of NCFSMs. We believe our work reveals the nature of decidability vs. undecidability of DDP.

The language concepts were first (as far as the authors know) introduced in the analysis of NCFSMs by K. Okumura [5]. The work in [5] concentrated mainly on establishing correspondence between the languages and the networks, in analogy with the traditional formal languages theory. However, the executable event sequences are context-sensitive even for many trivial networks. It appears that a direct use of the executable event sequences would not aid the analysis. This was our main motivation for introducing the concept of cover set for a particular property (DDP in this paper). Although we only discussed the decidability of DDP in this paper, the idea can also be applied to the decidability of detecting other properties such as unspecified receptions and unboundedness.

References

[1] G. Bochmann. Finite State Description of Communication Protocols. *Computer Networks*, Vol.2, 1978, pp.361–371.

[2] D. Brand and P. Zafiropulo. On Communicating Finite-state Machines. *JACM*, 30(2), 1983, pp.323–342.

[3] M. Gouda, E. Manning, and Y. T. Yu. On the Progress of Communication between Two Finite State Machines. *Information and Control*, 63(3), 1984, pp.308–320.

[4] M. Gouda, E. M. Gurari, Ten-Hwang Lai, and L. E. Rosier. On Deadlock Detection in Systems of Communicating Finite State Machines. *Computers and Artificial Intelligence*, Vol.6, 1987, No.3, pp.209–228.

[5] K. Okumura. Protocol Analysis from Language Structure. In *Protocol Specification, testing, and verification VIII*, S.Aggarwal and K.Sabnani (Editors), North-Holland, 1988, pp.113–124.

[6] Jan Pachl. Protocol Description and Analysis Based on a State Transition Model with Channel Expressions, In *Protocol Specification, testing, and verification VII*, H. Rubin and C. H. West (Editors), North-Holland, 1987, pp.207–219.

[7] W. Peng and S. Purushothaman. Analysis of Communicating Processes for Non-Progress. In *Proc of the 9th IEEE International Conference on Distributed Computing Systems*, June 1989, pp.280–287.

[8] Y. T. Yu and M. Gouda. Deadlock Detection for a Class of Communicating Finite State Machines. *IEEE Transactions on Communications*, Dec 1982, pp.2514–2518.

Branching Time Regular Temporal Logic for Model Checking with Linear Time Complexity

Kiyoharu Hamaguchi, Hiromi Hiraishi and Shuzo Yajima
Department of Information Science, Faculty of Engineering
Kyoto University, Kyoto, 606, Japan.

Abstract Firstly in this paper, we propose a branching time logic BRTL (Branching time Regular Temporal Logic) which has automata connectives as temporal operators. BRTL is more expressive than CTL proposed by Clarke et.al. and it is modest in terms of model checking, i.e. it has a model checking algorithm which runs in time proportional both to the size of a given Kripke structure and to the length of a given formula, as shown in the paper.

Secondly, in order to improve the succinctness of the temporal logic formulas, we introduce the mechanism of *substitutions to Boolean variables* and *references to the Boolean variables*. We propose EBRTL(Extended BRTL), i.e. BRTL with the substitution mechanism, and show examples of descriptions of some temporal properties. We develop its model checking algorithm whose time complexity is linear both in the size of a given Kripke structure and in the length of a given formula and exponential in the number of the Boolean variables used in the formula.

1 Introduction

In order to verify correctness of finite state systems, researchers developed various kinds of propositional temporal logics, applied model checking methods on Kripke structures and succeeded in verifying finite state systems of medium size[1, 2, 3].

CTL proposed in [1] has a model checking algorithm which runs in time $O(\text{Size}(S) \times \text{Len}(f))$ (Size(S) and Len(f) are the size of the given Kripke structure and the length of the given CTL formula respectively.) The expressive power of CTL is, however, restricted, because only $\mathcal{U}$(until) and X(next) with path quantifiers are allowed. ECTL[4] is an extension of CTL, which have *automata connectives*[5] as temporal operators and can express an arbitrary ω regular set. The model checking algorithm shown in [4] is exponential in the size of the number of the states of automata connectives, because Muller automata are used as their bases.

One of the two concerns of this paper is to develop a temporal logic system which is more tractable, in terms of model checking, than ECTL and more expressive than CTL. We propose a branching time logic BRTL (Branching Time Regular Temporal Logic) and show a model checking algorithm which runs in time $O(\text{Size}(S) \times \text{Len}(f))$, i.e. linear in the number of the states of automata connectives used in f.

The automata connectives of BRTL are based on restricted deterministic ω finite automata of Büchi type. Although this means that BRTL cannot express all ω regular sets, it can express the properties of repetition of some events as shown by examples.

The other concern is to improve the succinctness of the temporal logic formulas by introducing the mechanism of *substitutions to Boolean variables* and *references to the Boolean variables*. We propose EBRTL(Extended BRTL), i.e. BRTL with the above mechanism, and show examples of descriptions of some temporal properties. We develop its model checking algorithm whose time complexity is $O(\text{Size}(S) \times \text{Len}(f) \times 2^{|V|})$ ($|V|$ is the number of the different Boolean variables used in f).

In Section 2, BRTL is defined and its model checking algorithm is shown. In Section 3, BRTL is extended by introducing the mechanism stated above.

2 Branching Time Regular Temporal Logic

The truth values of propositional logic are represented by T(true) and F(false). X^ω and $X^\dagger$ represent sets of infinite and finite sequences of elements of a set X respectively. $|X|$ is the number of the elements of X. V_T and V_F are *tautology* and *invalid* formulas respectively.

2.1 Syntax and Semantics

Def. 2.1 *A logic-type deterministic ω finite automaton (ldo-fa)* $A = (Q, \Sigma, P, Br, \delta, q_0, F)$ is defined as follows.

Q and $\Sigma = \{T, F\}^n$ are sets of finite number of states and input symbols respectively. $P = \{p_1, \cdots, p_n\}$ is a set of atomic propositions. F is a set of accepting states and the elements of $Q - F$ are called rejecting states. q_0 is the initial state.

Let BF be a set of all propositional formulas constructed from the elements of P. $Br : Q \times Q \to BF$ is a partial function which satisfies the following two conditions.

We define $BF_q \overset{\text{def}}{=} \{f | \exists q'.Br(q, q') = f\}$, for each $q \in Q$.

For any $q \in Q$, (1) $f_1 \wedge f_2 \equiv V_F$ for any $f_1, f_2 \in BF_q$ and (2) $\bigvee_{f \in BF_q} f \equiv V_T$.

We define $\text{Edges}(A) \overset{\text{def}}{=} \{(q, q') | \exists f.Br(q, q') = f\}$.

$\delta : Q \times \Sigma \to Q$ is a state transition function such that, for $q, q' \in Q$ and $v \in \Sigma$, $\delta(q, v) = q' \Leftrightarrow f = Br(q, q')$ is defined and $f(v) = T$.

For $x \in \Sigma^\omega$, we define $Inf(x)$ as the set of the states through which A goes infinitely often. Then the set of words accepted by A is $\{x | Inf(x) \cap F \text{ is not empty }\}$ and is described by $|A|$. □

Although complementation of an ω finite automaton is not easy in general, even if it is deterministic, an automaton that satisfies the following Cond. 1 can be complemented easily, (i.e. by exchanging the accepting states and the rejecting states.)

Cond. 1 Let A be a ldo-fa $(Q, \Sigma, P, Br, \delta, q_0, F)$. There is no input sequence which transits A from a rejecting state $q_r \in (Q - F)$ to the state q_r via some accepting state $q_a \in F$. □

A logic-type deterministic ω finite automaton A which satisfies Cond. 1 is called *ldo-fa-1*.

It is easy to prove the following two lemmas.

Lem. 2.2 *For an ldo-fa-1 $A = (Q, \Sigma, P, Br, \delta, q_0, F)$, an ldo-fa-1 $\overline{A}$ which accepts $\Sigma^\omega - |A|$ is obtained by exchanging accepting states and rejecting states of A.*

Lem. 2.3 *For ldo-fa-1's $A_i = (Q_i, \Sigma, P, Br_i, \delta_i, q_{i0}, F_i)$ $(i = 1, 2)$, an ldo-fa-1 $A_1 | A_2$ which accepts $|A_1| \cup |A_2|$ is obtained as follows:*

- $Q = Q_1 \times Q_2 = \{(q_1, q_2) | q_1 \in Q_1, q_2 \in Q_2\}$

- $Br : Q \times Q \to BF$ is defined by $Br((q_1, q_2), (q_1', q_2')) = Br_1(q_1, q_1') \wedge Br_1(q_2, q_2')$, where $q_i, q_i' \in Q_i$ $(i = 1, 2)$ and both of $Br_1(q_1, q_1')$ and $Br_1(q_2, q_2')$ are defined.

- $q_0 = (q_{10}, q_{20})$.

- $F = \{(q_1, q_2) | q_1 \in F_1 \text{ or } q_2 \in F_2\}$.

□

The lemmas mean that the languages defined by ldo-fa-1's are closed under Boolean operations over sets.

Let AP be a set of atomic propositions. $S = \langle \Sigma, I, R, \Sigma_0 \rangle$ is called a Kripke structure, where Σ is a set of states, $I : \Sigma \to 2^{AP}$ is an assignment function, $R \subseteq \Sigma \times \Sigma$ is a *total* relation, $\Sigma_0 \subseteq \Sigma$ is a set of initial states. The size of a given Kripke structure, denoted by Size(S), is defined as $|\Sigma| + |R|$.

Def. 2.4 Syntax and Semantics

The syntax of BRTL(Branching Time Regular Temporal Logic) is as follows:

BRTL formulas: The set of all BRTL formulas are described by BF.

- $p \in AP$ is an BRTL formula.

- If f and g are BRTL formulas, then $\neg f$ and $f \vee g$ are BRTL formulas.

- If A is an automaton connective , then $\exists A$ is a BRTL formula.

Automata connectives

If A' is an ldo-fa-1, then $A'[AP/B]$ and $\neg A'[AP/B]$ are automata connectives, where $A'[AP/B]$ is obtained by replacing each atomic proposition $p_i \in AP$ $(1 \le i \le n)$ with $f_i \in B \subseteq BF$ corresponding to p_i simultaneously. The numbers of the elements in AP and B have to be equal.

The semantics of BRTL is defined on a Kripke structure $S = \langle \Sigma, I, R, \Sigma_0 \rangle$. $S, s \models f$ means that the BRTL formula f holds at the state s on S. In the following, $p \in AP$, f and g are BRTL formulas and A is an automaton connective.

- $S, s \models p$ iff $I(s) \ni p$

- $S, s \models f \vee g$ iff $S, s \models f$ or $S, s \models g$

- $S, s \models \neg f$ iff $S, s \not\models f$

- $S, s \models \exists A$ iff there exists an infinite sequence $\sigma = s_0 s_1 s_2 \cdots$ starting from s on S and a run (a sequence of states) $q_0 q_1 q_2 \cdots$ in Q such that, $S, s_i \models BR(q_i, q_{i+1})$ holds and all the states which appear infinitely in the run are in F.

- $S, s \models \exists \neg A$ iff $S, s \models \exists \overline{A}$

If $\forall s \in \Sigma_0. S, s \models f$, then we describe $S \models f$. □

The Boolean operators $\wedge$, $\equiv$ and $\Rightarrow$ are also used. Besides we define $\forall A \overset{\text{def}}{=} \neg \exists \neg A$ and $\forall \neg A \overset{\text{def}}{=} \neg \exists A$.

For BRTL formulas $f = \exists A$ or $\exists \neg A$, where $A = A'[AP/B]$ for some ldo-fa-1 A' and BRTL formulas B, the length of f, denoted as $\text{Len}(f)$ is defined as follows:

$$\text{Len}(A) \overset{\text{def}}{=} |Q| + |\text{Edges}(A')| + \sum_{f \in Br'(Q \times Q)} \text{Len}(f) + \sum_{g \in B} \text{Len}(g)$$

$Br'(Q \times Q)$ represents the set of propositional formulas labeled to transitions of A'. Intuitively the length means the length of the description of the ldo-fa-1 A' and the set of B.

2.2 Comparison with CTL

Def. 2.5 The truth value of a given CTL formula is defined on each state on a given Kripke structure.

While BRTL uses path quantifiers $\exists$ and automata connectives, $\forall X f$, $\exists X f$, $\forall [f_1 \mathcal{U} f_2]$ and $\exists [f_1 \mathcal{U} f_2]$ are allowed to be used[1].

The semantics of these formulas are as follows:

- $S, s \models \forall X f$ ($\exists X f$) iff $S, s_1 \models f$ for all (some) s_1 such that $(s, s_1) \in R$

- $S, s \models \forall [f_1 \mathcal{U} f_2]$ ($\exists [f_1 \mathcal{U} f_2]$) iff $S, s_i \models f_1$ ($i = 0, 1, 2, \cdots, n$) and $S, s_{n+1} \models f_2$ for all (some) infinite sequences $s = s_0 s_1 s_2 \cdots$ starting from s and an integer $n \geq 0$.

$\square$

Since the automata connectives are not easy to write, we introduce informally a description language like programming languages. For example, the automata connectives of Fig. 1 and Fig. 2 are described by the descriptions (1) and (2) in the following.

- *state:* [a] and [r] represent an accepting state and a rejecting state of an automaton connective respectively. [ac] and [re] are special states. [ac] ([re]) represents an accepting (rejecting) state from which the automaton never transits to the other states for any input.

- *branch:* if S_1 else if $S_2 \cdots$ else S_n endif represents a branch from a state. S_i has a higher *precedence* than S_{i+1}.

- *loop:* loop S endloop represents a loop from a state to the loop.

- *label and goto:* 'L:' is a label to be put on a state and goto L means the transition to the state.

If no transition is possible from a state, it is assumed to transit to a rejecting state. [x] [x] is interpreted as [x] V_T [x] (x = a, r, ac or re). The first state is assumed to be the initial state of the automaton.

With the above notations, we can define the temporal operators used in CTL. Let $Next(f)$ and $Until(f_1, f_2)$ be automata connectives such that

$$Next(f) \overset{\text{def}}{=} ([r][r]f[ac]) \tag{1}$$

$$Until(f, g) \overset{\text{def}}{=} ([r]\text{if loop } f_1 \wedge \neg f_2 \text{ endloop} \\ \text{else if } \neg f \wedge \neg f_2 \text{ [re] else if } f_2 \text{ [ac] endif }) \tag{2}$$

Then,

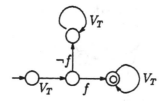

Figure 1: $Next(f)$

$\forall[f_1 \mathcal{U} f_2] \overset{\text{def}}{=} \forall Until(f_1, f_2)$ $\exists[f_1 \mathcal{U} f_2] \overset{\text{def}}{=} \exists Until(f_1, f_2)$
$\forall X f \overset{\text{def}}{=} \forall Next(f)$ $\exists X f \overset{\text{def}}{=} \exists Next(f)$

Temporal properties containing the concept of repetition can also be expressed. For example, an BRTL formula $f = \forall A_R$, where A_R is shown in Fig. 3, means "for a given $m \geq 2$, p is true on every state s_i $(i = km, k = 0, 1, \cdots)$".

It can be shown that this property cannot be expressed by any CTL formula by extending the proof by Wolper in [5] to Kripke structures.

For a Kripke structure $S = \langle \Sigma, I, R, \Sigma_0 \rangle$ and a state $s_0 \in \Sigma$,

$$Re_i(S, s_0) \overset{\text{def}}{=} \{s_i | (s_j, s_{j+1}) \in R, j = 0, 1, 2, \cdots i - 1\}$$

Lem. 2.6 *Let p be an atomic proposition. C_i is defined as a set of pairs of a Kripke structure $S = \langle \Sigma, I, R, \Sigma_0 \rangle$ and a state s, i.e. (S, s), satisfying the following conditions.*

1. $\forall s' \in Re_j(s).S, s' \models p$ $(j = 0, 1, \cdots, i)$,

2. $\forall s'' \in Re_{i+1}(s).S, s'' \models \neg p$,

3. $\forall s''' \in Re_j(s).S, s''' \models p$ $(j = i + 1, i + 2, \cdots)\}$

If CTL formula f has at most n 'X's , then, for any $(S_i, s_i) \in C_i$, $(S_{i'}, s_{i'}) \in C_{i'}$ $(i, i' > n)$,

$$S_i, s_i \models f \Leftrightarrow S_{i'}, s_{i'} \models f \tag{*}$$

In other words, for any $i > n$, the truth value of f is same for $(S_i, s_i) \in C_i$. □

Theo. 2.7 *Given an integer $m \geq 2$, $D \overset{\text{def}}{=} \{(S, s) | \forall s' \in Re_{km}. S, s' \models p, (k = 0, 1, \cdots)\}$, where S is a Kripke structure and s is a state on S.*

There exists no CTL formula f which satisfies $(S, s) \in D \Leftrightarrow S, s \models f$

(Proof) Let us assume that a CTL formula f satisfies the above condition and f has l 'X's.

Let $km - 1 \geq l$. From Lemma 2.6, $S, s \models f \Leftrightarrow S', s' \models f$ holds, for any $(S, s) \in C_{km}$ and any $(S', s') \in C_{km-1}$. Besides $(S, s) \in D$ and $(S', s') \notin D$. Thus there exists no CTL formula that can express the property. □

ECTL[4] has as automata connectives deterministic ω finite automata of Muller type, the class of which is equivalent to ω regular sets, as automata connectives. This means that there are some properties which cannot be expressed in BRTL, but can be expressed in ECTL.

From the above arguments, the relation among CTL, BRTL and ECTL in terms of expressive power is as follows:

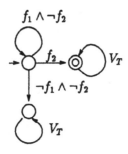

$$f_1 \wedge \neg f_2$$

Figure 2: $Until(f_1, f_2)$

$$CTL < BRTL < ECTL$$

2.3 A Model Checking Algorithm

A model checking algorithm is shown in this section. The algorithm is constructed, based on the next lemma.

Lem. 2.8 *The necessary and sufficient condition for $S, s \models \exists A$ is that there exists a finite sequence $(s_0, q_0)(s_1, q_1) \cdots (s_k, q_k)(s_{k+1}, q_{k+1}) \cdots (s_n, q_n)$ satisfying the following conditions:*

1. $(s_i, s_{i+1}) \in R \ (i = 0, 1, \cdots n - 1)$
2. $S, s_i \models Br(q_i, q_{i+1}) \ (i = 0, 1, \cdots n - 1)$
3. $s_k = s_n$ *and* $q_k = q_n$
4. $q_j \in F \ (j = k, k + 1, \cdots n)$

Algorithm 2.9 • Input: A Kripke structure S and an BRTL formula f

- Output: $S \models f$?
- Method:

1. The automata connectives of the form $\neg A$ occurred in f are transformed to $\overline{A}$ by exchanging the accepting states and rejecting states.

2. By the following (a)-(c), every subformula g of f is labeled to every state $s \in \Sigma$ if and only if $S, s \models g_i$, in a bottom up manner.

   The output is 'yes' if and only if f is labeled to every $s_0 \in \Sigma_0$.

   (a) When g is an atomic proposition, its truth value on each $s \in \Sigma$ is given.
   (b) When g is $h_1 \vee h_2$ or $\neg h$ for some BRTL formulas h_1, h_2 or h, the truth value of g on each state can be obtained by the truth values of h_1, h_2 or h which are already labeled.
   (c) When g is $\exists A$ for some automata connective A, the following (i)-(iv) are performed.

      i. A graph $G = (V, E)$ is constructed, where
         $V = \{(s, q) | s \in \Sigma, q \in Q\}$ and
         $E = \{((s, q), (s', q')) | (s, s') \in R \text{ and } S, s \models Br(q, q')\}$.

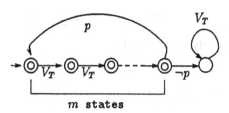

m states

Figure 3: An automata connective representing repetition

ii. The set of vertices of all the strongly connected components on the subgraph G' of G whose vertices are in $V' = \{(s, q_F)|q_F \in F\}$ is obtained. It is described as V_C

iii. The set of the vertices in G which can reach some vertex in V_C is obtained. It is described as V_R

iv. $\exists A$ is labeled to every $s \in \Sigma$ such that $(s, q_0) \in V_R$ (q_0 is the initial state of A).

□

Prop. 2.10 *Algorithm 2.9 runs in time $O(Size(S) \times Len(f))$.*

(Sketch of Proof) This is proved by the following facts:

- In 1, the transformation from $\neg A$ contained in f to $\overline{A}$ can be performed in time $O(\text{Len}(f))$.

- In 2.(b), if the truth values of h_1, h_2 or h are determined on each state $s \in \Sigma$, then the truth values of $h_1 \vee h_2$ and $\neg h$ can be determined in time $O(|\Sigma|)$.

- In 2.(c).i, the determination of $S, s \models Br(q, q')$ for all $s \in \Sigma$ and all $q, q' \in Q$ can be performed in time $O(Size(S) \times (\sum_{f \in Br'(Q \times Q)} \text{Len}(f) + \sum_{g \in B} \text{Len}(g)))$. Then the construction of the graph G costs $O(Size(S) \times (|Q| + |Edges(a)|))$ in time, thus in total, $O(Size(S) \times (\text{Len}(A)))$.

- In 2.(c).ii, iii and iv, the construction of strongly connected components and the calculation of the vertices reachable to the components can be performed in linear to the size of the graph. (Here the size is the sum of the numbers of edges and vertices.)

□

3 Temporal Logic with Substitutions and References

3.1 Succinctness of Descriptions

Some temporal properties expressed in the linear time temporal logic with □ (always) and ○ (next) as temporal operators cannot be described easily in CTL or BRTL. For example, an assertion for a sequence detector such that the value of a signal line z is 1 iff it recognizes the sequence 110 on a signal line x, is described in the linear time temporal logic as follows.

$$\Box((x \land \bigcirc(x \land \bigcirc\neg x)) \equiv \bigcirc\bigcirc z)$$

Using BRTL, this property is expressed as

$$\forall\Box\forall A,$$

where $\Box f$ is equivalent to $\neg Until(V_T, \neg f)$ and $A \overset{\text{def}}{=}$

```
[r]   if x₁ [r] {
          if x₁ [r] {
              if ¬x₁ ≡ z [ac]
              else if x₁ ∧ ¬z [ac]
              else [re]
          }
          else L:[r]  {
              if ¬z [ac]
              else [re]
          }
      else [r] Vₜ goto L
```

By allowing substitution to Boolean variables and references to them, the property can be expressed succinctly as

$$\forall\Box\forall(\; [r] \; V_T, v_1 := x \, [r] V_T, v_2 := x \; [r] \; (v_1 \land v_2 \land \neg x) \equiv z \; [ac] \;)$$

Here v_1 and v_2 are Boolean variables and $v_1 := x$ means that the value of x is substituted to v_1.

Def. 3.1 Syntax of EBRTL(Extended BRTL)

Let $V = \{v_1, v_2, \cdots, v_m\}$ be a set of Boolean variables. AP of BRTL contains V. EBRTL formulas are defined similarly to BRTL formulas.

Automata connectives are similar to those of BRTL except that

- Formulas labeled to the transitions are composed of EBRTL formulas.

- Substitutions $v_i := f$ (f is an EBRTL formula) can be labeled to the transitions. For a transition, substituting to a variable v_i is allowed only once.

More precisely, the function Br is redefined as follows:

$Br : Q \times Q \to BF \times SUB$ is a partial function which satisfies the following three conditions, where SUB is a class of sets of substitutions.

We define $BF_q \overset{\text{def}}{=} \{f | \exists q'.Br(q, q') = (f, sub), sub \in SUB\}$ for each $q \in Q$.

For any $q \in Q$, (1) $f_1 \land f_2 \equiv V_F$ for any $f_1, f_2 \in BF_q$, (2) $\bigvee_{f \in BF_q} f \equiv V_T$ and (3) each element $sub \in SUB$ has at most one substitution of the form $v_i := f$ for each $v_i \in V$. (sub can be empty.) $\qquad\Box$

Before defining the semantics, we show a sequence along which $\exists(\; [r] \; v_1 := x \; [r] \; v_2 := x \; [r] \; (v_1 \land v_2 \land \neg x) \equiv z \; [ac] \;)$ holds in Fig. 4. The tuple labeled to each state stores the values of v_1 and v_2. We have to give initial values of the Boolean variables. In the example, $v_1 = F$ and $v_2 = F$ are the initial values.

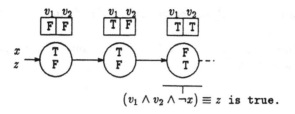

$$(v_1 \wedge v_2 \wedge \neg x) \equiv z \text{ is true.}$$

Figure 4: Intuitive Semantics of an EBRTL formula

Note that the change of the values of the Boolean variables caused by substitutions at a state occurs at its next state. This prevents oscillations of the values at each state.

Def. 3.2 Semantics

The semantics of EBRTL is defined for a Kripke structure S. When an EBRTL formula f holds at a state s on S with initial values of Boolean variables represented by a vector notation $\vec{v} = (v_1, v_2, \cdots, v_m)$ $(v_j = F$ or $T)$, we denote as $S, s, \vec{v} \models f$. $\vec{v}[j]$ represents the j-th component of the vector $\vec{v}$.

While the truth value of each atomic propositions except Boolean variables are determined for $(S, s, \vec{v})$ by $I(s)$, that of the Boolean variable v_j $j = 1, 2, \cdots, m$ is the value of the j-th component of $\vec{v_j}$.

Here we define the semantics for $S, s, \vec{v} \models \exists A$ (A is an automaton connective).

$S, s, \vec{v} \models \exists A$ iff there exists an infinite sequence $\tau = t_0 t_1 t_2 \cdots$ that satisfies the following conditions.

Here $t_i = (s_i, \vec{v_i}, q_i) \in \Sigma \times \{T, F\}^m \times Q$.

- $(s_i, s_{i+1}) \in R$

- Let $(f, sub) = Br(q_i, q_{i+1})$. If $(\vec{v_i}[j] := g) \in sub$, then $\vec{v_{i+1}}[j] = T$ iff $S, s_i, \vec{v_i} \models g$. Otherwise $\vec{v_{i+1}}[j] = \vec{v_i}[j]$.

- Let $(f, sub) = Br(q_i, q_{i+1})$. $S, s_i, \vec{v_i} \models f$.

□

3.2 A Model Checking Algorithm for EBRTL

The outline of a model checking algorithm of EBRTL is shown.

Algorithm 3.3

- Input: A Kripke structure S, an BRTL formula f and initial values of the Boolean variables $\vec{v}$

- Output: $S, s, \vec{v} \models f$ for all the initial states of S ?

- Method: We modify the model checking algorithm shown in Algorithm 2.9 as follows:

  - In constructing the graph G, tuples of the form $(s, q, \vec{v})$, where $s \in \Sigma$, $q \in Q$ and $\vec{v}$ is the vectors of the truth values for $v_j \in V$, are used instead of (s, q).

Prop. 3.4 *The model checking algorithm for EBRTL runs in time $O(Size(S) \times Len(f) \times 2^{|V|})$.*

(Sketch of Proof) This is proved by the fact that the number of the tuples of the form $(s, q, \vec{v})$ is $Size(S) \times Len(f) \times 2^{|V|}$. ☐

4 Conclusion

We proposed a temporal logic BRTL which is more expressive than CTL and is more tractable than ECTL in terms of model checking. Furthermore, we provided an extension of BRTL which has the mechanism of substitutions to Boolean variables and references to them.

The linear time complexity of the model checking algorithms of CTL or BRTL is obtained by excluding nondeterminism from their descriptions. CTL* or ETL can express, in a sense, the concept of nondeterminism by allowing formulas of the form $\forall(f \vee g)$ (f and g are the formulas without the path quantifiers $\forall$ and $\exists$) and $A \vee B$ (A and B are automata connectives based on nondeterministic finite automata) respectively.

The mechanism of substitutions and references to Boolean variables also introduces nondeterminism to BRTL, in a sense. In the example of the sequence detector, the branches of the automaton connective of BRTL formula are "folded" by using substitutions. This means that the substitution mechanism is another way to introduce nondeterminism.

Future problems are as follows:

- Determination of the expressive power of EBRTL.

- Determination of the complexity of model checking problem for EBRTL.

References

[1] E. M. Clarke, E. A. Emerson, and A. P. Sistla. Automatic Verification of Finite State Concurrent Systems Using Temporal Logic Specifications: A Practical Approach. In *10th ACM Symposium on Principles of Programming Languages*, pages 117–126, January 1983.

[2] M. C. Browne, E. M. Clarke, D. L. Dill, and B. Mishra. Automatic Verification of Sequential Circuits Using Temporal Logic. *IEEE Transactions on Computers*, C-35(12):1035–1044, December 1986.

[3] H. Hiraishi. Design Verification of Sequential Machines Based on a Model Checking Algorithm of ϵ-free Regular Temporal Logic. In *Computer Hardware Description Languages and their applications*, pages 249–263, June 1989.

[4] E. M. Clarke and O. Grümberg and R. P. Kurshan. A Synthesis of Two Approached for Verifying Finite State Concurrent Systems. Technical report, Carnegie Mellon University, 1987. manuscript.

[5] P. Wolper. Temporal Logic Can Be More Expressive. In *Proceedings of 22nd Annual Symposium on Foundations of Computer Science*, pages 340–348, 1981.

THE ALGEBRAIC FEEDBACK PRODUCT OF AUTOMATA(EXTENDED ABSTRACT).

VICTOR YODAIKEN
DEPARTMENT OF COMPUTER AND INFORMATION SCIENCE
UNIVERSITY OF MASSACHUSETTS
AMHERST, MASSACHUSETTS 01003
YODAIKEN@CS.UMASS.EDU

1. INTRODUCTION

Finite state automata provide a clear and faithful mathematical representation of digital devices and programs. Since large-scale digital systems are generally constructed by interconnecting less complex components, the problem of representing such systems in terms of interconnected automata has received a great deal of attention in the computer science literature, e.g., [COGB86, Har84, Kur85]. In this paper we describe an algebraic automata product, called the *feedback product* which can be employed to emulate the intricate control and component interconnection of a wide variety of digital systems[1]. A classical Moore machine or finite state transducer [HU79] provides a mathematical model of a discrete finite state mechanism with output. A *feedback product form transducer* provides a mathematical model of an interconnected system of finite state mechanisms. A product form transducer contains some number of *component* transducers representing sub-systems. Each component transducer is associated with a *feedback function*. When input is provided to the product form machine, each feedback function generates an input sequence for its associated transducer. The generated sequence depends on the input provided to the product machine, the global system state, and the outputs of some or all of the other components. Thus, each system transition corresponds to parallel computations in each of the components. Because the feedback functions can access the outputs of all the components, arbitrary (finite state) synchronization can be modeled.

- Generated sequences can be of differing lengths (or empty), so components may change state at varying rates.

This research was funded in part by the Office of Naval Research under contract N00014-85-K-0398.

[1] The feedback product used here is a variation of the general product described in [Gec86].

- A feedback product form transducer can be "multiplied-out" to obtain a classical transducer, we remain within the domain of finite state machine theory.
- Since components may, themselves, be in product form, we can model multi-layer systems.
- There is no need to make the alphabets of components and the product transducer overlapping or disjoint — the feedback functions, rather than the transition labels, define interactions.

The feedback product provides several advantages over inter-connection techniques previously described in the computer science literature. Among these advantages are: a direct (non-interleaved) model of concurrency, a model for systems containing components that change state at differing rates, and a natural model of encapsulation and information hiding. The feedback product also provides a formal framework for describing concurrent systems *without* making any assumptions about underlying computation environment: e.g., scheduling, storage, or communication mechanisms. We are interested in specification of operating systems, circuits, and real-time controllers. These systems are not implemented within environments that provide uniform, fixed communication primitives, and these systems may contain components which have radically different granularities of state change. We are hesitant to accept, for example, the paradigm of a concurrent algol-like programming language as fundamental. And we are dubious about the prospects of verifying the behavior of systems which implement scheduling, storage and communication mechanisms in terms of a formal framework which assumes the previous existence and character of these mechanisms.

Direct analysis and definition of product form automata would be rather awkward. In preference, we define a modal formal language based on the primitive recursive functions [Pet67, Goo57]. The modal primitive recursive (m.p.r.) functions are evaluated in the implicit context of a product form transducer and *trace* (sequence of transition symbols). The trace is intended to represent the sequence of state transitions which have driven the transducer from its initial state. We never explicitly construct either traces or transducers. Instead, we describe these objects in terms of the values that they confer on m.p.r. functions.

We say a transducer P *satisfies* a m.p.r. expression E iff E is non-zero in the context of (P, w) for every w that does not drive P to an undefined state. In other words, P satisfies E iff E is true in every reachable state of P. Every m.p.r. boolean expression is, thus, a specification of the class of transducers which satisfy it. We can show that some fairly simple extensions to the primitive recursive functions are sufficient to obtain a powerful and expressive language in which to describe product form transducers.

Related formalisms. In some respects, our work is closest to that of Clarke *et al* [CAS83, EH83] and Ostroff [OW87], who have also used modal formalisms to describe finite automata. Our work can be seen as providing an alternative semantics for temporal logic based formalism. We believe that this semantics solves some of the problems that temporal logic presents in terms of composition. Statecharts [Har84] are an extension of state diagrams to allow for description of concurrent and composed systems. While statecharts are quite expressive, they suffer from some of the same limits as traditional state diagrams, e.g., we cannot extend the statechart of a 2 bit shift register to obtain the statechart of a 3 bit shift register. Furthermore the formal semantics of statecharts is exceedingly complex.

Outline. The remainder of this extended abstract is in 2 sections. Section 2 defines product form automata, the m.p.r. functions, and an interpretation. In section 3 we show the expressive range of the m.p.r. functions and develop some proof methods.

<u>Notational Preliminaries</u>. Much of the body of this paper is concerned with sequences and paths. Let $\langle\rangle$ denote the empty sequence and let juxtaposition denote sequence concatenation: $\langle a_1, \dots, a_n \rangle \langle b_1, \dots, b_n \rangle = \langle a_1, .., a_n, b_1, \dots, b_n \rangle$. Let $a : u \stackrel{def}{=} \langle a \rangle u$ and let $u :: a \stackrel{def}{=} u \langle a \rangle$. Finally, let $length(w)$ be the length function.

2. FEEDBACK PRODUCTS AND MODAL FUNCTIONS.

2.1. Product form transducers. We define product form transducers inductively. First we define the class $\mathbf{P}_0$ of flat product form transducers — those with no factors. The elements of this class are essentially standard Moore machines [HU79]. The class $\mathbf{P}_{i+1}$ of transducers consists of those transducers containing at least one factor belonging to $\mathbf{P}_l$, and no factors of belonging to $\mathbf{P}_j$ for $j > l$. The class $\mathbf{P}$ is the union over all $\mathbf{P}_i$.

Each product form transducer contains a tuple of *factor* product form transducer and *feedback functions*: $(\phi_1, P_1, \dots, \phi_r, P_r)$. Factors are connected by making the input to each factor depend on the input to the product form automaton, the state of the product form automaton, and the outputs of all of the factors. When a product form automaton accepts a single input symbol, each factor is provided with a sequence of 0 or more input symbols, representing the parallel activity of all the components. Note that the transition and output functions can only use the output of the factors, and do not see the internal state of factors.

266

Figure 2.1. *A product form transducer.*

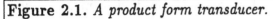

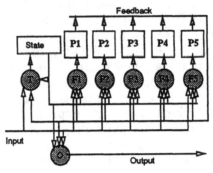

$\Gamma = 14$

$P1 - P5$ are factors, T is a transition function, and $F1 - F5$ are feedback functions.

Definition 2.1.   *The class of product form transducers*
The class $\boldsymbol{P}$ of product form transducers is the infinite union of the classes $\boldsymbol{P_i}$ for $i \geq 0$. The class $\boldsymbol{P_0}$ consists of all Moore machines. Each class $\boldsymbol{P_{l+1}}$ of product form transducers is the smallest set containing all tuples of the form:

$$P = (A, O, S, start, \lambda, \delta, (\phi_1, P_1, \dots, \phi_r, P_r))$$

- $A = \{1, \dots n\}$ is a finite transition alphabet, with 0 reserved for use as the null transition that leaves state unchanged,
- $O = \{1, \dots k\}$ is a finite output alphabet,
- $S = \{1, \dots h\}$ is a finite state set (called the top-level state set),
- $start \in S$ is a distinguished start state,
- Each P_i $(0 < i \leq r)$ is a product form transducer, with $l = max\{l' : (\exists 0 < i \leq r) P_i \in \boldsymbol{P}_{l'}\}$.
- $\lambda : S \times O.1 \dots \times O.r \to O$,
- $\delta : S \times O.1 \dots \times O.r \times A \to S$ is a transition function, with $\delta(s, o_1, \dots, o_r, 0) = s$.
- Each $\phi_i : S \times O.1 \dots \times O.r \times A \to (A.i)^*$ is a feedback function, with $\phi_i(s, o_1, \dots, o_r, 0) = \langle\rangle$.

As the product form transducer accepts input, its internal state changes according to δ, and the outputs of the factors change according to their own transition and output functions and the feedback functions of the product. A *configuration* is a tuple $(s, o_1, \ldots, o_r)$ representing a state of the product form transducer where the top-level state is s, and the output of each P_i is o_i. We let $\Delta(P, w)$ represent the configuration reached by following w fro the initial configuration. Let $\psi(P, w, i)$ denote the sequence generated for P_i by ϕ_i when the product transducer follows w from its initial state. Whenever, $\Delta(P_i, \psi(P, w, i))$ is undefined, we want $\Delta(P, w)$ to be also undefined. This means that any trace of a product form transducer will not drive any of the factors into undefined states. The language of traces $\mathcal{L}(P)$ is simply the set of traces that do not drive Δ to an undefined configuration.

2.2. The Modal Primitive Recursive Functions.

The class MPR of modal primitive recursive functions includes a small set of initial functions, and all those functions definable from the initial functions with a finite number of applications of an even smaller set of function construction rules. The m.p.r. functions form an extension to the class PR of primitive recursive functions [Pet67, Smo85]. Every p.r. function is also a m.p.r. functions. In fact, m.p.r. functions are simply p.r. functions with *hidden* parameters for a product form transducer and its trace. We first define the m.p.r. functions, and then give a partial definition of a map $\rho : MPR \to PR$ which makes the context dependencies explicit.

Definition 2.2. *The class of m.p.r. functions*
The class MPR of the modal primitive recursive functions consists of all the functions which can be generated after a finite number of steps using the following rules and the defining rules of the primitive recursive functions.[2].

(1) *Alphabet:* $f \stackrel{def}{=} Alphabet$,

(2) *Output function:* $f(x) \stackrel{def}{=} Out(x)$,

(3) *Component names:* $f \stackrel{def}{=} Components$,

(4) *Pumping number:* $f \stackrel{def}{=} pump\_number$,

(5) *Enabling:* $f(x) \stackrel{def}{=} enable(x)$,

(6) *Feedback:* $f(x, y) \stackrel{def}{=} effect(x, y)$,

(7) *Path offset:* $f(y, \vec{x}) \stackrel{def}{=} (after\, y)g(\vec{x})$,

(8) *Component selection* $f(y, \vec{x}) \stackrel{def}{=} (in\, n\, y)g(\vec{x})$.

---

[2]The form of this presentation is adapted from [Smo85].

Definition 2.3. *A fragment of the evaluation functional, ρ*
Let $P = ((A, O, S, start, \lambda, \delta, (\phi_1, P_1, \ldots \phi_r, P_r))$

> If $f \overset{def}{=} \textbf{pump\_number}$
> $$(\rho\, f)(P, w, x) \overset{def}{=} size(S) * \Pi_i\, size((\rho\, \textbf{pump\_number})(P_i))$$
> If $f(x) \overset{def}{=} \textbf{enable}(x)$
> $$(\rho\, f)(P, w, x) \overset{def}{=} \begin{cases} 1 & if\ w :: x \in \mathcal{L}(P) \\ 0 & otherwise. \end{cases}$$
> If $f(x) \overset{def}{=} \textbf{effect}(x, y)$
> $$(\rho\, f)(P, w, x, y,) \overset{def}{=} \begin{cases} \phi_y(\Delta(P, w), x) & if\ 0 < x \leq r; \\ () & otherwise. \end{cases}$$
> If $f(y, \vec{x}) \overset{def}{=} (\textbf{after}\, y)g(\vec{x})$
> $$(\rho f)(P, w, y, \vec{x}) \overset{def}{=} (\rho g)(P, wy, \vec{x})$$
> If $f(y, \vec{x}) \overset{def}{=} (\textbf{inn}\, y)g(\vec{x})$
> $$(\rho f)(P, w, y, \vec{x}) \overset{def}{=} \begin{cases} (\rho g)(P_y, \phi_y(st, w, y), \vec{x})) & if\ 0 < y \leq r; \\ 0 & otherwise. \end{cases}$$

Theorem 2.1. *The functional ρ is an effectively 1-1 map from the m.p.r. functions to the p.r. functions.*
Proof. *Note that (ρf) is obviously a p.r. function when f is an initial m.p.r. function. Proceeding by induction we can see that (ρf) must be p.r. for all functions $f \in MPR$. The function is "effectively" 1-1 because for every p.r. function f, there is a m.p.r. function, f itself, so that $f'(\vec{x}) = (\rho f)(P, w, \vec{x}) = f(\vec{x})$. That it, the map is 1-1 modulo hiding the extra, meaningless arguments.*

3. Proof System

Goodstein [Goo57] has described a purely syntactic proof system for p.r. functions, and this system is sound for m.p.r functions by the construction of ρ. We need , however, to provide for modal deductions as well as arithmetic deductions.

3.1. Elementary proof methods. We find it convenient to be able to treat expressions as un-named functions. That is, we might write $(\textbf{inn}\, c)(E \wedge E')$ to force evaluation of the entire expression $t \wedge t'$ within the context of component c. Similarly, we write $(\textbf{after}\, a)(E \wedge E')$ to force the evaluation of the expression $E \wedge E'$ in the state reached via an a transition. We can now list some simple results. The first result is that state independent functions can be "factored" out of the scope of **inn** and **after**.

Theorem 3.1. *Distributive law for modal functionals.*

$$(\textbf{after}\,u)(f(E_1,\ldots,E_n\}) = (\textbf{after}\,u)f((\textbf{after}\,u)E_1,\ldots,(\textbf{after}\,u)E_n)$$

and

$$(\textbf{inn}\,c)(f(E_1,\ldots,E_n)) = (\textbf{inn}\,c)f((\textbf{inn}\,c)E_1,\ldots,(\textbf{inn}\,c)E_n)$$

The second, trivial, result is that primitive recursive functions are state independent.

Theorem 3.2. *Distributive law for non-modal functions. If f is a primitive recursive function then,*

$$(\textbf{after}\,u)f(\vec{x}) = f(\vec{x}) = (\textbf{inn}\,c)f(\vec{x}).$$

A consequence of these two theorems is that many primitive recursive functionals can be factored out of expressions. For example:

$$(\text{after}\,u)\sum f(x) = \sum(\text{after}\,u)f(x).$$

A more interesting theorem allows *inversion* of expressions containing nested modal modifiers.

Theorem 3.3. *Inversion.*

$$(\textbf{after}\,u)(\textbf{inn}\,c)f(\vec{x}) = (\textbf{inn}\,c)(\textbf{after effect}(u,c))f(\vec{x})$$

This theorem states that the result of advancing the entire product state machine to the state reached by following u, and then evaluating $f(\vec{x})$ in the context of component c, is equal to the result of simply advancing component c by the sequence induced by u, and then evaluating $f(x)$.

The "compositional proof rules" that are the staple of process based formal methods are not necessary in m.p.r. arithmetic.

3.2. Modal Grammars . The most unconventional aspect of our proof theory, is the concept of syntactic proofs of finite state realizability. We will show that there are certain syntactic restrictions, so that if f meets the restrictions, then $f(\vec{m})$ defines exactly one (minimal) product form automaton. Thus, the construction of f is a proof that the system specified by $f(\vec{m})$ is realizable as a finite state machine. Functions of this restricted form are called *modal grammars*. A modal grammer is a symbolic, and highly compact definition of a family of deterministic finite state machines. Modal grammars provide both a formal basis for exact specifications, and a style of specification that we find intuitive. In this abstract, we do not have space to define modal grammars fully. But a modal grammar is a boolean function defined as a conjunction of clauses which describe the alphabet, component set, feedback, and enabling rules of a product form transducer. It follows from the existence of modal grammars and theorem 2.1 that there is a recursive procedure for deciding $\Box \mathcal{G} \to t$ for closed modal grammar $\mathcal{G}$ and arbitrary closed term t.

3.3. Temporal logic style functionals. We let $(\textbf{sometimes}\,u)f(n)$ be true iff $f(n) > 0$ *sometime* during the computation of u. Similarly, we let $(\textbf{always}\,u)f(n)$ be true iff $f(n) > 0$ every state visited by u. We write $v \prec u$ to denote that v is a prefix of u, i.e., $v \prec u \leftrightarrow (\exists z)u = vz$. Note that $u \prec u$ and $\langle\rangle \prec u$.

$$(\textbf{sometimes}\,u)f(\vec{x}) \overset{def}{=} (\exists v \prec u)(\textbf{after}\,v)f(\vec{x}) > 0$$
$$(\textbf{always}\,u)f(\vec{x}) \overset{def}{=} (\forall v \prec u)(\textbf{after}\,v)f(\vec{x}) > 0$$

Let $Paths(i)$ be the set of enabled paths of length i: $Paths(i) = \{u : \textbf{enable}(u) \wedge length(u) = i\}$. We can now state the theorem that makes m.p.r. analogs of temporal logic operators possible.

Theorem 3.4. *There is a total map* $plen : MPR \to MPR$ *so that:*

$$(\exists u)\ (\textbf{after}\,u)f(\vec{x}) > 0$$
$$\to (\exists u \in Paths((plen\ f)(\vec{x})))(\textbf{sometimes}\,u)f(\vec{x})$$
and
$$(\exists u \in Paths((plen\ f)(\vec{x})))(\textbf{always}\,u)f(x) > 0$$
$$\leftrightarrow (\forall i)(\exists u\,in\,Paths(i))(\textbf{always}\,u)f(x)$$

The implications of theorem 3.4 can be illustrated by considering the concepts of eventuality and henceforth. Suppose that we wish to show that $f(x) > 0$ is inevitable (eventual). Thus, we need to show that there is some i so that no enabled path of length i or more keeps $f(x) = 0$. The second part of the theorem tells us that $f(x)$ is inevitable iff there is no enabled path $u \in Paths((plen\ f)(x))$ so that $(\textbf{always}\,u)(f(x) = 0)$. Similarly, $f(x) > 0$ after every enabled path iff there is no enabled path $u \in Paths((plen\ f)(x))$ so that $(\textbf{sometimes}\,u)(f(x) = 0)$. Thus, we can define constructive versions of the temporal logic operators $\Box$ and $\Diamond$ and their existential counterparts. We can also define a m.p.r. analog of the temporal logic operator *until*. In the full paper, we show that the axioms of the UB Branching time temporal logic [BAPM83] are valid in m.p.r. arithmetic under our definitions.

271

REFERENCES

[BAPM83] M. Ben-Ari, A. Pnueli, and Z. Manna. The temporal logic of branching time. *Acta Informatica*, 20, 1983.

[Bou88] R. T. Boute. On the shortcomings of the axiomatic approach as presently used in computer science. In *Compeuro 88 Systems Design: Concepts Methods, and Tools*, 1988.

[CAS83] E. M. Clarke, Emerson A., and A.P. Sistla. Automatic verification of finite-state concurrent systems using temporal logic specifications: A practical approach. In *Proceedings of the 10th Annual Symposium on Principles of Programming Languages*, pages 117–119, 1983.

[COGB86] E.M. Clarke, M.C. O. Grumberb, and Browne. Reasoning about networks with many identical finite state processes. Technical Report cmu-cs-86-155, Carnegie-Mellon University, October 1986.

[EH83] E. A. Emerson and J. Y. Halpern. Sometimes and 'not never' revisited: on branching versus linear time temporal logic. *Journal of the ACM*, 33(1), January 1983.

[GC86] G. Gouda, M. and C. Chang. Proving liveness for networks of communicating finite state machines. *ACM Transactions on Programming Languages and Systems*, 8(1), January 1986.

[Gec86] Ferenc Gecseg. *Products of Automata*. Monographs in Theoretical Computer Science. Springer Verlag, 1986.

[Goo57] R. L. Goodstein. *Recursive Number Theory*. North Holland, Amsterdam, 1957.

[Har84] D. Harel. Statecharts: A visual formalism for complex systems. Technical report, Weizmann Institute, 1984.

[Har87] D. Harel. On the formal semantics of statecharts. In *Proceedings of the Symposium on Logic in Computer Science*, Ithaca, June 1987.

[HU79] John E. Hopcroft and Jeffry D. Ullman. *Introduction to Automata Theory, Languages, and Computation*. Addison-Welsey, Reading MA, 1979.

[Kro87] F. Kroger. *Temporal Logic of Programs*. EATCS Monographs on Theoretical Computer Science. Springer-Verlag, 1987.

[Kur85] Kurshan. Modelling concurrent processes. In *Proc of Symposia in Applied Mathematics.*, 1985.

[Moo64] E.F. Moore, editor. *Sequential Machines: Selected Papers*. Addison-Welsey, Reading MA, 1964.

[MP79] Z. Manna and A. Pnueli. The modal logic of programs. In *Proceedings of the 6th International Colloquium on Automata, Languages, and Programming*, volume 71 of *Lecture Notes in Computer Science*, New York, 1979. Springer-Verlag.

[OW87] J.S. Ostroff and W.M. Wonham. Modelling, specifying, and verifying real-time embedded computer systems. In *Symposium on Real-Time Systems*, Dec 1987.

[Pet67] Rozsa Peter. *Recursive functions*. Academic Press, 1967.

[Smo85] C. Smorynski. *Self-Reference and Modal Logic*. Springer-Verlag, 1985.

Synthesizing Processes and Schedulers from Temporal Specifications*

Howard Wong-Toi and David L. Dill
Department of Computer Science
Stanford University

Abstract

We examine two closely related problems: synthesizing processes to satisfy temporal specifications of reactive systems, and the synthesis of a scheduler to interact with and control a group of processes in order to meet a specification. Processes communicate through shared and distributed variables, either synchronously or asynchronously. In the finite state case, processes and specifications are arbitrary ω-regular languages, and both synthesis problems are solvable in doubly exponential time and space. The framework we present is flexible enough to incorporate dense real time into the model of concurrency, thereby allowing us to study the synthesis of real-time processes and schedulers. Real-time implementability and scheduling are also doubly exponential.

1   Introduction

A specification for a reactive process defines the desired ongoing behavior of the process along with its environment (the other processes with which it is connected). If the implementer of a process does not have control over the behavior of its environment, it may not be possible to meet the specification. For example, if a specification says that "x is always 1", and the environment can assign x the value 2, the specification is not implementable — the process can set x to 1, but cannot prevent it from taking other values.

The fundamental problem is to derive an implementation that always satisfies the specification in the face of adversarial behavior by the environment. Intuitively, an implementation is a winning strategy in the game against the environment. When the specification can be modeled as a finite automaton on infinite words, or a formula from a logic interpreted over infinite words, the synthesis problem is decidable. The decision procedures consist of reductions to the Church *solvability problem*, [Chu63, BL69] which can be solved using automata on infinite trees [Rab72].

Our processes have three different types of variables: shared, read-only and "distributed" (variables that can be read by other processes but written only by one). We solve the problem of process synthesis under both synchronous and asynchronous paral-

---

*Supported by NSF under grant MIP-8858807.

lel execution[1]. These results extend, unify, and simplify previous work [PR89a, PR89b, ALW89]. Next, we solve the problem of synthesizing a *scheduler*, which controls a collection of processes so as to satisfy a specification. For example, a critical region between processors could be enforced by a scheduling strategy instead of explicit synchronization within the processes. We also discuss ways to solve the same problem when there are additional constraints on the scheduler (such as processor availability). Finally, we show how these methods can also be applied to meet general real-time requirements.

For arbitrary untimed ω-regular processes and specifications the implementability problem is doubly exponential in the length of a temporal logical formula for the specification. Scheduling is doubly exponential in the number of processes as well, but only singly exponential in their size. If there is any implementation, the algorithm produces one of at most doubly exponential size. Real-time implementability and scheduling are also doubly exponential.

A more detailed presentation of this material appears in [WD91].

2 Background

In the last two years, several people have discovered the relevance of the solvability problem for determining the implementability of a linear temporal specification. In 1988, Dill considered it as a well-formedness condition on models of asynchronous circuits, but did not consider synthesis *per se* [Dil89]. Later, Pnueli and Rosner, [PR89a, PR89b], solved the problem of synthesizing concurrent reactive processes (in both synchronous and asynchronous models of concurrency) from a linear temporal logic specification. This solution was applicable to systems with only *distributed variables*. Shared variables were not considered.

Abadi, Lamport and Wolper did model shared variables, but their *computers* had the ability to observe *every* change made to the state, including all changes made by the environment [ALW89]. This is inappropriate for asynchronous parallelism, since in reality an implementation can base its actions only on the state changes that it can observe by reading variables. It does not have full observability of the entire system. Our implementations use only information directly inferable from their own actions.

We simplify considerably Pnueli and Rosner's solution for the asynchronous case. Their original solution involved transformations on several different representations of the specification. In our case, all transformations are performed on automata. The key simplifying idea is that the reductions used in solving implementability can be derived from projections that are used in the definitions of parallel composition.

3 Processes

We choose an action-based model of execution as opposed to a state-based approach because a process is always completely aware of its actions, but often ignorant of the

---

[1]By synchronous parallelism, we mean that processes run in "lock-step"; by asynchronous parallelism, we mean that processes run at arbitrary speeds. Elsewhere, these terms are used incompatibly to mean unbuffered and buffered communication between processes.

global state of the system. Our model of a single execution of a process is an infinite sequence of *sets* of actions (the reading and writing of variables). A process is assumed to be acting in an environment of other agents, with the only communication being indirectly through the values of the variables.

Suppose that $\mathcal{P}$ is an infinite set of primitive process names. Then every process P has a *type* describing what basic components it has and the variables it can access. Its type is a quadruple $\mathcal{T}_P = (Procs, \mathcal{D}_P, \mathcal{S}_P, \mathcal{E}_P)$, where $Procs$ is a subset of $\mathcal{P}$, $\mathcal{D}_P$ is a set of (owned) *distributed variables* which it can write and other processes can only read, $\mathcal{S}_P$ is a set of *shared variables* which may be written and read by P and other processes, and $\mathcal{E}_P$ is a set of *external* variables, which are read-only. We omit the subscripts when no confusion arises and abbreviate $\mathcal{D} \cup \mathcal{S} \cup \mathcal{E}$ by $\mathcal{V}$. $Procs$ is used to record which primitive process performs each action; this seems necessary for dealing with shared variables[2]. We assume all variables range over a common domain, Dom. When discussing finite-state processes, we assume Dom is finite.

The process P has a set of *primitive actions*, A_P, consisting of *reads* of the form $read(P_i, x, v)$ and *writes* of the form $write(P_i, x, v)$, where P_i is a primitive agent name in $\mathcal{P}$ and v is some value in the domain of the variable x. A *composite action* is a set of primitive actions. These correspond to primitive actions occurring *simultaneously*. The empty set of actions represents nothing happening and is called *skip*.

A *state*, s, is a function assigning a value to every variable in $\mathcal{V}$. An initial condition is a function mapping from $\mathcal{D}$ to $\mathcal{V}$. Let σ_i be the ith element of σ. An *action run* of P corresponds to a trace of actions that P may perform, in which P's distributed variables change only if P writes to them, but shared and external variables can change arbitrarily. Formally, an action run $a = a_1, a_2, \ldots$ of type $(Proc, \mathcal{D}, \mathcal{S}, \mathcal{E})$ is an infinite sequence of composite actions of P, satisfying the following consistency condition: there exists an infinite sequence of states $s = s_1, s_2, \ldots$ such that

(*read-values*) if $read(-, x, v)$ occurs in a_i, then x must have the value v in s_i,

(*write-values*) for every $x \in \mathcal{D}$, if a $write(-, x, v)$ occurs in a_i, x's value in s_{i+1} must be v' where some $write(-, x, v')$ occurs in a_i, and, if no $write(-, x, v)$ occurs in a_i, then $s_i(x) = s_{i+1}(x)$, and,

(*type*) whenever $write(-, x, v)$ occurs in a, x is in $\mathcal{D} \cup \mathcal{S}$.

Intuitively, if several writes to x occur simultaneously, x arbitrarily takes the value of one of them. An action run is consistent with an initial condition I if there is a satisfying sequence of states starting with state s_1 which agrees with I on $\mathcal{D}$.

The behavior of a process P is a triple $(\mathcal{T}_P, I_P, Runs_P)$, where $\mathcal{T}_P$ is its type, I_P its initial condition and $Runs_P$ (usually denoted $R(P)$) a set of action runs of type $\mathcal{T}_P$ consistent with its I. We sometimes use P as an abbreviation for $R(P)$.

3.1 Synchronous parallelism

The behaviors of two processes can be combined to yield the behavior that results when they run synchronously (in "lock step"). The runs of $P = P_1 \parallel P_2$ are essentially those

---

[2]The usual way of dealing with shared variables is have two labels for actions: π if performed by the process or ε if by the environment[BKP84]. But for scheduling, it helps to record exactly which component executed each action.

runs of the correct type which look like runs to both P_1 and P_2. P is defined only when $Proc_1$ and $Proc_2$ are disjoint, $\mathcal{D}_1$ and $\mathcal{D}_2 \cup \mathcal{S}_2$ are disjoint, and $\mathcal{D}_2$ and $\mathcal{D}_1 \cup \mathcal{S}_1$ are disjoint. Let del_Q be the projection on action runs that deletes all atomic actions made by agents in Q. Formally, $P = (\mathcal{T}_P, I_P, R(P))$ is defined as follows:

$\mathcal{T}_P = (Proc, \mathcal{D}, \mathcal{S}, \mathcal{E})$ where $Proc = Proc_1 \cup Proc_2$, $\mathcal{D} = \mathcal{D}_1 \cup \mathcal{D}_2$, $\mathcal{S} = \mathcal{S}_1 \cup \mathcal{S}_2$ and
$\mathcal{E} = (\mathcal{E}_1 \cup \mathcal{E}_2) - (\mathcal{D} \cup \mathcal{S}) = (\mathcal{E}_1 - (\mathcal{D}_2 \cup \mathcal{S}_2)) \cup (\mathcal{E}_2 - (\mathcal{D}_1 \cup \mathcal{S}_1))$.

I_P is the extension of both I_{P_1} and I_{P_2}, i.e. I_P has domain $\mathcal{D}$ and agrees with I_{P_1} on $\mathcal{D}_1$
and I_{P_2} on $\mathcal{D}_2$.

$R(P) = del_{P_2}^{-1}(R(P_1)) \cap del_{P_1}^{-1}(R(P_2))$
$= \{r \mid r \text{ is a consistent run of type } \mathcal{T}_P, del_{P_2}(r) \in R(P_1) \text{ and } del_{P_1}(r) \in R(P_2)\}$.

Because del_{P_2} removes all P_2 actions, it deletes everything unobservable to P_1, leaving exactly what P_1 observes. Thus $del_{P_2}(r)$ is in $R(P_1)$ precisely when P_1's view of the global system run r is a run of P_1.

3.2 Generalized parallelism

A run is in the parallel composition of P_1 and P_2 when it is observed as a run by each process. Process observability is represented by functions f_{P_i} mapping consistent runs of type $\mathcal{T}_{P_1 \| P_2}$ to sets of consistent runs of type $\mathcal{T}_{P_i}{}^3$. In synchronous parallelism, we had del_{P_2} for f_{P_1} and del_{P_1} for f_{P_2}.

$R(P_1 \| P_2) = f_{P_1}^{-1}(R(P_1)) \cap f_{P_2}^{-1}(R(P_2))$
$= \{r \mid r \text{ is a consistent run of type } \mathcal{T}_{P_1 \| P_2}, f_{P_1}(r) \subseteq R(P_1) \text{ and } f_{P_2}(r) \subseteq R(P_2)\}$

3.3 Asynchronous parallelism

We often want the behaviors of processes that run in parallel *asynchronously*. Inserting an arbitrary (finite) number of *skip* actions between the non-*skip* actions of a run of a process allows other processes to perform as many or as few actions as they like between the process's non-trivial actions. The *delskip* operation removes the *skip* actions of P from an action run. The *stuttering closure* of an action run a is $stut(a) = \{a' \mid delskip(a') = delskip(a)\} = delskip^{-1}(delskip(a))$, see [ALW89]. This definition is extended to sets of action runs. For convenience, we use $R_{AS}(P)$ to refer to $stut(R(P))$, the asynchronous runs, or asynchronous closure of P. Asynchronous parallelism, $\|_{AS}$, is defined as synchronous parallelism on the asynchronous closure of the processes. That is, $R(P_1 \|_{AS} P_2) = del_{P_2}^{-1}(R_{AS}(P_1)) \cap del_{P_1}^{-1}(R_{AS}(P_2))$. This is a special case of generalized parallelism, with $delskip^{-1} \circ delskip \circ del_{P_2}$ for f_{P_1}.

4 Implementability

A reactive system consists of processes which communicate among one another and that are intended to run forever. Suppose P is a reactive process where the only other agent is a completely unpredictable environment over which it has no control. The interaction between the implementation and the environment can be viewed as a game. Each player

---
[3] It may be that P_i's view of a run is not uniquely determined.

takes a turn extending a computation. The implementation wins the game if the resulting computation satisfies the specification. Moves in the game are actions, the *reads* and *writes* of variables. If the process always wins the game, it is an implementation of the specification.

A *strategy* is a special kind of process which decides its next *move* from its history of actions. Its moves are *read* requests of *individual* variables, and *write* requests of individual variables *with* the values to be written. Notice that a strategy is restricted to accessing only one variable at a time. Formally, a strategy S of type $(\{S\}, \mathcal{D}, \mathcal{S}, \mathcal{E})$ is a function $g_S : As^* \to Moves_S$, where $Moves_S = \{read(var) \mid var \in \mathcal{S} \cup \mathcal{E}\} \cup \{write(var, value) \mid var \in \mathcal{D} \cup \mathcal{S}, value \in Dom\}$. Because the adversarial environment can choose to return any value in response to a *read* action by the strategy, $R(S)$ is the set of *all* sequences of actions $a = a_1, a_2, \ldots$ such that

1. $a_{i+1} = read(S, var, value)$ for some $value \in Dom$ whenever $g_S(a_1, a_2, \ldots, a_i) = read(var)$, and

2. $a_{i+1} = write(S, var, value)$ whenever $g_S(a_1, a_2, \ldots, a_i) = write(var, value)$.

The strategy plays against an environment E of type $(\{E\}, \mathcal{E}, \mathcal{S}, \mathcal{D})$ whose actions runs are simply all the action runs for its type. It is an implementation of the specification *Spec* if every run of the combined system is in the specification, i.e. $R(S \parallel E) \subseteq Spec$. In this case it is a winning strategy for the game.

4.1 The Read-Write Game

Most of the games in this paper are games of partial information. The implementation cannot directly observe the entire system; the only information it knows about the environment has been inferred from reading the environment's variables. A solution to the implementability problem proceeds in two steps: reducing the game of partial information to one of complete information, and then solving the game of complete information, the *read-write game*. Runs in this game are traces of strategy actions only, and therefore reflect exactly what the strategy observes. A read-write game is stated as: Given a set of action runs W of type $(\{S\}, \mathcal{D}, \mathcal{S}, \mathcal{E})$, is there a strategy of the same type whose action runs are all in W, i.e. $R(S) \subseteq W$?

If the problem is finite state (the behaviors of the processes and the specification can be represented as ω-regular sets), the read-write game is decidable. This result follows almost immediately from the decidability of the Church-Büchi solvability problem, [BL69, Rab72].

Theorem 1 *If A is a nondeterministic Büchi automaton (NBA) defining Spec, a set of action runs for S over k variables, and A has n states, then the read-write game for Spec is decidable in $2^{O[(kn)^2 \cdot log(kn)]}$ time. If there is a winning strategy for Spec, the algorithm produces one of size $2^{O[(kn)^2 \cdot log(kn)]}$.*

Proof: The solution is a variant on the algorithm for "synchronous implementability" in [PR89a]. We construct from A a tree automaton that accepts exactly the trees that correspond to winning strategies. Hence to solve the game, we need only test for emptiness of the tree automaton [EJ88]. □

4.2 Cores

It is usual for a process to have only partial observability of the entire system; it cannot observe all the environment's actions. Suppose the environment owns the variables x_1 and x_2. While the process is reading x_1, the environment may be changing the value of x_2. In an asynchronous system the environment may even change x_2 an arbitrary number of times without the process noticing. Suppose two global system runs, r_1 and r_2, both look like the action run r to the implementation. The implementation only observes r and cannot distinguish whether r_1 or r_2 actually occurred. Thus a strategy or implementation should only generate action runs for which *all* corresponding global runs lie in the specification.

We assume the strategy has observability function f_S. The core of a global specification *Spec*, *core(Spec)*, is the set of (local) strategy action runs whose inverse images under f_S are all in the specification, i.e. $core(Spec) = \{a \mid f_S^{-1}(a) \subseteq Spec\}$.

Lemma 1 *Let Spec a global specification and S be a strategy, interacting with the environment under observability function f_S. The following are equivalent:*
1) S is an implementation of Spec,
2) $R(S \parallel E) \subseteq Spec$,
3) $f_S^{-1}(R(S)) \subseteq Spec$,
4) $R(S) \subseteq core(Spec)$,
5) S wins the read-write game over core(Spec). □

Therefore implementability can be constructively reduced to the read-write game, provided we have an algorithm for finding the core of a specification. Since $core(Spec) = \{r \mid f_S^{-1}(r) \subseteq Spec\}$, its complement is $\{r \mid f_S^{-1}(r) \cap \overline{Spec} \neq \emptyset\}$. Thus a run r is in the complement of the core if it has a preimage under f_S in $\overline{Spec}$. Hence $\overline{core(Spec)} = f_S(\overline{Spec})$. The core can be found by complementation and substitution.

4.3 Synchronous Implementability

Until now, we have not considered any particular representation of the specification. We want to express ω-regular properties of variables ranging over *finite non-boolean* domains. Rather than defining a temporal logic with finite-valued variables, for convenience we express properties in *propositional* temporal logic (PTL). We encode each bit-value of a variable as a separate proposition.

Lemma 2 *Let ϕ be a PTL formula for Spec. Then there is an algorithm to find a deterministic Rabin automaton A accepting the $core_S(Spec)$.*

Proof: We proceed in three steps. From ϕ construct a NBA B for $\overline{Spec}$, ([WVS83]). Find a NBA B' for $\overline{core_S(Spec)}$, (apply the homomorphism $f_s = del_E$). Complement and determinize B' to get the desired Rabin automaton A ([EJ89]). □

Theorem 2 *Let ϕ be a PTL formula of length n_0 for the property Spec. There is a $2^{O[(k \cdot |Dom|)^3 \cdot 2^{(3n_0)}]}$ algorithm for testing the synchronous implementability of Spec. If Spec is implementable, it yields a $2^{O[(k \cdot |Dom|)^3 \cdot 2^{(3n_0)}]}$ folded representation of a strategy for Spec.*

Proof: Because of Lemma 1, we can follow the algorithm of Lemma 2, then solve the read-write game. The complexity result follows from analysis of the tableau procedure of [WVS83], the simultaneous complementation and determinization algorithm of [EJ89] and the test for emptiness of [EJ88]. □

4.4 Asynchronous Implementability

The analogous results from the last section all hold; the observability function is merely changed from del_E to $f_{AS} = delskip^{-1} \circ delskip \circ del_E$. Performing the substitution for $delskip$ is similar to way ϵ-transitions are removed from finite state automata.

Theorem 3 *Let ϕ be a PTL formula of length n_0, defining the set Spec . The asynchronous implementability of Spec is solvable with complexity $2^{O[(k \cdot |Dom|)^3 \cdot 2^{(3n_0)}]}$.* □

5 Scheduling

The ideas in the previous section can be used to find a *scheduler*, which is a process designed to provide external coordination for other processes. Such a situation may arise in the fields of distributed computing, networking, hardware design and operating systems. For example, we may be given information about the processes to be run and asked to schedule them to ensure mutual exclusion of their critical sections or guarantee some form of liveness. The scheduling problem is stated as follows: For processes $P_1, \ldots, P_n$ and specification $Spec$, is there a strategy S such that $P_1 \parallel \cdots \parallel P_n \parallel S$ satisfies $Spec$?

Let $P_1 \parallel \cdots \parallel P_n$ be collectively called P. We assume $P \parallel S$ is a system with no external variables. A truly external environment can always be modeled as another primitive process within P. By definition, $R(P \parallel S) = f_P^{-1}(R(P)) \cap f_S^{-1}(R(S))$. Thus, $P \parallel S$ satisfies $Spec$ iff $f_S^{-1}(R(S)) \subseteq \overline{f_P^{-1}(R(P))} \cup Spec$. This problem is essentially implementability with a modified specification, $Spec' = \overline{f_P^{-1}(R(P))} \cup Spec$, (see Lemma 1). Intuitively, $Spec'$ asserts that if the environment behaves like the processes we need to schedule, then the original specification $Spec$ must hold.

5.1 The Finite State Case

For both synchronous and asynchronous parallelism, the processes can be effectively composed and $Spec'$ computed, which implies that the scheduling problem is decidable. The complexity is the same as that for implementability except for an extra multiplicative factor of p^n for the composition of the processes.

Theorem 4 (Scheduling) *Assume the system has k variables over a finite domain Dom. Let $\{P_1, \ldots, P_n\}$ be a set of processes, where each P_i is defined by a NBA A_i of size $\leq p$. Let Spec, a property of action runs with agents $P_1, \ldots, P_n$ and S, be defined by a PTL formula of length n_0. Then there is a $2^{O[(k \cdot |Dom| \cdot p^n)^3 \cdot 2^{(3n_0)}]}$ algorithm to solve the scheduling problem, giving a $2^{O[(k \cdot |Dom| \cdot p^n)^3 \cdot 2^{(3n_0)}]}$ state scheduler, if any scheduler exists.* □

6 Transformations on specifications

Most specifications are only expected to be met when the processes or the environment behave in a certain way. For example, a mutual exclusion program may only guarantee progress for each process requesting entry to its critical section if every process has a terminating critical section, $Term\_Crit \Rightarrow Prog$. An environmental assumption can be handled simply by adding it as an antecedent to the specification.

Additional requirements on the scheduler can be added to specifications. For example, suppose we have 10 processes to schedule, but only 5 processors available. Then we must disallow any scheduler that permits more than 5 processes to execute at any time. To handle this, the scheduler needs substantial control over the processes. One way to achieve this control is to add a new boolean *signal variable* $\{go_i\}$ for each process P_i. These new variables are owned by the scheduler. The process constraints C_P would assert that no P_i ever makes a non-*skip* action when the signal go_i is false. We then add to the scheduler the constraint "no more than 5 go_i signals are true at any time".

A constraint on the scheduler C_{sched} can be enforced by changing the specification $Spec$ to $C_{sched} \cap Spec$. The same technique can be used for more general resource constraints, for example, "processes 5 and 6 can never run on processors A and B at the same time."

7 Real-Time Processes and Specifications

Synchronous parallelism requires all processes to advance at the same time, while asynchronous parallelism is completely speed independent. It is often the case that processes are not interacting synchronously, but nor are they completely asynchronous — something is known about timing in the system. Knowledge of the relative timing of events may make it easier to implement a specification.

We consider *dense time*, where events may occur at arbitrarily close times; time is interpreted over the nonnegative real numbers, R. The execution of concurrent real-time processes is modeled by *timed traces* [AD90], which are infinite sequences together with times at which each event occurs. We use the *timed Büchi automata (TBA)* of [AD90] to model specifications and processes.

7.1 Real-Time Implementability

The methodology for solving the implementability and scheduling problems for untimed processes and specifications also works when timing is introduced. Using the techniques of section 6, timing assumptions on the environment and timing requirements on the implementation may all be written into the specification. Thus a strategy has no timing information itself and may be synthesized from the core of the specification, exactly as above for the untimed case. To solve real-time implementability then, we must show how to derive the core from a timed specification.

Timed processes are the natural extension of untimed processes, in that runs are now *timed* traces. Let $Untime$ be the function that removes the timing information from a timed trace. The observability function for P_1 in $P_1 \parallel_t P_2$ is given by $f_{P_1} = Untime \circ delskip^{-1} \circ delskip \circ del_{P_2}$. A strategy S is a timed implementation of $Spec$ iff $Untime^{-1}(S) \parallel_t Untime^{-1}(E) \subseteq Spec$. It is clear that S implements $Spec$ if it wins the read-write game over $core_t(Spec) = \{r \mid f_S^{-1}(r) \subseteq Spec\}$.

Lemma 3 *Given a TBA $A' = \langle \Sigma, S, s_0, F, C, \delta \rangle$ for $\overline{Spec}$, there is an $O(|A|^3)$ algorithm to find a NBA B for $\overline{core(Spec)}$. B has size $O(|C|! \cdot |\Sigma| \cdot |S|^2 \cdot 2^d)$, where d is the number* *of bits in the binary encoding of A''s timing constants.* □

Because TBAs are not closed under complementation, we allow only specifications given as *deterministic* timed automata (DTA), which have at most one run for any timed trace. However DTA with Büchi acceptance conditions are not as expressive as DTA with Muller, Streett and Rabin acceptance condition. All these deterministic automata can be complemented, yielding TBA with polynomially many states[4]. Timing conditions on the implementation and the environment can also be included as antecedents to the original specification. An implementation of $Spec' = \overline{T_{imp} \cap T_e} \cup Spec$ implements $Spec$ under the timing assumptions T_{imp} and T_e.

Theorem 5 (Real-time Implementability) *Let A_{imp} and A_e be TBA for timing requirements T_{imp} and T_e on the implementation and the environment respectively. Let A_{Spec} be a DTA[4] for the specification. Then the real-time implementability problem is solvable in time exponential in the number of variables and number of states of the automata but doubly exponential in the number of bits to encode the timing constants.* □

7.2 Real-time Scheduling

The problem of scheduling a collection of activities to meet hard real-time deadlines has been studied extensively [CS88]. Previous work in this area considers scheduling a set of tasks (which may or may not be known in advance) to meet completion deadlines, usually on a single processor. These problems are usually at least NP-hard.

Here we are scheduling *reactive* processes which communicate and coordinate among themselves. This problem is not easily modelled as a collection of tasks to be scheduled. Furthermore, the expressiveness of ω-regular languages enables us to handle a far wider range of timing properties than the simple meeting of completion deadlines. For example, restrictions can be enforced on the ordering of completion times and the differences between completion times. Our specifications include, for example, "if y is set to 1 sufficiently often, it will eventually always be reset to 0 within 2 seconds".

We deal only with static scheduling, where the characteristics of the processes are known beforehand. Specifications must be given as *deterministic* TBA, but the processes to be scheduled and the timing constraints on the scheduler and the environment may be arbitrary TBA. Real-time scheduling then reduces to real-time implementability over the property "if the environment behaves like the processes, and all components in the system satisfy their timing constraints, then the specification is met".

Theorem 6 (Real-time Scheduling) *Let $\{P_1, \ldots, P_n\}$ be a set of processes with each P_i defined by the TBA A_i of size $\leq p$. Let $Spec$ be given by a DTBA A_{Spec} of size n_{Spec}, and let T_{imp} and T_e be timing constraints on the implementation and the environment defined by TBA A_{imp} and A_e of sizes n_{imp} and n_e respectively. Let c be the total number of clocks and d the number of bits in the encoding of all the timing constants. The complexity of real-time scheduling is $2^{O(\lfloor c! \cdot p^n \cdot k \cdot |Dom|(n_{imp} \cdot n_e \cdot n_{Spec})^2 \cdot 2^d \rfloor^3)}$.* □

---

[4] The number of acceptance pairs of a Rabin automaton must be logarithmic in the number of states.

8 Conclusion

The idea of finding strategies for games is not only applicable to the specific problem of synthesizing an implementation from a temporal specification, but can also be applied to other problems such as scheduling. Although the doubly exponential algorithms presented here may prove to be useful for very small systems, easier special cases of the problem will have to be discovered for real practicality. The problem of synthesizing a scheduler to meet a specification closely resembles the *supervisory control problem* in the study of discrete event systems, which historically has fallen in the domain of control theory. The relations between these two problems should be studied in greater detail.

Acknowledgements

We thank Amir Pnueli for a helpful discussion, and Elizabeth Wolf for careful reading of a draft of the paper.

References

[AD90] R. Alur and D. Dill, "Automata for modeling real-time systems", ICALP 1990.

[ALW89] M. Abadi, L. Lamport, P. Wolper, "Realizable and unrealizable specifications of reactive systems", *International Colloquium on Automata, Languages, and Programming*, Lecture Notes in Computer Science, 1989, Springer-Verlag

[BKP84] H. Barringer, R. Kuiper and A. Pnueli, "Now you can compose temporal logic specifications", *Proceedings of the ACM Symposium on Theory of Computing*, 1984, pp. 51–63.

[BL69] J. R. Büchi and L. H. Landweber, "Solving sequential conditions by finite-state strategies", Transactions of the American Mathematical Society, 138, 1969, pp. 295–311.

[CS88] S-C. Cheng and J. A. Stankovic, "Scheduling algorithms for hard real-time systems – a brief survey", in *Hard Real-Time Systems*, IEEE Press, 1988, pp. 150–173.

[Chu63] A. Church, "Logic, arithmetic, and automata", in *Proceedings of the International Congress of Mathematicians, 1962*, Institut Mittag-Leffler, 1963, pp. 23–35.

[Dil89] D.L. Dill, "Trace Theory for Automatic Hierarchical Verification of Speed-Independent Circuits", MIT Press, 1989.

[EJ88] E.A. Emerson, C.S. Jutla, "The complexity of tree automata and logics of programs", *Proc. of the 29th IEEE Symp. on Foundations of Computer Science*, 1988, pp. 328–337.

[EJ89] E.A. Emerson, C.S. Jutla, "On simultaneously determinizing and complementing ω-automata", *Pro. of the Symp. on Logic in Computer Science*, 1989, pp. 333–342.

[PR89a] A. Pnueli, R. Rosner, "On the synthesis of a reactive module", *Proc. 16th ACM Symp. Principles of Programming Languages*, 1989, pp. 179–190.

[PR89b] A. Pnueli, R. Rosner, "On the synthesis of an asynchronous reactive module", *International Colloquium on Automata, Languages, and Programming*, Lecture Notes in Computer Science Vol 372, 1989, Springer-Verlag

[Rab72] M. O. Rabin, "Automata on infinite objects and church's problem", *Regional Conference Series in Mathematics*, Vol. 13, American Mathematical Society, 1972.

[WVS83] P. Wolper, M.Y. Vardi, A.P. Sistla, "Reasoning about infinite computation paths", *Proc. of the 24th IEEE Symp. on Foundations of Computer Science*, 1983, pp. 185–194.

[WD91] H. Wong-Toi and D.L. Dill, "Synthesizing processes and schedulers from temporal specifications", *Computer-Aided Verification (Proc. CAV90 Workshop)*, DIMACS Series in Discrete Mathematics and Theoretical Computer Science Vol. 3 (American Mathematical Society, 1991).

TASK-DRIVEN SUPERVISORY CONTROL
OF DISCRETE EVENT SYSTEMS

C.H. Golaszewski and R.P. Kurshan

AT&T Bell Laboratories, Murray Hill, NJ 07974

Abstract

The supervisory control framework formulated by Ramadge and Wonham is extended to allow the synthesis of supervisors which control a given system to perform an arbitrary ω-regular task specified by a nondeterministic Büchi automaton. To this end, the supervisory control paradigm is applied to R.P. Kurshan's L-processes which provide a convenient model for nonterminating discrete event system behaviors. Necessary and sufficient conditions for the existence of supervisors are derived and the synthesis of supervisors is discussed.

1. Introduction

One of the goals of supervisory control for discrete event systems is the synthesis of supervisors which guarantee that the closed-loop (or supervised) system performs certain prescribed tasks. In this paper a task is assumed to be specified by an ω-regular language defined by a Büchi automaton. The given discrete event system is not constrained to be completely observable, i.e., it may be required to synthesize a supervisor for the system which cannot observe all system events and which may not possess complete state information about the system.

To motivate our study, consider a communication system consisting of a sender A, a receiver B and an unreliable channel C which nondeterministically loses packets. A standard problem is to design a supervisor (communication protocol) that ensures that a packet sent by A is eventually received by B, under the assumption that the channel eventually delivers some message which was transmitted by A. The channel can delay packets arbitrarily long, the loss of a message is unobservable and all components operate asynchronously. Thus, based on the past observations of the system trajectory it cannot be decided whether a packet which was not received was actually lost or will be received at some point in the future. Consequently, a supervisor which is to guarantee packet delivery must at some point cause retransmission of that message, irrespective of the event of packet loss. The timing of this retransmission is not critical, and for efficiency may be adjusted with consideration of system component delay. Therefore, it is natural to model the supervisory event which causes retransmission as a nondeterministic control action. However, in order for the supervisor to fulfill its required task, it will often be required to satisfy certain obligations eventually. These considerations lead in a natural way to the definition of the supervisor as a nondeterministic automaton with acceptance conditions for infinite sequences.

The major new contribution of this paper is a synthesis procedure for Büchi supervisors. These can be understood as supervisors that for certain states of the system may choose nondeterministically between several control actions. The choice is not forever arbitrary, however, as we require that eventually some particular sequence of control actions is selected; this ultimate occurrence of sequences of control actions is characterized by automata acceptance conditions. Allowing nondeterminism in the supervisor permits to defer certain timing and sequencing decisions to a lower level of system design, while at the same time providing "place holders" during the high-level supervisor synthesis. It was already illustrated with the previous example that such an extension is needed for the synthesis of communication protocols. Lacking this feature, the RW-Model cannot be applied directly to the synthesis of, say, communication protocols as they are conventionally modeled. For example, Cieslak et al. had to include the structure of the alternating bit protocol in the model of the open-loop system in order to derive their solution in [2]. Clearly, a general synthesis procedure should not require much knowledge about the solution. This suggests that the basic RW-Model lacks certain features that are necessary to synthesize a supervisor for this class of systems.

The present investigation is by no means the first attempt to extend the RW-Model so that it can be applied to a larger class of supervisory control problems. For example, recently it was realized by Thistle and Wonham, and Ramadge that one important aspect absent from the basic model is the notion of periodic and asymptotic behaviors of discrete event systems. This led to proposals for extensions of the RW-Model to include infinite sequences ([8,10]). These approaches are very similar in spirit in that they both extend the notion of string controllability of the basic model to the prefix set of (infinite) sequences. The extended model was used to pose and solve several interesting supervisory control problems ([3,8]). However, these extensions do not address the problem of achieving a desired sequential closed-loop behavior which is neither realizable ([10]) nor topologically closed ([8]). Simply stated, this framework applies only to the case of transition structures which are structurally incapable of generating infinite sequences incompatible with the given task. We show in this paper that this is too restrictive in the context of communication protocols. Instead, by defining the supervisor as a nondeterministic L-process, we may place constraints on the eventuality of the closed-loop behavior to exclude arbitrarily long but finite periods of behavior, which, if continued forever, would constitute a failure of the system.

Some of our research overlaps with work done by Brave et al. in [1] and Özveren et al. in [6]; our independently derived results differ from the work of both of these groups, though. Their research is focused on the synthesis of deterministic supervisors which guarantee that the closed-loop system visits a given subset of states infinitely often. In contrast, we use a formal framework for the infinite behavior of discrete event systems, which includes Büchi supervisors and L-processes, to pose and solve supervisory synthesis problems for ω-regular tasks.

The organization of the paper is as follows. In Section 2 the discrete event system model is reviewed. Section 3 introduces the control-theoretic framework used in the paper. Section 4 discusses the existence and synthesis of supervisors for fully observable systems and in Section 5 these results are extended to the general case.

2. L-Processes

In this paper, the logical behavior of discrete event systems is modeled by *L-processes*. The reason is that L-processes are a class of automata with a simple (polynomial) product and the property that the language of the product system is the intersection of the languages of the subsystems. We will review only the basic definitions; for a detailed presentation we refer to [4].

Let $L = \langle \mathbf{L}, +, *, \sim, 0, 1 \rangle$ be a Boolean algebra over the set $\mathbf{L}$. A Boolean algebra L admits a partial order $\leq$ defined by $x \leq y$ if and only if $x * y = x$. An *atom* of L is a minimal element with respect to this order.

Let $L_1, \ldots, L_k$ be subalgebras of L. Define their product

$$\prod_{i=1}^{k} L_i = \left\{ \sum_{j \in J} x_{1j} * \ldots * x_{kj} \mid x_{ij} \in L_i \text{ for } i = 1, \ldots, k, j \in J \text{ and } J \text{ is finite} \right\}$$

Clearly, $\prod L_i$ is a subalgebra of L. $L_1, \ldots, L_k$ are *independent* if

$$0 \neq x_i \in L_i \text{ for } i = 1, \ldots, k \Rightarrow x_1 * \ldots * x_k \neq 0.$$

Subalgebras will be used later to describe the concurrent operation of several discrete event processes.

Let V be a non-empty set. The map $M : V \times V \to L$ is an *L-matrix* with state space $V(M)$; the elements of $V(M)$ are the *states* of M. It is *deterministic* if for all $u, v, w \in V(M)$, $v \neq w \Rightarrow M(u, v) * M(u, w) = 0$, and *complete* if for all $v \in V(M)$

$$\sum_{w \in V(M)} M(v, w) = 1.$$

Let M and N be L-matrices over the same Boolean algebra L. Their *tensor product* $M \otimes N$ is defined by

$$(M \otimes N)((v, w), (v', w')) = M(v, v') * N(w, w').$$

Given an L-matrix M, a sequence $s \in L^\omega$ and a sequence of states $w \in V(M)^\omega$, we call w a *state trajectory* of s starting at q if $w_0 = q$ and $s_i * M(w_i, w_{i+1}) \neq 0$ for all i.

An *L-automaton* is a quadruple $\Gamma = (M_\Gamma, I(\Gamma), R(\Gamma), Z(\Gamma))$, where M_Γ, the *state transition matrix*, is a complete L-matrix, $\emptyset \neq I(\Gamma) \subseteq V(M_\Gamma)$ are the *initial states*, $U(\Gamma) \subseteq V(M_\Gamma)$ are the *recurring states* (Büchi states), and $Z(\Gamma) \subseteq 2^{V(M_\Gamma)}$ is the *cycle set*.

Denote the set of all sequences of atoms of L by $S(L)^\omega$. A sequence $s \in S(L)^\omega$ is Γ-*cyclic* if for some $N > 0$ and some $C \in Z(\Gamma)$, the corresponding state trajectory w satisfies $w_i \in C$ for all $i > N$; it is Γ-*recurring* if $w_i \in U(\Gamma)$ for infinitely many i. A sequence $s \in S(L)^\omega$ is a *tape* of Γ if it is Γ-cyclic or Γ-recurring. The *behavior* (or language) $\mathcal{L}(\Gamma)$ accepted by Γ is the set of tapes of Γ.

An *L-process* G is a quintuple $G = (M_G, S_G, I(G), U(G), Z(G))$, where $M_G, I(G)$, $U(G)$, and $Z(G)$ are as in the definition of an L-automaton in the previous section. However, M_G is not required to be complete. Finally, S_G is the nondeterministic *selection function* $S_G : V(G) \to 2^L$. We restrict attention to the case where $S_G(q)$ is an atom for every $q \in V(G)$.

A selection $x \in S_G(q)$ enables the state transition $q \to p$ if $x * M_G(q, p) \neq 0$. We also require that

$$\sum_{p \in V(G)} M_G(q,p) \leq \sum_{x \in S_G(q)} x. \qquad (2.1)$$

If equality holds in (2.1) for all $q \in V(G)$, and if $0 \notin S_G(q)$, then G is said to be *lockup-free*. (This is similar to the notion of nonblocking in [8].)

In contrast to the definition of a tape for an L-automaton, an element s of $S(L)^\omega$ is a tape for G if and only if it is not G-cyclic and not G-recurring. If $s \in S(L)^\omega$ is not a tape for G because it is G-cyclic or G-recurring, then s is said to be *excepted* from the language of G. The *behavior* (or language) $\mathcal{L}(G)$ generated by an L-process G is the set of all tapes of G.

The concurrent operation of several discrete event processes is modeled as follows: each individual component system is represented by an L_i-process G_i over a Boolean algebra L_i. The algebras L_i can be interpreted as subalgebras of a Boolean algebra L which is given by the product $L = \prod_i L_i$. This procedure associates with each L_i-process a corresponding L-process.

Given L-processes $G_1, \ldots, G_k$ define their *synchronous product* to be

$$G = \bigotimes_{i=1}^{k} G_i = (M_G, S_G, I(G), U(G), Z(G)) =$$

$$= \left(\bigotimes_{i=1}^{k} M_{G_i}, \prod_{i=1}^{k} S_{G_i}, I(G_1) \times \ldots \times I(G_k), \bigcup_{i=1}^{k} \pi_{G_i}^{-1} U(G_i), \bigcup_{i=1}^{k} \pi_{G_i}^{-1} Z(G_i) \right),$$

where π_G denotes the projection $G \times H \to G$ and

$$\left(\prod_{i=1}^{k} S_{G_i} \right)(q_1, \ldots, q_k) = \{x_1 * \ldots * x_k | x_i \in S_{G_i}(q_i), i = 1, \ldots, k\}.$$

The model described so far applies to the concurrent operation of synchronous discrete event processes, i.e., state transitions occur simultaneously in the component processes. To model partially and completely asynchronous selections we introduce two additional selections called *wait* and *pause*, respectively. A process can select *pause* independently of the selections of the other processes and, for example, use *pause* to self-loop for an undetermined amount of "time." The *wait* selection is used to model the case where the possible state transitions of a process depend on the selections of other processes. Until this process can proceed it remains at the current state (in a self-loop) by selecting *wait*.

Consider two alphabets Σ_1 and Σ_2, and let $\phi \colon \Sigma_1 \to \Sigma_2$ be an arbitrary map. We can extend ϕ to a *language homomorphism* $\Sigma_1^\omega \to \Sigma_2^\omega$ in the usual fashion by setting $\phi(s) = \phi(s_1)\phi(s_2) \ldots$ for all $s \in \Sigma_1^\omega$.

Let G_1 be an L_1-process over the alphabet Σ_1 and G_2 be an L_2-process over the alphabet Σ_2. The pair $\Phi = (\varphi, \phi)$, with $\varphi \colon V(G_1) \to V(G_2)$ and ϕ a language homomorphism, is said to be a *process homomorphism* if

$$\sigma \in M_1(v, w) \Rightarrow \phi(\sigma) \in M_2(\varphi(v), \varphi(w)). \qquad (2.2)$$

Now let G be an L-process, not necessarily deterministic. Define a binary relation Ψ on $V(G) \times V(G)$ according to

(1) for all $q \in V(G)$, $(q, q) \in \Psi$, and
(2) $(p, q) \in \Psi$ if there exists $(u, v) \in \Psi$ such that $\sigma \in M(u, p)$ and $\sigma \in M(v, q)$ for some $\sigma \in \Sigma$.

A pair of states $(p, q) \in \Psi$ is said to be *indistinguishable*. Note that in general Ψ is not an equivalence relation.

For two relations Ψ_1 and Ψ_2, a partial order is defined in the usual fashion:

$$\Psi_1 \le \Psi_2 \quad \Leftrightarrow \quad ((p, q) \in \Psi_1 \Rightarrow (p, q) \in \Psi_2).$$

Partial or incomplete observations of the tapes of G are modeled by letting the language homomorphism ϕ be a projection onto a subset of the alphabet. In this case, $\ker(\phi) = \{\sigma \in \Sigma : \phi(\sigma) = 1\}$ induces a binary relation on $V(G) \times V(G)$ via (2.2).

3. Tasks and Supervisors

For a given L-process G with behavior $\mathcal{L}(G)$ we can specify its admissible behavior $\mathcal{L}$. In this paper $\mathcal{L}$ is assumed to be an ω-regular language. Thus a task can be represented by a deterministic L-automaton T. A standard verification problem is to decide if $\mathcal{L}(G) \subseteq \mathcal{L}$. Suppose this condition fails to hold, but there exists some L-process C such that $\mathcal{L}(G \otimes C) \subseteq \mathcal{L}$. In situations where G is independent from its environment this cannot happen. However, it is often the case that some (but not necessarily all) of the state transitions of G depend on shared or controllable variables which can be accessed by other processes. Thus, the language $L(G)$ can be restricted to a subset by external *control selections*.

We call the L-process C in (3.1), if it exists, a *Büchi supervisor* (or supervisor) for G with respect to the *task* $\mathcal{L}$. Note that we permit a supervisor to be an L-process with nondeterministic selection function. Hence, certain control selections (or sequences of control selections) can be required to occur eventually. For example, in the model of a communication protocol an eventual control action may represent the fact that the protocol will retransmit a window after a time-out has occurred.

Let $h : V(G) \to 2^{V(C)}$ be the map induced by $G \otimes C$, which for each tape in $\mathcal{L}(G \otimes C)$ assigns to the present state q of G the corresponding state(s) $h(q)$ of C. A state transition $M(q, p)$ of G is *disabled*, i.e., prevented from occurring, if

$$S_G(q) * S_C(h(q)) * M(q, p) = 0;$$

otherwise $M(q, p)$ is *enabled* by C.

Define an equivalence relation Θ on $V(G) \times V(G)$ according to

$$\Theta = \{(q, p) : \quad S_C(h(q)) = S_C(h(p))\}$$

Let $B \subseteq V(G)$ be an arbitrary subset. Say that B is *controlled-invariant* if there exists a supervisor C such that every state-trajectory of the process $G \otimes C$, once it enters B, remains indefinitely inside B. A state $q \in V(G)$ is B-steerable, if there exists a supervisor C such that for all $p \in V(G)$, $S_C(h(q)) * M_G(q,p) = 1$, implies that p is either B-steerable or an element of B.

Now consider two tasks T_1 and T_2 and suppose C_1 and C_2 are two supervisors with

$$\mathcal{L}(G \otimes C_1) = \mathcal{L}_1 \subseteq \mathcal{L}(T_1) \quad \text{and} \quad \mathcal{L}(G \otimes C_2) = \mathcal{L}_2 \subseteq \mathcal{L}(T_2).$$

It is natural to require that supervisors which control the same system do not share control variables. From Proposition 4.9 in [6] it follows that if G, C_1 and C_2 are independent L-processes, i.e., there exist independent subalgebras L_G, L_1 and L_2 such that

$$S_G = \bigcup_{q \in V(G)} S_G(q) \subseteq L_G, \quad S_{C_1} = \bigcup_{q \in V(C_1)} S_{C_1}(q) \subseteq L_1$$

and

$$S_{C_2} = \bigcup_{q \in V(C_2)} S_{C_2}(q) \subseteq L_2,$$

then

$$\mathcal{L}(G \otimes C_1 \otimes C_2) = \mathcal{L}_1 \cap \mathcal{L}_2,$$

and the resulting product system is lockup-free.

Now suppose we are given an L-process G, a task T and a supervisor C satisfying $\mathcal{L}(G \otimes C) \subseteq \mathcal{L}(T)$, i.e., G is controllable relative to T. Furthermore suppose that for some reason (e.g. distributed implementation) we desire to partition T into subtasks $T_1, \ldots, T_n$ with $\mathcal{L}(T) = \mathcal{L}(T_1) \cap \ldots \cap \mathcal{L}(T_n)$. Since there is a supervisor for T there also exists a supervisor C_i which implements task T_i. However, if there are two tasks T_i and T_j such that any pair of supervisors C_i and C_j, which implement the tasks, is dependent, then it is impossible to realize T through the given subtasks $T_1, \ldots, T_n$.

In certain cases it is possible to eliminate the dependency between supervisors by introducing new variables. For example, let $x \in S_{C_1} \cap S_{C_2}$ be a "shared" control variable and introduce two variables $x_1 \notin S_G \cup S_{C_2}$ and $x_2 \notin S_G \cup S_{C_1}$ (possibly by extending L) and set $x = f(x_1, x_2)$ where f is an appropriately chosen Boolean function. At the present time it is not clear when such an f exists; this question is currently under investigation.

4. Existence of Supervisors: Deterministic Case

Consider an L-process G with deterministic resolution and a task T. Our goal is to derive necessary and sufficient conditions for the existence of a lockup-free supervisor C such that $\mathcal{L}(G \otimes C) \subseteq \mathcal{L}(T)$. Intuitively, such a supervisor C exists if all state trajectories which originate in $I(G)$ and reach a void cycle of uncontrollable events, can be disabled by appropriately selected control actions.

Recall that any tape of $\mathcal{L}(G)$ has to be T-cyclic or T-recurring to be a tape for $\mathcal{L}(T)$. Since the two acceptance conditions are independent of each other, they are investigated separately in the next two sections.

4.1 Largest Invariant Recurrent Set

The case of recurrent sets is similar to the problem considered in [1] and [6] in the context of the RW-Model for supervisory control. The following algorithm computes the largest invariant recurrent set via a fixpoint iteration:

Algorithm 1:

$$K_1 := \pi_G(V(G \otimes T));$$
$$H_1 := \pi_G(U(G \otimes T));$$
REPEAT
$$\quad K_{i+1} := \{q \in K_i : q \text{ is } H_i - \text{steerable }\};$$
$$\quad H_{i+1} := H_i \cap K_{i+1};$$
UNTIL $H_{i+1} = H_i$;

Denote the (unique) fixpoint of the algorithm by $H(G) = \lim_{i \to \infty} K_i$. Since we consider only finite state systems it is clear that the algorithm converges in a finite number of steps.

Proposition 4.1. *The set $H(G)$ is controlled invariant.*

Proof. Every state in $H(G)$ is steerable to some other state in $H(G)$. ◇

Proposition 4.2. *Suppose $I(G) \subseteq H(G)$. Then there exists a lockup-free supervisor C such that $\mathcal{L}(G \otimes C) \subseteq \mathcal{L}(T)$.*

Proof. Follows from control invariance and the fact that the states of every cycle which does not intersect $U(G \otimes T)$ are $U(G \otimes T) \cap H(G)$-steerable. ◇

Proposition 4.3. *Suppose $Z(G \otimes T) = \emptyset$. Then there exists a lockup-free supervisor C such that $\mathcal{L}(G \otimes C) \subseteq \mathcal{L}(T)$ if and only if $I(G) \subseteq H(G)$.*

Proof. (if.) This is the statement of Proposition 4.2.

(only if.) Suppose there is a supervisor such that $\mathcal{L}(G \otimes C) \subseteq \mathcal{L}(T)$. Since $Z(G \otimes T) = \emptyset$ this implies that C can control G in such a way that every state trajectory intersects the set $U(G \otimes T)$ infinitely often, irrespective of the initial state. But now it follows from the construction of the set $H(G)$ that $I(G) \subseteq H(G)$. ◇

The last proposition implies that, in the absence of cycle sets, Algorithm 1 produces the largest set of initial states for which a lockup-free supervisor with the desired properties exists.

4.2 Cycle Sets

We proceed in a similar fashion to derive conditions on the set of initial states with respect to the cycle set.

Algorithm 2:

FOR ALL $K \in Z(G \otimes T)$ DO

$\quad$ $A(K) :=$ largest controlled invariant set contained in $\pi_G(K)$;

$\quad$ $H(K) := \{q \in V(G) : q \text{ is } A(K) - \text{steerable }\}$;

The computation of the sets $A(K)$ can be done with a variant of Algorithm 1:

$V_1 := \pi_G(K)$;
$H_1 := \pi_G(K)$;
REPEAT
$\quad$ $V_{i+1} := \{q \in V_i : q \text{ is } H_i - \text{steerable }\}$;
$\quad$ $H_{i+1} := H_i \cap V_{i+1}$;
UNTIL $H_{i+1} = H_i$;

For each $K \in Z(G \otimes T)$ set $A(K) = \lim_{i \to \infty} V_i$, i.e., $A(K)$ is the fixpoint of the above procedure.

The following two propositions summarize the properties of the sets $H(K)$:

Proposition 4.4. *Suppose $I(G) \subseteq \bigcup_{K \in Z(G \otimes T)} H(K)$. Then there exists a lockup-free supervisor C such that $\mathcal{L}(G \otimes C) \subseteq \mathcal{L}(T)$.*

Proof. We can assume without loss that there is only one set $H(K)$. Consider the set $D = H(K) - A(K)$. Since each state in D is $A(K)$-steerable it follows that the states on every cycle in D not contained in $Z(G \otimes T)$ are $A(K)$-steerable. This implies that the required supervisor C exists. $\diamond$

Proposition 4.5. *Suppose $U(G \otimes T) = \emptyset$. Then there exists a lockup-free supervisor C such that $\mathcal{L}(G \otimes C) \subseteq \mathcal{L}(T)$ if and only if $I(G) \subseteq \bigcup_{K \in Z(G \otimes T)} H(K)$.*

Proof. (if.) This is the statement of Proposition 4.4.

(only if.) The argument is identical to the one used for the proof of Proposition 4.3. $\diamond$

4.3 Necessary and Sufficient Conditions

The sufficient conditions of the two previous sections can now be combined to yield necessary and sufficient conditions for the existence of a lockup-free supervisor. Let G be an L-process and T be a task. With the notation from above define

$$B = \bigcup_{K \in Z(G \otimes T)} H(K) \cup H(G).$$

Theorem 4.1. *There exists a lockup-free supervisor C such that $\mathcal{L}(G \otimes C) \subseteq \mathcal{L}(T)$ if and only if $I(G) \subseteq B$.*

Proof. (if). Follows from the previous propositions.

(only if). Suppose $I(G) \nsubseteq B$. Then there exists a sequence $s \in S(G)^\omega$ starting at an initial state in $I(G) - B$ which, independent of any control actions chosen, eventually reaches a cycle which does not intersect $U(G \otimes T)$ and is not contained in $Z(G \otimes T)$. Hence, there cannot exist a supervisor C with the desired property. $\diamond$

4.4 Supervisory Synthesis

Based on the above theorem the synthesis of a supervisor C for the task T is straight-forward. We already presented algorithms (Algorithm 1, Algorithm 2) which compute the sets $H(K)$ for $K \in Z(G \otimes T)$, and $H(G)$. There is a standard procedure in supervisory control to synthesize a lockup-free controller for a controlled invariant set (see [8]). So, for each of the above sets we can compute a supervisor, say C_i. The supervisor $C' = C_1 \otimes C_2 \otimes \ldots$, forces G to remain inside the set of "good" states B, provided $I(G) \subseteq B$. Now extend C' to a supervisor C by excepting all states which correspond to void cycles, i.e., C eventually "interrupts" every cycle in $H(G)$ and $H(K)$, $K \in Z(G \otimes T)$, which does not intersect $U(G \otimes T)$ and is not contained in $Z(G \otimes T)$. Then the language of $G \otimes C$ satisfies $\mathcal{L}(G \otimes C) \subseteq \mathcal{L}(T)$.

5. Existence of Supervisors: Nondeterministic Case

We now consider the case where G is permitted to have nondeterministic resolution; everything else is like in the previous section. In the present setting, L-processes with nondeterministic resolution arise from ones with deterministic resolution and partial observations. As discussed previously, partial observations are modeled by a language homomorphism and give rise to a relation Ψ on the state space of G.

It is clear that Algorithms 1 and 2 can still be used to compute the sets $H(G)$ and $H(K)$. However, the existence of supervisors does no longer follow immediately. For the moment assume that there is a supervisor C with $\mathcal{L}(G \otimes C) \subseteq \mathcal{L}(T)$, which by some unspecified means is capable of exact state observations, independent of Ψ. In this case a necessary and sufficient condition for the existence of an "ordinary" supervisor C' with $\mathcal{L}(G \otimes C') = \mathcal{L}(G \otimes C)$, whose resolution is limited to $\ker(\Psi)$, is that

$$\Psi \leq \Theta. \tag{5.1}$$

This means that whenever two states are indistinguishable the respective control actions must be identical.

We contend that the requirement (5.1) is artificial and too severe for many applications; for example, it is rarely satisfied in communication systems or distributed systems. On the other hand, if (5.1) fails to hold, then clearly a mechanism is needed that allows the supervisor to glean the necessary state information from G. Mathematically, we can always change the system through *cylindrification* and satisfy (5.1). Roughly, this amounts to refining Ψ by adding new selections to transition conditions, i.e., making G_Φ more deterministic. In communication systems cylindrification is achieved through the introduction of sequence numbers or time stamps; more generallly, cylindrification can be interpreted as introducing tags for events. However, to give a generally applicable practical solution appears difficult and remains an open problem for future research.

6. Conclusion

A modified supervisory control framework based on the models proposed by Ramadge and Wonham ([7]), and by Kurshan ([4]) was presented. We introduced the notion of a Büchi supervisor to implement arbitrary ω-regular behaviors by a feedback control. In contrast to work along similar lines by other researchers, these behaviors are allowed to violate the conditions of realizability in [10] and topological closure in [8]. We briefly discussed the decomposition of complex tasks into subtasks, a topic which will receive more attention in a future paper. Conditions for the existence of supervisors that lead to feedback implementations of prespecified ω-regular tasks were derived for the case of L-processes with deterministic state transition matrix.

Future research will address the questions of distributed control, task decomposition and computational aspects.

References

[1] I. Brave and M. Heymann, "On Stabilization of Discrete-Event Processes", *Proceedings of the 28th Conference on Decision and Control*, Tampa, Florida, 1989.

[2] R. Cieslak, C. Desclaux, A. Fawaz, and P. Varaiya, "Supervisory Control of Discrete Event Processes with Partial Observations", *IEEE Transactions on Automatic Control*, Vol. 33(3), pp. 249-260, March 1988.

[3] C.H. Golaszewski and P.J. Ramadge, "Mutual Exclusion Problems for Discrete Event Systems with Shared Events", *Proceedings of the 27th Conference on Decision and Control*, pp. 234-239, Austin, Texas, 1988.

[4] R.P. Kurshan, "Analysis of Discrete Event Coordination", Lecture Notes on Computer Science 430, pp. 414-453, Springer Verlag, 1990.

[5] R.P. Kurshan and C.H. Golaszewski, "An Automaton-Based Approach to the Synthesis of Communication Protocols", in preparation.

[6] C.M. Özveren and A.S. Willsky, "Output Stabilizability of Discrete Event Dynamic Systems", Discrete-Event Dynamical Systems", *Proceedings of the 28th Conference on Decision and Control*, Tampa, Florida, 1989.

[7] P.J. Ramadge and W.M. Wonham, "Supervisory Control of a Class of Discrete-Event Processes", *SIAM J. Control and Optimization*, Vol. 25(1), pp. 206-230, January 1987.

[8] P.J. Ramadge, "Some Tractable Supervisory Control Problems for Discrete Event Systems Modeled by Büchi Automata", *IEEE Transactions on Automatic Control*, Vol. 34(1), pp. 10-19, January 1989.

[9] P.J. Ramadge, "Observability of Discrete Event Systems", *Proceedings of the 25th Conference on Decision and Control*, Athens, Greece, 1986.

[10] J.G. Thistle and W.M. Wonham, "On the Synthesis of Supervisors Subject to ω-Language Specifications", *22nd Annual Conference on Information Sciences and Systems*, Princeton NJ, pp.440-444, March 1988.

A Proof Lattice-Based Technique for Analyzing Liveness of Resource Controllers

Ugo Buy

Department of Electrical Engineering
and Computer Science
University of Illinois
Chicago, Illinois 60610
E-mail: buy@uicbert.eecs.uic.edu

Robert Moll

Department of Computer
and Information Science
University of Massachusetts
Amherst, Massachusetts 01003
E-mail: moll@cs.umass.edu

1 Introduction and Background

The automatic synthesis of programs is one of the most challenging activities in computational logic. Deductive approaches to the synthesis of sequential programs have met with only limited success, in part because of the relative inadequacy of proposed specification languages for such programs. That is, specifications often turn out to be less enlightening and more difficult to construct than the code they specify. This is not the state of affairs with synchronization code. Desired program properties, such as liveness or freedom from deadlock, are relatively easy to include in a specification but are quite difficult to discern in code. Indeed, it is precisely this feature of concurrent programming that makes automated program synthesis an attractive goal. In this paper we discuss an aspect of the automatic synthesis of synchronization code for asynchronous processes.

Our synthesis approach conforms to the following paradigm: 1) first a specification is written in a nonconstructive specification language; 2) that specification is analyzed in an attempt to establish that crucial concurrency properties are respected; and 3) if the concurrency properties of the specifications are established, then Ada code is generated.

Here we report on the most difficult part of this process: establishing that a program specification has a crucial concurrency property, namely liveness. (Other related aspects of our proposed approach, such as safety properties and deadlock avoidance, are discussed elsewhere [1]). By liveness we mean a condition that must be satisfied at some point during the execution of a program. Liveness contrasts with a safety property, a condition that must hold continuously throughout program execution.

Underlying our approach to synthesis is a model of a concurrent program in which processes communicate by accessing and modifying shared resources. This *resource-oriented model* uses temporal logic for specification and analysis. By a *process* we mean an abstract state machine defining an active computation. A process computation generally requires allocation of shared resources. By a *synchronizer* we mean an abstract state machine that defines a set of encapsulated resources [8]. A synchronizer specification is expressed in terms of the safety and liveness properties that the synchronizer must satisfy. Our approach to program specifications is similar to that presented in [8], which, however, does not address the issue of liveness. Synchronizer liveness is the main focus of this paper.

We believe our approach to synthesis and, in particular, our approach to liveness analysis can be automated for two reasons. First, by identifying indiscernible synchronizer computation states, we drastically reduce the size of the resulting synchronizer state space graph. Second, by using this graph to drive the formal analysis of program specifications, we limit the need for general theorem proving capabilities. At present a partial implementation has been completed.

This paper is organized as follows. Synchronizer specifications are described in section 2. A graph model of synchronizer behavior is discussed in section 3. A proof lattice-based method for liveness analysis is given in section 4 along with the analysis of the familiar readers/writers example. Some conclusions are presented in section 5.

2 Synchronizer Specifications

Our specification language is built around the concept of the state of the resources encapsulated in a synchronizer. Specifications are formulated in terms of a set of state variables and a set of operations that can access and modify those variables. Linear-time temporal logic is used to define the semantics of the constructs appearing in a synchronizer specification. Thus, a synchronizer specification consists of a set of clauses defining the state transitions performed by each synchronizer operation. A specification also gives the safety and liveness conditions governing the execution of synchronizer operations. An example, the specification of synchronizer *buffer* from the readers/writers problem, is given in Figure 1. In this problem a set of reader and a set of writer processes must access a shared memory area simultaneously. Synchronizer *buffer* controls access and modifications to the shared memory on behalf of the reader and writer processes.

The control_resources clause of the specification defines resources *read_permission* and *write_permission* for synchronizer *buffer*. The clause asserts that, in order for a reader (or a writer) process to access the shared memory, the resource *read_permission* (or *write_permission*) must be allocated first. Allocation is performed by operations *start_read* and *start_write*. Operations *end_read* and *end_write* perform the corresponding deallocations.

The state_variables clause defines the type and initial value of the variables *reader_#* and *writer_#* used by the synchronizer to represent the state of the encapsulated resources. These two variables track the total number of reader and writer processes that are in the synchronizer at any point in time. The operations clause declares the operations that

Synchronizer Buffer
Control_resources write_permission **allocated_by** start_write;
deallocated_by end_write;
read_permission **allocated_by** start_read;
deallocated_by end_read;

State_variables
reader_#: 0..10 **initially** 0 ;
writer_#: 0..1 **initially** 0 ;

Operations
start_read, start_write, end_read, end_write ;

Operation_preconditions
$\forall s \in$ start_write: (reader_# = 0) $\wedge$ (writer_# = 0)
$\forall s \in$ start_read: (writer_# = 0)

Operation_state_changes
$\forall s \in$ start_read: reader_# $\leftarrow$ reader_# + 1
$\forall s \in$ start_write: writer_# $\leftarrow$ 1
$\forall e \in$ end_read: reader_# $\leftarrow$ reader_# - 1
$\forall e \in$ end_write: writer_# $\leftarrow$ 0

Operation_priorities
start_write > start_read

Liveness
$\forall s \in$ {start_write, start_read} s@request $\rightarrow$ $\Diamond$s@terminate

Figure 1: Specification of synchronizer *buffer* in the readers-writers problem

units outside synchronizer *buffer* can invoke to perform allocation and deallocation of the resources encapsulated in this synchronizer. Whenever an operation is invoked, a model of operation execution is used in which the invocation goes through a sequence of three phases. In the *request* phase the operation has been invoked and is waiting to be selected for execution; in the *service* phase the operation has been selected for execution and will be executed next; and in the *terminate* phase the operation has completed execution. For a given invocation o of a synchronizer operation O, the propositions *o@request*, *o@service* and *o@terminate* are true when o is in the corresponding phase and false otherwise.

The **operation_preconditions** clause specifies the conditions to be satisfied in order for each operation request to be executed. Temporal operators are not allowed in this clause. A request for operation *start_write* can be serviced only if the synchronizer state satisfies:

$$(reader\text{-}\# = 0) \wedge (writer\text{-}\# = 0)$$

The **operation_state_changes** clause specifies the changes in the value of the state variables resulting from the execution of a given operation. Whenever a request for operation *start_read* is executed, the value of the state variable *reader_#* is incremented. The operation priorities clause defines the order in which instances of invocations of different operations are to be considered for service. In synchronizer *buffer*, operation *start_write* has higher priority than operation *start_read*.

Finally, the liveness clause specifies that invocations of operations *start_read* and *start_write* must eventually be serviced. As with most temporal logic systems, the formula $\Box p$ asserts that predicate p is true in the current state and any subsequent state. The formula $\Diamond p$ asserts that p is true in the current state or in some subsequent state.

Linear time temporal logic underlies the computational model of a synchronizer specification. Thus, time is viewed as a totally ordered sequence of *time instants*. Each time instant corresponds to a well-defined synchronizer state (i.e. the assignment of a value to every synchronizer state variable). In general a synchronizer operation O can be invoked an arbitrary number of times by the units contained in the concurrent program. Each operation invocation o can be in the *request* phase for an arbitrary (and possibly indefinite) number of consecutive time instants, while waiting for the appropriate preconditions and priority conditions to become true. When these conditions become true, invocation o is *enabled*, in which case the predicate *enabled(o)* is true.

When an invoked operation o is selected for execution, o goes into the *service* phase for exactly one time instant. At the following instant operation o is in the *terminate* phase. At this instant the synchronizer state variables have been modified to reflect the changes caused by the execution of o. Operation execution in a synchronizer is strictly sequential: only one operation invocation can be in the service phase at any time instant. Once an operation request reaches the *terminate* phase it remains in that phase forever.

3   Graph Model of Synchronizer Behavior

We model synchronizer behavior using an augmented finite state machine called a Reduced State Transition Graph or RSTG. This model underlies the proof lattice mechanism we introduce to establish liveness, and our proof lattice inferences are aimed at identifying valid transition sequences between the states in an RSTG.

The RSTG of a synchronizer is a graph $G = (N, E)$. Nodes N in this graph represent sets of synchronizer states and arcs E represent transitions resulting from the execution of enabled operation requests. Each node N_i is associated with a set of synchronizer states X_i, and each edge E_i is associated with a synchronizer operation O_i and a predicate P_i on the state of the synchronizer, subject to the following conditions. First, the states associated with the same node are indiscernible in that they enable the same operations. Second, if E_i is an edge labeled (P_i, O_i) leading from node N_j to N_k, if operation O_i is executed when the synchronizer is in one of the states associated with N_j, and if predicate P_i is true, then the synchronizer is in a state associated with N_k after O_i has been executed. Third, state sets associated with distinct nodes are disjoint.

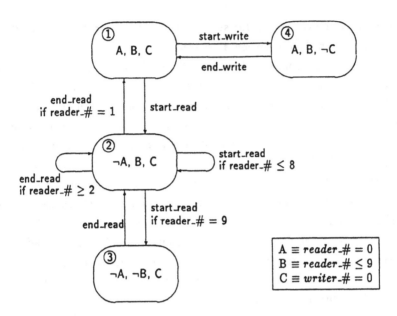

Figure 2: RSTG of synchronizer *buffer*

Construction of an RSTG is carried out in two phases. First, the nodes in the graph are defined. Each node corresponds to a combination of truth values of the predicates appearing in a program specification. Then edges are defined by identifying the operations that are enabled at each node and by tracking the state transitions caused by the execution of each enabled operation. The RSTG construction is discussed in detail in [1].

The RSTG for synchronizer *buffer* is shown if Figure 2, along with the predicate truth values corresponding to each node. Node 1 corresponds to the initial synchronizer state. Operations *start_read* and *start_write* are enabled in this state, as shown by the edges leaving node 1. Node 2 represents the synchronizer states in which resource *read_permission* has been allocated to a number of reader processes that is less than the maximum number of readers allowed. Consequently both *start_read* and *end_read* are enabled in the states corresponding to this node. The execution of the former operation leads to node 3 when the variable *reader_#* is one short of the maximum reader number; otherwise it leaves the synchronizer in a state still contained in node 2. Likewise, the execution of operation *end_read* can either lead to node 1 or back to node 2, depending on whether variable *reader_#* is one or greater than one. Node 3 corresponds to the synchronizer state in which *reader_#* equals the maximum number of readers. Only operation *end_read* is enabled in this state, leading to state 2. Finally, node 4 represents the state in which a writer process is in the synchronizer. Only operation *end_write* is enabled in this state.

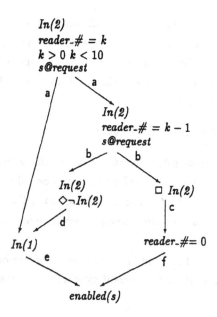

Figure 3: Proof lattice for operation *start_write* for RSTG node 2

4 Proof Lattice-based Analysis

We now describe our deductive method for establishing the following liveness condition:

$$\forall o \in O \ \Box(o@request \rightarrow \Diamond \ o@service) \tag{1}$$

We prove liveness using the notion of a *proof lattice*. Proof lattices were introduced in [5] to prove first-order formulas. They were used in [7] to analyze a subset of first-order temporal logic. A proof lattice is a finite directed acyclic graph in which each node is labeled with an assertion and such that:

1. There is a single entry node with no incoming edges

2. There is a single exit node with no outgoing edges

3. If a proof lattice node labeled P has outgoing arcs to nodes labeled R_1, R_2, ... R_n, the following formula holds for the synchronizer:

$$\Box(P \rightarrow \Diamond(R_1 \vee R_2 \vee ... \vee R_n))$$

Our approach to liveness is based on several assumptions. First we assume that the implementation of a synchronizer uses a *fair scheduler*, that is, a scheduler that does not ignore an operation invocation that is enabled infinitely often:

$$\Box\Diamond enabled(o) \rightarrow \Diamond o@service \tag{2}$$

Second, we assume that a resource allocation is always followed by the corresponding re-source deallocation. This assumption is realistic because deadlock prevention is addressed independently of the liveness analysis described here [1]. Thus, if operations A and D allocate and deallocate a given resource, any invocation a of operation A is eventually followed by an invocation d of operation D:

$$\Box(a@request \rightarrow \Diamond d@request)$$

Third, we make use of the notion of *conformity* between a state variable x and a resource R. We say x is *conformal* with R if the value of x reflects the number of open allocations of R (i.e. the number of executions of the operation that allocates R that have not been followed by matching deallocations). In the *buffer* example variable *reader_#* is conformal with resource *read_permission*.

The following theorem establishes the validity of the proof lattice-based approach [1]. The predicate $In(N_i)$ indicates whether the current synchronizer state belongs to an RSTG node N_i.

Fundamental Theorem

Given an operation O defined in synchronizer S and an RSTG for S, if for every node N in the RSTG a proof lattice can be built whose entry node is labeled $In(N) \wedge o@request$ and whose exit node is labeled *enabled(o)*, then the following assertion is true for every synchronizer state

$$\Box(o@request \rightarrow \Box\Diamond enabled(o))$$

The conclusion of the theorem is that, under the stated hypotheses, every invocation o of operation O is enabled infinitely often. In the presence of the fair scheduler assumption, this guarantees that the liveness condition (1) above is satisfied. Thus, every invocation of operation O is eventually serviced. Consequently the liveness analysis of synchronizer operations can be performed by constructing a suitably labeled proof lattice for each node N in the RSTG of the synchronizer.

Proof lattice construction rules fall into three categories. Rules in the first group tie proof lattice deductions to RSTG transitions. For example, if an RSTG node N has arcs leading to nodes $N_1, \ldots, N_k$, a proof lattice node L labeled $In(N)$ has descendants labeled $In(N_1), \ldots, In(N_k)$.

Rules in the second group are based on axioms of linear-time temporal logic. For example, suppose there is a proof lattice node L labeled $\Box p$. In this case, predicate p can be added to the label of any descendant of node L, based on these two axioms of temporal logic: $(\Box p \rightarrow p)$ and $(p \rightarrow \Diamond p)$.

Rules in the third group reflect the behavior of resource allocation and deallocation induced by our synchronizer model. As an example, consider the *allocation completion* rule. Suppose that a state variable x of synchronizer S is conformal with resource R, which is allocated and deallocated by operations A and D, respectively. Then a proof lattice node L labeled $\Box\neg enabled(A)$ has a descendant labeled $(x = \bar{x})$, where $\bar{x}$ is the initial value of x. The validity of this rule is proved by use of the assumption of resource deallocation. Since operation A is never enabled, eventually all executions of this operation will be followed by executions of operation D. Roughly speaking, the effects of each execution of operation A on variable x will eventually be "undone" by the corresponding execution of operation D. Consequently, variable x will eventually return to its initial value.

A schematic description of a proof lattice for operation *start_write* is given in Figure 3. This proof lattice is aimed at proving that the operation is eventually serviced, assuming that it is invoked when synchronizer *buffer* is in a state corresponding to RSTG node 2. Thus, the entry node is labeled by the predicates $In(2)$ and *s@request*, assuming s is an invocation of operation *start_write*.

Step (a) is performed by applying the first set of construction rules to the entry node in the proof lattice. Note that four edges leave node 2 in the RSTG; however, the edges labeled *start_read* are temporarily disabled, because this operation cannot be serviced due to the pending invocation of higher-priority operation *start_write*. Consequently, only two edges are created in the proof lattice, corresponding to the RSTG edges labeled *end_read*.

Step (b) is performed by applying the second set of rules and temporal logic axiom $(\Box p \vee \Diamond\neg p)$, where p is instantiated to the predicate $In(2)$. So the formula $(In(2) \rightarrow (\Box In(2) \vee \Diamond\neg In(2)))$ is also valid. Consequently, the node under consideration has two descendants, one for each of the disjuncts in the temporal logic axiom.

Step (c) is performed by noting that *end_read* is enabled when $In(2)$ is true, whereas *start_read* is not enabled. Moreover, these two operations deallocate and allocate resource *read_permission*, respectively. Since variable *reader_#* is conformal with this resource, the allocation completion rule can be applied to produce a descendant labeled $(reader\_\# = 0)$. This predicate is in contradiction with the earlier assumption $\Box In(2)$. As a result, a rule leading directly to the exit node, whose label is *enabled(s)*, can be applied in the following proof step (f).

Step (d) is performed by applying a construction rule from the first group. This rule is similar to the rule applied in step (a); however, here the label of a proof lattice node has the additional predicate $\Diamond\neg In(2)$. The execution of an enabled operation can lead from node 2 to node 1 or node 3; however, the edge leading to node 3 is disabled due to the priority specification. Consequently, the formula $(\neg In(2) \rightarrow \Diamond In(1))$ is valid and a descendant labeled $In(1)$ is created for the proof lattice node under consideration. Finally, a rule leading to the exit node can be applied in the step (e), concluding the proof lattice construction.

The above example has shown the liveness of operation *start_write*, assuming this operation is invoked when synchronizer *buffer* is in a given state subset. Similar proofs have been made for the other synchronizer operations and states, in an effort to complete the liveness analysis of synchronizer *buffer*. These proofs have been generally successful. However, difficulties have arisen in the proof of liveness for operation *start_read*, which has a lower priority than operation *start_write* in the specification of synchronizer *buffer*. Because of this priority condition invocations of *start_read* may "starve", due to the continuous presence of *start_write* invocations. To avoid this phenomenon, the specification of synchronizer *buffer* could be modified to allow invocations of *start_read* and *start_write* to be placed in the same FIFO queue. In this case, all the proof lattices for synchronizer *buffer* can be constructed successfully, thus establishing the liveness of this synchronizer.

Our approach to liveness analysis has been applied successfully to a variety of traditional examples in the domain of concurrent programs, such as the producers/consumers, the sleepy barber [4] and a simplified version of a memory controller. While the full expressive power of this approach is still under active investigation, a prototype that partially automates the proof lattice construction shown here has been implemented [1].

Several features of our approach have made an implementation feasible. First, we have wedded our proof lattice machinery to the RSTG construction. This means that proof flow is directed by RSTG state transitions, thus controlling the size of the proof search space. Second, the absence of loop constructs in operation state changes means that loop invariants need not be generated during proof lattice construction. Third, our assumptions about the behavior of resource allocation and deallocation simplify proof lattice construction.

5 Conclusions

In this paper we have briefly sketched the workings of a proposed automatic synthesis system for synchronization code and we have described one particular crucial feature, a deductive system for establishing the liveness of synchronizer operations.

We believe our method is automatable, and is even potentially practical, for several reasons. First of all, the RSTG construction is relatively straightforward. This is crucial because the RSTG reduces the excessive state information present in system behavior to a small, tractable body. Second, the absence of temporal operators in the specification of operation preconditions means that existing theorem provers can be used in the construction of the RSTG of a synchronizer. Third, our liveness analysis is rather stylized, thus sidestepping the need to use the full deductive power of a temporal logic system during proof lattice construction. Finally, proof lattice construction is facilitated by the absence of loop constructs in operation state transitions, by the absence of temporal operators in operation preconditions, and by the stated assumptions about the behavior of resource allocation and deallocation. As a result, only a limited subset of temporal logic has been required to build the proof lattices in the example set considered so far.

We believe that the practicality of the proposed approach is enhanced by the structure of program specifications. First, our approach to synchronizer specification is close to the way people think about resources (i.e. in terms of operations performing resource allocations and deallocations). Second, high-level support is provided, so that specifications can include such features as operation preconditions and priorities. Third, these properties are clearly separated in a program specification. We believe that these aspects of our approach compare favorably with previous work that does not support the specification of these properties explicitly [2, 6].

References

[1] U. Buy. *Automatic Synthesis of Resource Sharing Concurrent Programs*, PhD Dissertation, Computer Science Department, University of Massachusetts, Amherst, MA, September 1990.

[2] E. M. Clarke, E. A. Emerson. Design and Synthesis of Synchronization Skeletons Using Branching-Time Temporal Logic, *Lecture Notes in Computer Science 131*, Springer-Verlag, New York, 1981.

[3] E. W. Dijkstra. Guarded Commands, Nondeterminacy and Formal Derivation of Programs, *Communications of the Association for Computing Machinery, 18*, 8, August 1975.

[4] E. W. Dijkstra. *Cooperating Sequential Processes*, Technical Report EWD-123, Technological University, Eindhoven, The Netherlands, 1965.

[5] L. Lamport. Proving the Correctness of Multiprocess Programs, *IEEE Trans. on Software Engineering, SE-3*, 2, March 1977.

[6] Z. Manna, P. Wolper. Synthesis of Communicating Processes from Temporal Logic Specifications, *ACM Transactions on Programming Languages and Systems, 6*, 1, January 1984.

[7] S. S. Owicki, L. Lamport. Proving Liveness Properties of Concurrent Programs, *ACM Transactions on Programming Languages and Systems, 4*, 3, July 1982.

[8] K. Ramamritham. Synthesizing Code for Resource Controllers, *IEEE Transactions on Software Engineering, SE-11*, 8, August 1985.

Verification of a Multiprocessor Cache Protocol using Simulation Relations and Higher-Order Logic (Summary)

Paul Loewenstein* (Mitsubishi Electronics America) and
David L. Dill†(Stanford University)

Abstract

We present a formal verification method for concurrent systems. The technique is to show a correspondence between state machines representing an implementation and specification behaviour. The correspondence is called a *simulation relation*, and is particularly well-suited for theorem-provers. Since the method does not rely on enumerating all the states, it can be applied to systems with an infinite or unknown number of states. This substantially expands the class of hardware designs that can be formally verified. The method is illustrated by proving the correctness of a particularly subtle example which is likely to be of increasing importance: a directory-based multiprocessor cache protocol. The proof is carried out using the HOL ("higher-order logic") theorem-prover.

1 Introduction

It has been shown that formal verification of certain types of non-trivial hardware designs using theorem-provers is possible. Indeed, several complete (albeit simple) processors have been formally verified [4, 3, 10, 17] using theorem-proving. But before theorem-proving techniques can be applied to most hardware designs, the problem of modelling and verifying concurrent behaviour must be solved. Unforseen interactions among concurrent systems are one of the major source of bugs in hardware designs today. Significantly, architectural features that depend on concurrent interaction have been omitted from almost all formally verified processor designs. Such features include caches, pipelining, interrupts, and page mapping. Concurrency is crucial for high performance, so its use will continue to increase. In particular, multiprocessor architectures are certain to be a major source of complicated concurrent hardware designs. This paper describes a successful attempt to verify a high-level concurrent hardware design.

---

*Work performed at Stanford University while visiting scientist for National Semiconductor Corporation.

†Supported by NSF grant number MIP-8858807 and DARPA contract number N00014-87-K-0828.

The cache coherence protocol of our example is *directory-based*. The main idea behind directory-based cache coherence protocols is to maintain a partial model of cache contents in a central directory (or directories). When, for example, an address is written for the first time in one cache, the directory can send "invalidate" messages to the specific caches which have data present at the same address. The advantage is that cache invalidations do not have to be broadcast, eliminating a major bottleneck in large multi-processors.

The protocol involves the interaction of several concurrent agents, namely the individual cache controllers and the directory. In general, models of concurrent systems are fundamentally different from, say models of functional programs. It is not sufficient to model a system as an input/output mapping on data-values or memory states. A model must also capture the *sequence* of states during a computation, because the interactions with other systems depends crucially on the intermediate states of the computation.

We use a well-known method for verifying concurrent systems based on *simulation relations*. The implementation (system behaviour) and specification (desired behaviour) are both represented as state graphs. A simulation relation establishes a correspondence between implementation states and specification states. For the example of the paper, we use *successive refinement*. We transform a higher-level description into an implementation in several steps. A simulation relation is used to show that the automaton after the step implements the automaton before the step.

Previously, there have been proposals for logics and methods for reasoning about concurrent systems too numerous to mention. However, one promising class of methods is to establish a correspondence between two automata (the correspondence can be a single- or multi-valued function, or a relation) [13, 1, 12, 19, 11, 16]. Although some impressive examples have been done by hand these methods have generally not been combined with automatic theorem-proving (the one other example of which we are aware is due to Nipkow [18]).

The most successful methods for verifying concurrent hardware are *state enumeration methods*. In essence, these methods *automatically* find all of the reachable states in a state graph representing the behaviour of a circuit For example, in the *model-checking* approach, the state graph is compared with *temporal logic* formulas, which make assertions about the execution paths through the system [2, 7]. Other techniques compare formal languages of finite automata [5, 6].

Of course, the major limitation of state enumeration methods is that they potentially need to inspect all of the states of the system. They all suffer from the *state explosion problem* — the number of states grows exponentially with the size of the system being verified. They are also generally not good for systems that are partially described. For example, in our description of a cache system, we never specify the number of processes, size of memory, size of data values, or many other parameters of the system. Hence, we have verified the correctness of all the systems that can be obtained by giving values to these parameters. A state enumeration approach would require not only that values be assigned to these parameters, but that the values be very small integers. In some case, one can show that a parameterized design is correct for all parameter values by verifying a few designs with particular values for the parameters. However, these techniques are either highly restricted or it requires human insight to extend the proof from special cases to the general case.

Theorem proving and state enumeration are complementary methods — each is appropriate for a particular class of problems. Eventually, we hope that a combination of both approaches will make hardware design verification economically feasible (and routine).

We carry out proofs using the HOL theorem-prover ("Higher-Order Logic") [9, 8]. HOL uses a very expressive logic, in which variables can range over functions as well as values (unlike first-order logic). Our approach is to express everything in HOL: indeed, the implementation and specification state graphs are given as logical formulas. There are several advantages to using HOL. The generality of HOL allowed us to prove theorems not only about hardware, but about mathematics, including the soundness of the method itself. Also, the HOL type system allows proofs about objects without knowing their types. This results in generic results that can easily be reused.

Notation

For concreteness, we have included some of the theorems and formulas used to describe state graphs. For brevity, we have omitted most of them (for a computer file containing everything, including the proofs, please contact one of the authors). We have modified the notation somewhat to improve readability for readers who are not very familiar with HOL (for a detailed description of the HOL logic, see [8]).

HOL logic reads very much like ordinary predicate logic with a few differences. Variables and constants can range over functions or values, depending on their type. Types are left implicit in this paper; they are explained in the text unless they are obvious from context. Function application is indicated by juxtaposition: $f\,x$ denotes function f applied to x. Often multiple argument functions are *curried*; $g\,x\,y$ denotes a function g applied to x, the result (again a function) is then applied to y. Informally one can consider g to be merely a multiple argument function. The (x, y) notation is used for pairing (cartesian product). When pairs are used as function arguments, as in $f(x, y)$, the notation appears conventional. Sometimes, (as in this paper), currying and pairing are mixed, as in $g(u, v)(x, y)$. Logical implication is indicated by the $\supset$ operator.

Proven theorems are indicated by the symbol $\vdash$. These have been derived using the mechanical theorem prover, HOL[9], and should be correct apart from errors in typographic transcription. Defined constants are printed in sans serif typeface.

Let $x = y$ in $z[x]$ is syntactic sugar for a term semantically identical to $(\lambda x. z[x])y$, which reduces to $z[y]$. It saves repeating a long term y which may have multiple occurrences in $z[y]$.

We have adopted some special conventions for describing sequences and automata. Variables denoting sequences, or functions of time, are emboldened (e), as opposed to other variables (e). When referring to next-state relations, unprimed variables (s) refer to the current state, and primed variables (s') to the next state. The initial state predicate of an automaton Name, is always Name_P. The next-state relation is Name_N and any invariant is Name_I.

This paper gives an overview of the proof. More details are available in a longer version which appears in [15]. In the remainder of the paper, the next section describes the state-graph model of behaviour and the definition of simulation relations. The third section outlines our proof strategy, which transforms a high-level description of the system through several intermediate representations to the final implementation. The final section discusses the appropriate use of the verification method and future work.

2 The Automaton Model

We regard a digital system as a concurrent process with externally visible variables (corresponding to wires, or bundles of wires) and invisible internal state. A *trace* is an infinite externally-visible execution history of the system. The *behaviour* of a system is the set of all possible traces that it can exhibit. This set can be related to a state graph, consisting of of a set of states, a subset of those representing possible initial states, and a next-state relation.

Each state is denoted by a pair (e, s), where e is the visible (or external) part of the state and s is the hidden, or internal, part. A state graph is defined by a pair of predicates, (P, N): P is a unary predicate that defines the the initial states, and N is a binary next-state relation: $N(e, s)(e', s')$ holds when (e', s') is a possible successor state to (e, s).

A binary predicate (Labelled State Automaton) $\mathsf{LSA}(P, N)e$ is defined in HOL. It states that e is a trace of the automaton defined by P (initial states) and N (next state relation). In more detail, e is a trace of the automaton if there exists a sequence of hidden states, s, such that the first state satisfies P, and N holds between every state and the next state in the sequence:

$$\mathsf{LSA}(P, N)e = (\exists s.P(e\,0, s\,0) \wedge (\forall t.N(e\,t, s\,t)(e(t+1), s(t+1)))) \tag{1}$$

2.1 Relating two automata

One automaton *implements* another if every trace of the first is an trace of the second. A powerful method for showing that this relation holds is to find a *simulation relation* between them. A simulation relation is a three-way relation $R\,e\,s_1\,s_2$, where (e, s_1) is a state of the implementation graph and (e, s_2) is a state of the specification graph (the visible components must be the same in both states). It must satisfy two properties: first, every initial state of the implementation is related by R to some initial state of the specification; and, second, whenever $R\,e\,s_1\,s_2$ and $N_1(e, s_1)(e', s'_1)$ both hold, there exists some specification state (e', s'_2) such that $N_2(e, s_2)(e', s'_2)$ and $R\,e'\,s'_1\,s'_2$ both hold.

To show that one automaton is an implementation of another, it is sufficient to produce a simulation relation from the first to the second. This fact is proven as a theorem in HOL[1]

$$
\begin{aligned}
\vdash \forall &P_1 N_1 P_2 N_2. \\
&(\exists R. \\
&(\forall e s_1.P_1(e, s_1) \supset (\exists s_2.P_2(e, s_2) \wedge R\,e\,s_1\,s_2))\wedge \\
&(\forall e e' s_1 s'_1 s_2.R\,e\,s_1\,s_2 \wedge N_1(e, s_1)(e', s'_1) \supset \\
&\quad (\exists s'_2.R\,e'\,s'_1\,s'_2 \wedge N_2(e, s_2)(e', s'_2)))) \supset \\
&(\forall e.\, \mathsf{LSA}(P_1, N_1)e \supset \mathsf{LSA}(P_2, N_2)e)
\end{aligned}
\tag{2}
$$

The types of e and s_1 and s_2 are left open; the theorem therefore holds for finite, countable and uncountable-state automata.

This is a soundness result for our verification method: to prove that one automaton implements another, it is sufficient to supply a simulation relation. The converse would

---

[1]This theorem was derived from the corresponding theorem for a labelled *transition* automaton presented with proof in [14].

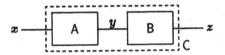

Figure 1: Composition of automata

be that whenever one graph is an implementation of another, there exists a simulation relation. This property (completeness) does not hold. Nevertheless, simulation relations have been adequate for our purposes. Abadi and Lamport have shown that a similar method can be made complete if additional variables are added to the implementation to record the past and predict the future (history[2] and prophecy variables), a result that is equally applicable to our method, should the addition of variables prove necessary [1].

An *invariant* is a predicate on individual states. An invariant is satisfied by a state graph if every state that is reachable from an initial state satisfies the invariant. An important method for simplifying proofs is to prove strong invariants before proving more complicated properties. Once an invariant has been proven on a state graph, the invariant can be incorporated into the next-state relation of original state graph without changing the behaviour of the system (this just makes explicit a property that is already implicitly true). This transformation often substantially reduces the amount of case analysis that must be used in proofs of additional properties of the state graph. The correctness of this transformation (although obvious) has been proven in HOL by specialisation of theorem 2.

2.2 Composition and Hiding

Figure 1 shows a parallel composition of two circuits, A and B. Given definitions of the automata for A and B, we can write the initial state predicate of C as

$$C\_P((x, z), s_1, s_2, y) = A\_P((x, y), s_1) \land B\_P((y, z), s_2), \tag{3}$$

where A_P and B_P are the initial state predicates of the individual machines. Note that y has moved from being a component of the *label* of the individual machine state, to being a component of the *hidden state* of the composed machine (composition without hiding can be defined be putting y with x and z, if desired). The next-state of C relation is defined similarly.

$$\begin{aligned} C\_N((x, z), s_1, s_2, y)((x', z'), s'_1, s'_2, y') = \\ A\_N((x, y), s_1)((x', y'), s'_1) \land B\_N((y, z), s_2)((y', z'), s'_2) \end{aligned} \tag{4}$$

Although it is possible to prove a general schema as a theorem, we found it simpler to merely prove each composition individually by expanding the LSA definition.

3 Highlights of the Specification and Proof

This section gives a very high-level overview of the proof. A preliminary task was to define a common master-slave communication channel MS, which is used at every interface where

---

[2]Our simulation relation approach does not need history variables.

data is read or written. In the figures, communication delay is represented by circled arrows, with the arrow pointing from master to slave.

The actual proof refines an specification of memory behavior, which we call a *multi-port memory* to a memory system consisting of a cache for each processor, a directory, and a main memory (which is also a multi-port memory). The first step is to refine the multi-port memory into what we call a *cached multi-port memory*. The cached multi-port memory is not directly implementable as specified — it exists only to simplify the proof. The second part of the proof is to refine a component of the cached multi-port memory into a composition of two smaller components. Finally, the directory-based system is formed by merging several of the components into a single directory. This "step" is just an application of composition and hiding. We now discuss these steps in more detail.

3.1 Multi-port Memory

We first define the behaviour of an multi-port memory with no cache, MP_Mem. MP_Mem serves two purposes: MP_Mem is a part of the implementation because a cached memory is just a multi-port memory with a cache in front; MP_Mem is also the specification of the *desired* behaviour of the cached memory (a correctly functioning cached memory behaves just like an uncached memory).

The per-port components of the multi-port memory are modelled as functions from a port address, q, to a value. The type of q is left open so that the proofs are valid for any number of ports. This is possible because none of the consistency proof uses induction; the only induction is over time in the derivation of theorem 2.

3.2 Refining the Multi-Port Memory to a Cached Multi-port memory

We first refine the (uncached) multi-port memory to a simple cached multi-port memory. This involves defining the cached memory and proving that it implements the multi-port memory.

The cached multi-port memory consists of three types of components: a multi-port memory for primary storage, a cache Port_Cache (there is one for each port) and unit EXA which goes between the memory and port caches, whose primary purpose is to enforce mutually-exclusive access by the port caches to the multi-port memory.

Port_Cache looks very much like a single processor cache, except that it has a "window", s_d, on part of its hidden state. This window is used by EXA both to obtain information about the internal state of the Port_Caches, and to constrain the behaviour of the Port_Caches. It is very important to note that we have provided no explicit mechanism for this two-way transfer of information between Port_Cache and EXA (it is therefore not indicated on figure 2 as a circled arrow). This is an example of a generally useful trick: partitioning the logical description of a design in ways that are not physically implementable. In this case, it makes the proof much easier, and no harm is done because the next step will refine the unimplementable description into an implementable one by repartitioning the system. EXA uses s_d to decide when it can grant a cache exclusive access, and also to block access directly.

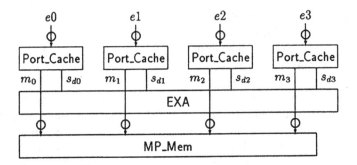

Figure 2: Modified caches with exclusive access control for a multi-port memory

It is then proved, by the use of a simulation relation, that the cached multi-port memory implements the uncached multi-port memory.

3.3 Splitting the port cache

The second part of the proof refines the Port_Cache into a composition of two halves: Cache, which is the final (per-port) cache specification, and Port_Dir, which will eventually end up being part of the directory. The two halves communicate via two MS communication channels, one of which carries directory-to-cache requests, and the other carries cache-to-directory requests. It is then proved, by using a simulation relation, that Port_Dir coupled to Cache implements Port_Cache. Despite the fact that this can be carried out on one port in isolation, it is by far the largest part of the total multiport correctness proof. The final system is shown in figure 3.

The directory consists of the EXA and the collection of Port_Dir, which is regarded as a single unit. The division between EXA and Port_Dir in the intermediate representation was only a convenient fiction which may have nothing to do with the actual implementation of the directory. The s_d state is now completely enclosed in the directory, so the communication between the different components is entirely via master-slave channels. So, if the individual components are implementable (as we believe they are), the entire system is also implementable.

At this point, we cease refining the specification. If we wanted to carry it further, MP_Mem could be implemented with a memory request arbiter and a single port memory, in which case the arbiter would also become part of the directory. One could also use another level of caching at each processor.

4 Conclusion

Before this project we thought that one could perform verification by proceeding top-down from a single specification, deciding on specifications of components and proving that the assembly of components conforms to the specification. Then repeat the process on the components, thus following a "verification tree" all the way to the primitive components.

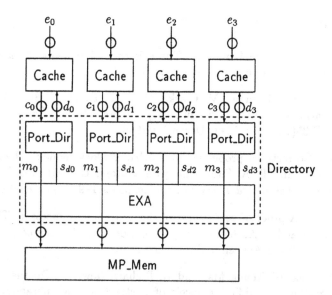

Figure 3: Configuration after splitting the caches

Things are not that simple. For example, modelling the directory state as a memory *per cache* requires *merging*, not division, as the proof proceeds down the abstraction hierarchy. The design hierarchy becomes a much more dynamic concept, different compositions being used to derive different properties.

The principle inspiration in the proofs is the finding of the simulation relation, and choosing invariants that are sufficiently strong. The rest of the proof is tedious and boring. To make this a routine way to approach engineering design we need a much improved user interface, much better interactive response time, and better automatic proof procedures than HOL currently offers. The re-running (without interaction) of the proofs in this paper takes about 20 hours on an unloaded Sun 3/60.

The automaton theory used in this paper was derived in a general-purpose theorem prover. The theory required to reason about other aspects of the design (such as Temporal logic for deriving liveness properties), can also be derived in the same environment, thus allowing more complete proof inside a single environment.

By using a suitable powerful logic can prove the correctness of systems with an arbitrarily large state-space, provided that the system is structured to allow a tractable case analysis.

References

[1] Martín Abadi and Leslie Lamport. The existence of refinement mappings. SRC Report 29, Digital Equipment Corporation, 1988.

[2] Michael C. Browne, Edmund M. Clarke, David L. Dill, and Bud Mishra. Automatic verification of sequential circuits using temporal logic. *IEEE Transactions on Computers*, C-35(12):1035–1044, December 1986.

[3] Avra Cohn. Correctness properties of the Viper block model: The second level. In G. Birtwistle and P. A. Subrahmanyam, editors, *Current Trends in Hardware Verification and Automated Theorem Proving*, pages 1–91. Springer-Verlag, 1989.

[4] Avra. J. Cohn. A proof of correctness of the Viper microprocessor: The first level. In G. Birtwistle and P. A. Subrahmanyam, editors, *VLSI Specification, Verification and Synthesis*. Kluwer Academic Publishers, 1988.

[5] Srinivas Devadas, Hi Keung Ma, and A. Richard Newton. On the verification of sequential machines at differing levels of abstraction. Memorandum UCB/ERL M86/93, University of California, Berkeley, 1986.

[6] David L. Dill. *Trace Theory for Automatic Hierarchical Verification of Speed-independent Circuits.* MIT Press, 1989.

[7] Masahirio Fujita, Hedehiko Tanaka, and Tohru Moto-oka. Verification with prolog and temporal logic. In T. Uehara and M. Barbacci, editors, *IFIP Sixth Computer Hardware Description Lanquages and their applications*, pages 103–114. North Holland Publishing Company, 1983.

[8] Mike Gordon. HOL: A machine oriented formulation of higher-order logic. Technical Report 68, University of Cambridge Computer Laboratory, 1985.

[9] Mike Gordon. HOL: A proof generating system for higher-order logic. In G. Birtwistle and P. A. Subrahmanyam, editors, *VLSI Specification, Verification and Synthesis*. Kluwer Academic Publishers, 1988.

[10] W. A. Hunt, Jr. The mechanical verification of a microprocessor design. In D. Borrione, editor, *From HDL Descriptions to Guaranteed Correct Circuit Designs*. North Holland, 1987.

[11] Nils Klarlund and Fred B. Schneider. Verifying safety properties using nondeterministic infinite-state automata. Technical Report TR 89-1037, Cornell University Computer Science Department, 1989.

[12] R. P. Kurshan. Analysis of discrete event coordination. In J.W. de Bakker, W.-P. De Roever, and G. Rozenberg, editors, *Stepwise Refinement of Distributed Systems*. Springer-Verlag, 1990.

[13] Simon S. Lam and A. Udaya Shankar. Protocol verification via projections. *IEEE transactions on software engineering*, SE-10(2):137–151, July 1984.

[14] Paul Loewenstein. Reasoning about state-machines in higher-order logic. In M. Leeser and G. Brown, editors, *Hardware Specification, Verification and Synthesis: Mathematical Aspects*. Springer Verlag, 1990.

[15] Paul Loewenstein and David L. Dill. Verification of a multiprocessor cache protocol using simulation relations and higher-order logic. In *Computer-Aided Verification (Proceedings of the CAV90 Workshop)*, volume 3 of *DIMACS Series in Discrete Mathematics and Theoretical Computer Science*. American Mathematical Society, 1991.

[16] Nancy A. Lynch and Mark R. Tuttle. Hierarchical correctness proofs for distributed algorithms. Technical Report TR-387, MIT Laboratory for Computer Science, 1987.

[17] Paliath Narendran and Jonathan Stillman. Formal verification of the sobel image processing chip. In G. Birtwistle and P. A. Subrahmanyam, editors, *Current Trends in Hardware Verification and Automated Theorem Proving*, pages 92–107. Springer-Verlag, 1989.

[18] Tobias Nipkow. Formal verification of data type refinement. In J.W. de Bakker, W.-P. De Roever, and G. Rozenberg, editors, *Stepwise Refinement of Distributed Systems*. Springer-Verlag, 1990.

[19] A. P. Sistla. A complete proof system for proving correctness of non-deterministic safety specifications. Technical Report TC-0060-8-89-378, GTE Laboratories, 1989.

Computer Assistance for Program Refinement

D.A. Carrington
Key Centre for Software Technology
Department of Computer Science
University of Queensland

K.A. Robinson
Department of Computer Science
University of New South Wales

Abstract

This paper explores the role for mechanised support for refining specifications to executable programs. The goal of refinement is to achieve the translation from specification to implementation without the introducing errors. The refinement calculus provides a set of rules for developing procedural programs from abstract specifications. A prototype editor for the refinement calculus is described that was constructed using the Cornell Synthesizer Generator. Based on our experiences, desirable features for future tools are suggested.

1 Introduction

Software development requires a process to transform an abstract specification into an executable substitute, that is, a computer program that exhibits the required behaviour. The transformation process is often referred to as refinement and normally requires a sequence of steps. These steps represent the design decisions leading to the choice of algorithms and of data representations. Formalizing the refinement process is important to establish the correctness of the product. A program is *correct* if it is a satisfactory implementation of the original specification. Correctness may be established by testing (almost always infeasible), by constructing a program proof (usually difficult if attempted after the program has been constructed), or by using a refinement strategy that ensures that each step in the development process maintains the original requirements.

The starting process for refinement is a specification. A major function of a specification is to assist communication between the developer and the client with the objective that both parties understand the intended behaviour of the product to be created.

There is growing acceptance of the benefits of formal specification methods that avoid the ambiguity of natural language. The objective of the specification is to define *what is required* of the product and not *how it is to be achieved*. A specification language should encourage precision, conciseness, and abstraction. It is also desirable to be able to manipulate the specification and prove properties or consequences of the specification. In this way, confidence in the specification as an accurate reflection of the requirements can be established. Unless the specification is expressed in a formal notation, it is not possible to have a systematic method of verifying that an implementation is correct.

Mathematical specification techniques using predicate logic, such as the Vienna Definition Method (VDM) [1,7] and Z [6,16] describe program behaviour as a relation between the input and output states. This allows us to capture the essential behavioural characteristics of a system or algorithm without concern for the implementation. Such specifications have the desirable properties listed above but are not suitable for expressing the concrete algorithms and data structures of the executable program. VDM also incorporates refinement techniques and is one of the few sustained research efforts to tackle this problem.

Having produced a specification that is considered to meet the informal requirements, we want to be able to refine the specification by changing the data structures and introducing programming language constructs until we have an implementation: an executable substitute. The aim of a formal refinement methodology is to ensure that at all times the refinement is in some essential way *consistent* with the initial specification. There are considerable advantages from using uniform notation throughout the refinement process. At the very least, it avoids the difficulties associated with converting between notations. There are two possible approaches: adding programming language elements to a specification language, or adding specification constructs to a programming language. Whichever approach is taken, it is crucial to have a formal semantics that encompasses the extended language. The refinement calculus method adopts the approach of adding specification to a programming language with the extended language defined using Dijkstra's weakest pre-condition semantics [3]. Section Two provides an overview of the refinement calculus and the specification statement on which it is based.

As well as ensuring that the final program is a correct implementation of the initial specification, the refinement steps also provide an excellent design history by documenting the choices made during the development. In that sense, the refinement calculus provides a more rigorous version of the development style that Knuth [8] calls *literate programming*. In section Three, we consider suitable roles for tools to assist in the application of the refinement calculus.

Section Four reviews a prototype refinement editor, a tool for developing programs from specifications based on the rules of the refinement calculus. Based on our experiences with this tool, we make some suggestions for future tools.

2 The refinement calculus

The refinement calculus developed by Morgan and Robinson [11,12,9] provides a rigorous technique for deriving an executable program from an abstract specification. The essential ingredients of the calculus are

- a programming language (with a formal semantics) that we use as our target language. We presume that this language can be executed on a computer. We follow Dijkstra and use both a guarded command language and a weakest pre-condition semantics. It should be noted that any programming language could be used, provided that it is given a weakest pre-condition semantics. In our work we denote the weakest pre-condition of S with respect to R by the functional application of S to R, i.e.

$$S\ R \quad \triangleq \quad \mathrm{wp}(S, R)$$

- a specification construct that is abstract *i.e.* it must allow the specification of programs without being restricted by implementation concerns that are present in any *real* programming language. Our formal semantics must include the specification construct.

- a formal definition of the notion of specification transformation, that we call *refinement*.

2.1 A specification construct

The specification statement [10] captures the essential behaviour of a program or part of a program, without concern for implementation issues. It is denoted

$$\vec{w} : \left[Pre \ / \ Post \right]$$

and denotes a program that, executed in a state satisfying *Pre*, will terminate in a state satisfying *Post* while modifying only variables in $\vec{w}$ (where $\vec{w}$ is a subset of the program variables $\vec{v}$). *Pre* is a predicate in $\vec{v}$ while *Post* is a predicate in $\vec{v}$ and $\vec{v}_0$ where the decorated variables $\vec{v}_0$ refer to the corresponding values in the initial state.

The weakest pre-condition for $\vec{w} : \left[Pre \ / \ Post \right]$ to terminate in a state satisfying R must be at least as strong as *Pre* and every program state that satisfies *Post* must also satisfy R. The formal definition is

$$\vec{w} : \left[Pre \ / \ Post \right] R \quad \triangleq \quad Pre \wedge \forall \left(\vec{w} \ . \ Post \Rightarrow R \right)_{[\vec{v}_0 \setminus \vec{v}]}$$

where $E_{[\vec{x} \setminus \vec{e}]}$ is the expression E with all free occurrences of each of the variables in $\vec{x}$ replaced by the respective expression in $\vec{e}$. We assume that the $\vec{v}_0$ are local to *Post* and hence are not contained in R; otherwise, a further renaming is required to avoid confusion.

Two syntactic variations are

- $\vec{w} : \left[Post \right]$ where the implicit pre-condition is $\exists \left(\vec{w} \ . \ Post \right)_{[\vec{v}_0 \setminus \vec{v}]}$, the weakest condition for the existence of a program state satisfying *Post*.

- $\vec{w} : \left[Pre \ / \ I \ / \ Post \right]$ where I is a predicate in $\vec{v}$ that is invariant, *i.e.* it is true in both the initial and the final state. It is a shorthand for $\vec{w} : \left[Pre \wedge I \ / \ I \wedge Post \right]$.

We introduce an example that is used subsequently to illustrate some refinement rules. The example is very simple and is chosen to demonstrate the principles involved. We wish to implement Min(b, N, m) that computes the minimum value in the array $b[0..N-1]$, passing back the result in m. This is represented by the following specification statement:[1]

$$\text{Min}(b, N, m) \quad \triangleq \quad m : \left[N > 0 \ / \ Post \right] \ where \ Post \quad \triangleq \quad \begin{array}{l} \exists \left(i : 0..N-1 \ . \ m = b[i] \right) \\ \forall \left(i : 0..N-1 \ . \ m \le b[i] \right) \end{array}$$

---

[1] We use vertical stacking of conjuncts to denote conjunction.

2.2 Refinement

With the refinement calculus, the development of a program is a process of replacing specification statements by other programming constructs using the rules of the calculus to perform the transformations. To ensure that the resulting program is a valid implementation of the initial specification, we require an ordering between specifications that captures the idea that one specification (S_i) may be replaced by another (S_j) in any context. This ordering is denoted $S_i \sqsubseteq S_j$. If in all initial states in which S_i does not abort, the set of final states of S_j is contained in the set of final states of S_i, then we say S_i is refined by S_j. The formal definition uses weakest pre-conditions:

$$S_i \sqsubseteq S_j \text{ iff } (S_i \ R) \Rightarrow (S_j \ R) \text{ for all } R$$

Based on this concept of refinement, we can depict the design process as the sequence

$$S_0 \sqsubseteq S_1 \sqsubseteq \cdots \sqsubseteq S_n$$

where S_0 is the initial specification and S_n is our executable program.

2.3 Refinement example

The rules of the refinement calculus express valid transformations of specifications. Many of the rules have an associated applicability condition that specifies when the rule may be used. Rather than attempting to supply a comprehensive set of rules, we give an example refinement. Consult [9] or [11] for exposition of the refinement rules.

A little contemplation of the specification statement for Min suggests that we need an additional variable to control the computation. The construct $[\![\text{ var } \vec{i} \ . \ S]\!]$ introduces new variables $\vec{i}^2$ into the program state inside the block delimited by the symbols $[\![$ and $]\!]$. Applying the appropriate rule to our example, we get

$$m: \left[N > 0 \ / \ Post \right] \quad \sqsubseteq \quad [\![\text{var } j \ . \ m, j: \left[N > 0 \ / \ Post \right]]\!] \quad \left\{ \begin{matrix} Variable \\ introduction \end{matrix} \right\}$$

Introducing an intermediate predicate we refine the inner specification to a sequential composition.

$$m, j: \left[N > 0 \ / \ Post \right] \quad \sqsubseteq \quad m, j: \left[N > 0 \ / \ I \right] ; m, j: \left[I \ / \ Post \right] \quad \left\{ \begin{matrix} Sequential \\ de\text{-}composition \end{matrix} \right\}$$

$$\text{where } I \ \triangleq \ \begin{pmatrix} 0 < j \leq N \\ \exists \left(i : 0..j - 1 \ . \ m = b[i] \right) \\ \forall \left(i : 0..j - 1 \ . \ m \leq b[i] \right) \end{pmatrix}$$

then $m, j: \left[N > 0 \ / \ I \right] \sqsubseteq m, j := b[0], 1 \quad \{ Assignment \ introduction \}$.

We intend to refine the second component to a loop, so we have to choose a set of guards $\{ B_i \}$ and a variant function. We choose a single guard $(j \neq N)$ observing that

$$I \wedge j = N \Rightarrow Post$$

---

[2]In this example, and in the discussion in general, we ignore the type of variables. Introducing typed variables adds a set membership constraint to the pre- and post-conditions.

and $N - j$ as our variant function to establish termination. Then

$$m,j: \left[I \;/\; Post \right]$$

$$\sqsubseteq \mathbf{do}\, j \neq N \rightarrow m,j: \left[j \neq N \;/\; I \;/\; N - j < N - j_0 \right] \mathbf{od} \quad \{Do\ introduction\}$$

It is obvious that the simplified postcondition, $j > j_0$, of the imbedded specification can be satisfied by incrementing j, so we use the weakest prespecification rule to refine to a sequential composition in which the second component is a specification that increments j.

$$m,j: \left[j \neq N \;/\; I \;/\; j > j_0 \right]$$

$$\sqsubseteq m,j: \left[j \neq N \wedge I \;/\; \left(\genfrac{}{}{0pt}{}{I}{j > j_0} \right)_{[j \setminus j+1]} \right]; j: \left[j = j_0 + 1 \right] \quad \left\{ \genfrac{}{}{0pt}{}{Weakest}{prespecification} \right\}$$

The second component refines to the assignment $j := j + 1$ and the first component is simplified, refined by deleting the variable j from the window and then reorganized to reveal an invariant component as follows:

$$m,j: \left[j \neq N \wedge I \;/\; \left(\genfrac{}{}{0pt}{}{I}{j > j_0} \right)_{[j \setminus j+1]} \right]$$

$$= m,j: \left[j \neq N \wedge I \;/\; \begin{array}{c} 0 < j + 1 \leq N \\ \exists \left(i : 0..j \cdot m = b[i] \right) \\ \forall \left(i : 0..j \cdot m \leq b[i] \right) \\ j + 1 > j_0 \end{array} \right]$$

$$\sqsubseteq m: \left[j \neq N \wedge I \;/\; \begin{array}{c} 0 < j + 1 \leq N \\ \exists \left(i : 0..j \cdot m = b[i] \right) \\ \forall \left(i : 0..j \cdot m \leq b[i] \right) \end{array} \right] \quad \{Restrict\ variables\}$$

$$= m: \left[\exists \left(i : 0..j - 1 \cdot m = b[i] \right) \;/\; \begin{array}{c} 0 < j < N \\ \forall \left(i : 0..j - 1 \cdot m \leq b[i] \right) \end{array} \;/\; \begin{array}{c} \exists \left(i : 0..j \cdot m = b[i] \right) \\ m \leq b[j] \end{array} \right]$$

The final specification above is now refined four ways:

1. to the guarded command $m \leq b[j] \rightarrow \mathbf{skip}$ $\quad \{Skip\ introduction\}$;
2. to the guarded command $m \geq b[j] \rightarrow m := b[j]$ $\quad \{Guarded\ command\ introduction\}$;
3. to the union of the two guarded commands $\quad \{Union\ introduction\}$;
4. to an if-statement containing that union $\quad \{If\ introduction\}$.

Assembling the refinements we get the following program:

$$\begin{aligned}
&[\mathbf{var}\ j\ .\ m, j := b[0], 1;\\
&\quad\ \mathbf{do}\ j \neq N \rightarrow\\
&\qquad\qquad \mathbf{if}\ m \leq b[j] \rightarrow \mathbf{skip}\\
&\qquad\qquad []\ m \geq b[j] \rightarrow m := b[j]\\
&\qquad\qquad \mathbf{fi};\\
&\qquad\qquad j := j + 1\\
&\quad\ \mathbf{od}\\
&]
\end{aligned}$$

The preceding example dealt exclusively with *procedural* refinement. The refinement calculus also contains rules for *data* refinement where abstract types in a program are replaced by more concrete types. For more details see [12].

3 Tool roles

The refinement calculus is intended to improve the quality of the software development process. However applying the calculus involves many small steps where each step may require the computation of a new specification and/or the checking of an applicability condition associated with a rule. While most of these are very straight-forward, there is scope for errors to be introduced by the number of steps.

From this observation we envisage two major roles for refinement calculus tools:

- correctly applying the rules of the refinement calculus as directed by the tool user, and

- recording the sequence of refinement steps in a software development.

The first role requires establishing that the applicability condition associated with each rule is satisfied whenever it is used and performing the computations of the calculus. We do not expect to automate software development; our aim is to provide systematic support. Thus the human designer retains the initiative in the design process with the tool playing the *careful assistant* role with responsibility for confirming the viability of each refinement step. Since the proof obligation of each step is a lemma in the first order predicate calculus, we cannot expect automatic confirmation of all steps. While some are easy to prove, many rely on domain knowledge. Some form of interactive proof editor seems necessary. This approach to tools for software development is not new; Floyd [4] predicted it and many people since have adopted this style for co-operative interaction. Perhaps the best known work is the programmer's apprentice project at MIT [15,17].

The recorded design history based on the refinement steps is a valuable artifact of the design process. A tool simplifies the task of capturing this information systematically and accurately. The design history can serve several purposes:

- to explain and document the current design

- for future modification by incremental changes to part of the refinement sequence

- for re-use in different contexts

4 A prototype refinement editor

A prototype refinement editor [2] has been built as an interactive tool for the refinement calculus. It was constructed using the Synthesizer Generator [13,14] which generates language-based editors from descriptions of an abstract syntax and display representation. Attribute grammars are used to incorporate context-sensitive constraints and incremental re-evaluation algorithms to ensure that a consistent document exists after each editing operation.

The development of a prototype was simplified by using parts of Bill Pugh's pv demonstration editor supplied with the Synthesizer Generator distribution. The pv editor is a preliminary program verification editor for Dijkstra's guarded command language. The editor generates verification conditions and incorporates an interactive simplifier for proving them. Using pv, programs are entered statement by statement with the verification conditions confirming the correctness of the program.

Our refinement editor manipulates similar objects but the development style is quite different. From pv we have been able to use the basic structures for predicates, predicate simplification and expression substitution for computing weakest pre-conditions. We also chose the same domain of discourse: integer arithmetic with either simple or array variables. The prototype is restricted to procedural refinement.

The abstract syntax for the refinement editor represents the refinement process as a tree with the initial specification as the root node and each edge corresponding to a refinement step. When the refinement is complete, the leaf nodes correspond to constructs of the executable programming language. Proof subtrees are attached to nodes to verify the applicability condition of each refinement rule. Some of these can be immediately simplified but there is the capability for user interaction for more difficult cases.

The refinement rules of the calculus are provided as transformation commands that act on selected nodes in the tree by extending the tree with an additional refinement step. The Synthesizer Generator displays a sub-window of valid transformation commands whose contents depend on the current edit selection.

The editor uses two display unparsing schemes to present two views of the program under development:

- the primary view is a linear exposition of the refinement steps based on a top-down depth-first traversal of the tree.

- the alternative view is a consolidation of the refinement tree to extract the "final" program. This displays only the leaf nodes. Parts of the program that are not fully refined are displayed as specification statements.

The editor user can dynamically choose either view for independent subtrees so partial compression or expansion is possible.

4.1 Experiences

While our first refinement editor is very rudimentary, it has provided considerable insight into the challenge of building support tools for the refinement calculus. Basing the support tool on the editing paradigm has been a good idea since it encourages an exploratory style where the user investigates refinements, deleting those that are not successful. We have used it to explore the refinement of many small programming examples and we are enthusiastic about the potential of the approach.

As an illustration, after using the editor to develop the Min example, we investigated how easily it could be modified to compute Max instead. The necessary editing changes were to

1. the initial specification (which then propagated through the complete refinement tree),
2. the guards within the if command(or equivalently the guarded statements), and
3. the intermediate predicate used to create the first sequential composition.

The last step was required because this predicate was entered by hand rather than by computation based on the goal post-condition. It would certainly be feasible to incorporate rules to transform predicates in this manner. This example highlights the benefits of recording the design steps and having the capability to incrementally modify individual steps and observe the consequences. Entering information once and then propagating it wherever it is required is convenient compared to performing refinements by hand where there is potential for transcription errors.

There are, of course, shortcomings with the prototype in the scope of problems that can be tackled and with performance. The Synthesizer Generator provides a realistic prototype very quickly and effectively but we encountered some limitations in the flexibility of the user interface.

5 Future tools

In future research into tools for the refinement calculus we plan to investigate the following issues:

Notation If the use of the predicate calculus is to be successful, we need to abbreviate our pre- and post-conditions. The normal way to do this is by defining names (possibly parameterized) for subcomponents of the predicates. Any more substantial tool needs this capability but must also be able to manipulate predicates that contain such abbreviations without requiring complete expansion. The work of Griffin [5] is of interest here.

Interactive proof techniques A realistic refinement editor will rely heavily on some form of interactive theorem prover for first-order predicates. The ability to manipulate predicates is crucial to the viability of a refinement tool. Also necessary are methods to establish and use theories about

- program objects — both abstract data types used in specifications such as sets and sequences and the concrete data types of our programming language.
- problem domain knowledge

User Interface The user interface for a refinement tool is extremely important. How to present the developing refinement sequence and provide facilities for navigating though and modifying that sequence is a significant challenge. Some of the problems are similar to those addressed by hypertext systems and we are interested to see if we can use results from that research. The user of a refinement tool needs to focus on different concerns at different stages of the refinement task. Once a refinement step has been established by verifying the applicability of the rule, the proof is normally of little concern and need not be displayed. By comparison, during the proof construction, we are unlikely to be interested in the refinement path leading to this proof.

Data refinement Extending our work to include data refinement will require a slightly different approach. Instead of focusing on the refinement of a single specification, data refinement is distributed over a larger scope and requires the transformation of each component of the scope. The transformations are driven by an abstraction invariant that relates the concrete variables to the abstract ones.

Storage and retrieval We see a need to go beyond a monolithic design document for each program. If the potential for reusing the design history is to be achieved, more flexible methods of storage and (particularly) retrieval are required. The possible application of object-oriented data base technology will be investigated.

Calculus rules The refinement calculus rules are not an inviolate set. There is a common core of rules which will be augmented and extended by new rules as experience with the calculus grows. Future tools should make it easy to add new rules and package derived rules (combinations of more basic rules). This provides potential difficulties as a refinement sequence is dependent on the rules used in its development. It may be necessary to store the rules used with each refinement sequence.

6 Conclusions

The development process is at least as important as the product, the executable software, since it serves to document the design and provides a method for checking the correctness of the program (with respect to the initial specification). The refinement calculus is a formal method for refining specifications. Applying the calculus effectively requires computer-based support to manage the many steps from the initial abstract specification to the final executable program. In the paper we have summarized our experiences with tool support for the refinement calculus and made suggestions about goals for future tools.

7 Acknowledgements

We gratefully acknowledge the influence and inspiration of the research on specification at the Programming Research Group, Oxford University. The refinement calculus was developed during Ken Robinson's study leave at PRG in 1985/86. Special thanks are owed to Carroll Morgan for his collaboration on the development of the refinement calculus.

This research is partially supported by an Australian Research Council grant.

References

[1] D. Bjorner and Jones C.B. *Formal Specification and Software Development.* Prentice-Hall International, 1982.

[2] D.A. Carrington and K.A. Robinson. A prototype program refinement editor. In *Australian Software Engineering Conference*, pages 45–63. ACS, 1988.

[3] Edsgar W. Dijkstra. *A Discipline of Programming.* Prentice-Hall, 1976.

[4] R.W. Floyd. Toward interactive design of correct programs. In *IFIP*, pages 7–10, 1971.

[5] T.G. Griffin. *Notational Definition and Top-Down Refinement for Interactive Proof Development Systems.* PhD thesis, Dept. of Computer Science, Cornell University, 1988.

[6] I.J. Hayes, editor. *Specification Case Studies.* Prentice-Hall International, 1987.

[7] C.B. Jones. *Systematic Software Development using VDM.* Prentice-Hall International, 1986.

[8] D.E. Knuth. Literate programming. *The Computer Journal*, 27(2):97–111, 1984.

[9] Carroll Morgan. *Programming from Specifications.* International Series in Computer Science. Prentice-Hall, 1990.

[10] C.C. Morgan. The specification statement. *ACM Transactions on Programming Languages and Systems*, 10(3):403–419, July 1988.

[11] C.C. Morgan and K.A. Robinson. Specification statements and refinement. *IBM Journal of Research and Development*, 31(5):546–555, September 1987.

[12] C.C. Morgan, K.A. Robinson, and P. Gardiner. On the refinement calculus. Technical Report PRG-70, Programming Research Group, 8-11 Keble Road, Oxford OX1 3QD, UK, 1988.

[13] T.W. Reps and T. Teitelbaum. *The Synthesizer Generator.* Springer-Verlag, 1989.

[14] T.W. Reps and T. Teitelbaum. *The Synthesizer Generator Reference Manual.* Springer-Verlag, third edition, 1989.

[15] C. Rich and H.E. Shrobe. Initial report on a LISP programmer's apprentice. *IEEE Transaction on Software Engineering*, 4(6):456–467, 1978.

[16] J.M. Spivey. *The Z Notation: A Reference Manual.* Prentice-Hall International, 1989.

[17] R.C. Waters. The programmer's apprentice: Knowledge based program editing. *IEEE Transaction on Software Engineering*, 8(1):1–11, 1982.

Program Verification by Symbolic Execution of Hyperfinite Ideal Machines (Extended Abstract)

James M. Morris and Mark Howard*
Odyssey Research Associates
301A Harris B. Dates Drive
Ithaca, New York 14850-1313
jmorris%oravax.uucp@cu-arpa.cs.cornell.edu

1 Introduction

This paper describes Ariel, a program verification system based on the symbolic execution of hyperfinite ideal machines. By *ideal machine* we mean a transition system that gives an operational semantics for the language in which the programs we wish to verify are written. The terms of a transition system can be thought of as the states of an ideal (mathematical as opposed to electronic) machine, tailored to the language in question, which executes the program to be verified. The meaning of the program is identified with the execution traces of the transition system or ideal machine. A program is verified by showing that its traces satisfy its specification.

Our transition systems are defined in an applicative programming language for which we have a theorem prover. This applicative language, called Caliban, is lazy, higher-order, and polymorphic. The theorem prover, called Clio, provides a meta-language for making assertions about the meanings of Caliban terms. One can write a Caliban expression whose meaning is the trace of a (sub-)program beginning in some initial state. The pre- and postconditions that the initial and final states of such a trace respectively are to satisfy are written in Clio's meta-language. A (sub-)program is verified by using Clio to show that its precondition implies its postcondition. Clio's proof rule for implication attempts to prove the postcondition under assumption of the precondition. As Clio's prover is basically a term rewriting engine, proving the postcondition is basically a matter of expanding the symbols in the postcondition until no more expansion is possible. Here two properties of Caliban are important. First, Caliban is lazy. It is possible to write Caliban terms that represent non-terminating computations. Only those components of a term (state) of the transition system needed to determine that term's (state's) successor are actually evaluated. This is important for verifying sub-programs independently of their invocations. Secondly, Caliban expressions are executable. This means that much

*Funded by the U.S. Air Force, RADC contract F30602-86-C-0116 and the STARS program contract BOA #3695.STARS-043.

of the calculation involved in determining the next state of an Ariel transition system can actually be carried out automatically by Clio, which has the effect of reducing the amount of "bookkeeping" that must be dealt with in an Ariel verification. Ariel can execute programs symbolically. This provides us with the means to define a trace that represents all those traces in which the program's variables have values satisfying its precondition.

A hyperfinite ideal machine is a transition system in which some of the components of terms are defined using constructs from non-standard analysis. The original version of Ariel was designed primarily for the verification of numerical programs written in a subset of C. We formulated a notion of correctness of numerical programs, called *asymptotic correctness*, using concepts from non-standard analysis (see [5] for a report on the theory of asymptotic correctness). This led to the formulation of non-standard models of computation which have proved to be of general utility in program verification.

The theory underlying Ariel uses non-standard analysis to formalize intuitive notions of what it means for a numerical program to compute a correct result. Numerical programs are not usually handled by verification systems because a sufficiently abstract yet tractable model of real arithmetic is difficult to design. When one attempts to model the amount of error introduced into a numerical calculation by finite arithmetic, one quickly gets bogged down in the intricacies of numerical analysis. The sort of verification we're interested in starts out with a program that purportedly implements some numerical method. We provide the means for showing that the program computes the desired result when run on a sufficiently accurate machine, i.e. that it's *logically* correct. Logical correctness turns out to be equivalent to asymptotic correctness. Asymptotic correctness captures programming errors that don't involve the *magnitude* of the error in computed results. Simple examples of such errors are : the sign of some variable is wrong, an expression uses the wrong operator, etc. A more subtle logical error is the following: comparing the signs of two quantities for equality by multiplying them and checking whether the product is positive. If the quantities are of different sign but small enough that the multiplication underflows to zero, the sign comparison will be incorrect.

2   The Use of Non-standard Analysis in Ariel

A program is asymptotically correct if, given *any* measure of the accuracy of the result to be computed, there *exists* a machine which will compute the result to the degree of accuracy given. Note first that this is weaker than the sort of properties the numerical analysts deal with. Numerical analysis tries to answer questions about how much accuracy in the machine operations and (most importantly) in the representation of the input data is required in order that the program compute a result that satisfies a given accuracy requirement. The problem with numerical analysis is its difficulty: questions of the sort just mentioned are hard to answer in general.

A program that is asymptotically correct will not contain errors of a logical nature. So, if a program has been specified to implement some numerical method, the theory of asymptotic correctness can be used to show that the program implements the method, as long as the specification doesn't require of the results some particular accuracy.

Why not just posit an ideal computing device that has the mathematical reals as a data type? The operations of such a machine would then deliver exact results when operating on exact data. The trouble is no worldly machine has such a data type.

Moreover, most numerical methods only compute exact results in the limit. Therefore, their (straightforward) implementations would be non-terminating on such a machine.

What we would like to model is an ideal machine whose representational power and accuracy are limited, but without having to say to what degree. Intuitively, we would like to say things about the model like "its operations compute results which are close to ideal when their operands are not too big", which is the translation , in our general setting, of the idea that the operations of a worldly machine compute results of a certain accuracy when their operands are not so large as to cause overflow.

Analysis is the mathematics of approximation. It formalizes, in a precise way, our intuitive notions of closeness or approximation. Consider *continuity*. Intuitively, a real function f is continuous if, whenever two values x and y in its domain are close (approximately equal), then $f(x)$ and $f(y)$ are close. Conventional analysis formalizes this intuitive notion by paraphrasing it in terms of the epsilon-delta argument we all learned in elementary calculus: f is continuous at x if, for all $\epsilon > 0$, there is a $\delta > 0$ such that $| f(y) - f(x) | < \epsilon$ for all y for which $| y - x | < \delta$. Non-standard analysis gives a more direct formalization by idealizing "close" by the non-standard notion of *infinitely close*([3] is a general introduction to non-standard analysis). The non-standard formalization of continuity is: f is continuous at a if, whenever x is infinitely close to a, $f(x)$ is infinitely close to $f(a)$. The direct way in which non-standard analysis expresses intuitive notions makes it easier to understand than conventional analysis and more easily implemented in theorem provers. Non-standard analysis can be expressed naturally in terms of operations rather than predicates and with fewer quantifiers, which makes it especially compatible with term-rewrite systems like Ariel's theorem prover.

Non-standard analysis is formed by adding the new undefined predicate *standard* to the usual language of mathematics. Other formal counterparts of informal concepts of non-standard analysis are defined from the predicate *standard*. A real number is *infinite*, corresponding to the informal notion "large", if its absolute value is larger than any standard real number; a real number is *infinitesimal*, i.e. "small", if its absolute value is smaller than any standard positive real number. Two real numbers are *infinitely close*, i.e. "close", if their difference is infinitesimal.

Concepts that can be defined without use of the predicate *standard* are said to be *internal*. These are the concepts of conventional mathematics. Those which cannot be defined without use of *standard* are called *external*, because they formalize notions which are outside conventional mathematics. The point of non-standard analysis is that it provides precise rules for manipulating the external notions. The rules of most interest to us here are:

1. there exists a real number which is non-standard,

2. the collection of standard real numbers satisfies the same internal properties as the collection of all real numbers.

Rule 2 is called the *transfer principal*; it is central to the theory underlying Ariel.

Numerical programs are useful in spite of the fact that they only compute approximations to ideal results, because the results they compute (one hopes) are close enough to ideal, i.e. they compute "good" approximations. For a given program, how good the approximation is depends on the accuracy of the machine on which its executed. If the program is correct, we should get increasingly accurate results, as we run it on increasingly accurate machines and, "in the limit", we should get exact results. This is what it means for a program to be asymptotically correct. Using the standard formalization of continuity as a guide, we might paraphrase asymptotic correctness in terms of an epsilon-delta argument: for any desired degree of accuracy ϵ of the result of running a program P, there exists a degree of accuracy δ of representational power and arithmetic operations such that, whenever P is run on a machine M of accuracy δ, the results will be within ϵ of the ideal ones. Such a formalization would be as unwieldly to reason with as-it is to state. Instead, we take the non-standard formalization of continuity as our guide. We let the machine M of the preceeding paraphrase be *hyperfinite* — the non-standard version of finite. In particular, its real data type is a hyperfinite subset of the non-standard reals. M's real data type will contain all the standard real numbers plus nonstandard reals infinitely close to each standard real. The non-standard values close to a standard real number x correspond to sequences of reals that converge to x. In our theory such sequences are the approximations computed by a program when run on machines of increasing accuracy. We then restate asymptotic correctness: a program P is asymptotically correct if it delivers results that are infinitely close to ideal when run a hyperfinite machine with finite data. This restatement is justified by the transfer principle. Let $\mathcal{M} =< M_0, M_1, M_2, \ldots >$ be a sequence of ideal machines of increasing accuracy and $\mathcal{R} =< r_0, r_1, r_2, \ldots >$ be the results of running P on the members of $\mathcal{M}$. If executing P on a hyperfinite member M_ω of $\mathcal{M}$ produces a result r_ω which is infinitely close to ideal (let r denote the ideal result), then for any standard real value ϵ, the following property is true of $\mathcal{R}$ ($^*\mathbf{N}$ is the non-standard natural numbers)

$$\Phi(\epsilon, {}^*\mathbf{N}) \leftrightarrow \exists n \in {}^*\mathbf{N} \, \forall m \in {}^*\mathbf{N}(m \geq n \rightarrow | \, r - r_m \, | < \epsilon)$$

By transfer, $\Phi(\epsilon, \mathbf{N})$ is true. Since ϵ is arbitrary, $\Phi(\epsilon, \mathbf{N})$ is asymptotic correctness.

3 Clio

Ariel is built around a theorem proving system called Clio. Clio is designed for proving properties of expressions written in a lazy higher-order polymorphic applicative language called Caliban. Caliban is similar to Haskell and Miranda. Theorems about Caliban expressions are stated in Clio's metalanguage, which is essentially first-order logic with Caliban expressions as terms. Clio's prover is based on term rewriting augmented with other proof techniques such as structural induction, case splitting over variables of constructed type, fixed point induction, etc.

3.1 Caliban

Caliban is so similar to the functional languages mentioned above that we needn't describe much of it here. We mention in passing that a lazy applicative language like Caliban is an apt notation for specifying operational semantics. Caliban's constructed types and definitions by pattern matching provide the means for creating very concise descriptions of the components of ideal machines. In addition, the fact that Caliban type constructor functions are taken to be non-strict means that infinite objects, such as the trace of a non-terminating computation, can be defined.

Since it was designed to be a specification language, Caliban also provides a means for working with arbitrary sets. A set together with a bottom element becomes a flat domain called a *sort*. A sort is introduced by declaring its name along with the names and signatures of the functions on it. Properties of the sort and functions are specified by means of axioms. The functions are assumed to be strict and total. The non-standard real numbers we work with in Ariel are represented as a sort.

3.2 Clio's Metalanguage

Clio's metalanguage is essentially the language of first order predicate calculus with Caliban expressions as terms and bounded quantifiers that range over Caliban types. For Ariel's purposes, atomic formulas are of the form $A = B$, where A and B are Caliban expressions enclosed in back quotes. The usual logical connectives are available. Both universal and existential quantifiers are available. All quantifiers are bounded, i.e. they are of the form $(x :: \tau)$ or $(Ex :: \tau)$ where τ is a Caliban type. Metalanguage predicates may be defined.

3.3 Clio's Prover

Clio's prover is basically a term rewriting engine. Whenever an assertion is proved or assumed it is made into a collection of rewrite rules which are added to Clio's rule base. Rules are used in two ways, as conditional rewrite rules and in a limited form of resolution (unit clause resolution when the result is a ground clause). Clio's rule manager also allows the user, interactively, to instantiate existing rules and to expand them (replace symbols by their definitions).

The basic strategy used by the prover is to skolemize an assertion to be proved and then reduce it to conjunctive normal form, i.e. a conjunction of clauses each of which is a disjunction. Each disjunct is a (possibly negated) atomic formula. The prover then attempts to prove each atomic formula by rewriting its left and right sides to the same normal form.

The prover's basic strategy is augmented by a number of additional proof techniques. If the assertion is a conjunction, Clio will break it up into its constituent conjuncts, each of which will have to be proved. If the assertion is an implication, Clio will assume the antecedent and add the corresponding rules to the rule base and then present the consequent to the user to prove (or try to prove it automatically). Existential assertions may be proved by supplying a witness to replace the quantified variable and then proving the resulting assertion. Proof by contradiction may also be used. Universal sentences may be proved either by generalization or structural induction (if the type of the variable of quantification is a constructed type). Disjunctions are proved by proving one of the equivalent assertions: $\neg B \Rightarrow A$ or $\neg A \Rightarrow B$, if the original assertion was $A \vee B$. Negations are proved by first pushing the negation inside the assertion. Then the usual suite of techniques is made available. A case split is available; one may case split both on Caliban and metalanguage expressions.

4   The Ariel Formalization of Numerical Computation

We formalize numerical computations as execution traces of an abstract interpreter. We do so for a number of reasons. First, we intend to expand our semantic model to include concurrent execution and the verification techniques for concurrent programs we find most attractive are based on operational models. Second, we wanted to model, to a certain extent, the possibility that a program was executed on a distributed system comprised of machines with different arithmetic, e.g. different rounding rules. On such a system the time at which a value was computed becomes significant — the approximate value resulting from evaluation of a given expression in the same context would, in principle, be different for different times. Two such values would differ only by an infinitesimal, but they wouldn't be the same. Third, an operational formalization seems most compatible with the concepts of non-standard analysis we use to model real arithmetic. The problem has to do with ensuring that one does not attempt to prove external properties of an execution trace by induction or by means of a proof rule whose soundness depends on an induction on the length of traces. An example of the latter is the theorem about while-loops that says that if I is an invariant of the loop body when the loop guard is true, then I and the negation of the loop guard is true when the loop terminates (if it does).

The semantic model used by Ariel is a transition system along the lines of G. Plotkin's work on structured operational semantics [4]. Ariel is based on an abstract interpreter, i.e. a transition system whose configurations consist of a syntactic representation of a program and some data that represent the current state of the program's execution. A program is represented by its abstract syntax tree coded as a Caliban constructed type. A program's static semantics are checked by the parsing program that builds its abstract syntax tree. The static semantic property most pertinent to execution is the resolution of ambiguous identifier references. In an abstract syntax tree an applied occurrence of a program variable is replaced by the abstract syntax tree for the declaration at the binding occurrence of the identifier determined by the scope rules of the language in which the program is expressed. Clio actually implements this replacement as a reference, so the size of the abstract syntax tree does not grow.

A state of the interpreter consists of a stack of procedure activation records called a *context*. Each activation record consists of a representation of which procedure corresponds to the activation record, where control is in that procedure, a partially evaluated expression, and a local store. As an expression is evaluated, its constituent parts are replaced by the results to which they evaluate. A store is a function from abstract locations to machine values (represented by its graph — a list of location-value pairs). A context is an element of a Caliban constructed type.

A next-state function is defined in Caliban. Traces of the abstract interpreter are defined by iterating the next-state function starting in some initial state. Thus, the meaning of a program or a piece of a program can be thought of as a function from initial states to traces. The next-state function may be iterated a fixed number of times or until some point in the program is reached depending on what one wants to verify. Since Caliban is a lazy language, just those parts of a trace required to prove a program's specification will actually be computed.

To illustrate how proofs of asymptotic correctness are mechanized, we must first describe the formalizationm of real arithmetic. The natural numbers are a built-in Caliban type, NAT. They are modelled as a constructed type generated by the zero-ary constructor ZERO and the strict unary constructor SUCC:

$$NAT ::= Zero \mid Succ \ !NAT$$

(The ! indicates strictness in an argument.) Operations on elements of NAT are defined by recursion in the usual way. Caliban also contains built-in types of truth values, BOOL, and characters, CHAR. Ariel uses the latter two Caliban types for its universes of truth values and characters, respectively. Ariel's integers are constructed from NAT. The Caliban definition is

$$INT ::= Intzero \mid Pos \ !NAT \mid Neg \ !NAT$$

Note that Pos Zero is the first positive integer, namely 1, and analogously for Neg Zero. The successor and predecessor functions on INT are defined in terms of the successor function on NAT in the obvious way.

The non-standard reals are formalized as a Caliban sort. Recall that a Caliban sort is essentially a set (made into a flat domain by appending $\perp$ below all the set's elements) plus functions on the set which are constrained to be strict and total. Sorts are incorporated into the Caliban domain by making their elements new atomic elements of the latter. The definition of the non-standard reals as a Caliban sort is as follows.

```
REAL :: sort including Int_to_real !INT
           with
rstp :: REAL -> REAL
                rfinite :: REAL -> BOOL
                rst :: REAL -> BOOL
                rzero :: REAL
                rplus :: REAL->REAL->REAL
                rminus:: REAL -> REAL -> REAL
                rneg :: REAL -> REAL
                rmult :: REAL -> REAL -> REAL
                rrecip :: REAL -> REAL
                rless :: REAL -> REAL -> BOOL
rclose :: REAL -> REAL -> BOOL
```

The including Int_to_real !INT clause indicates here that the sort contains all the integers plus additional elements; Int_to_real may be thought of as an injection function which also may be used as a constructor. It may be used in definitions by pattern matching, for example.

The functions on REAL are declared to satisfy a set of axioms — the usual axioms for the reals, minus the least-upper-bound axiom (which property the non-standard reals lack).

The axioms for non-standard analysis are based on the standard-part map, rstp, and the predicates rfinite, rst, and rclose (finite, standard and close, respectively). Ideally, there should be a single, polymorphic, undefined predicate *standard*, which should commute with applications of standard functions. This would yield a completely general transfer principle. However, such a predicate could not, in general, be continuous on non-flat domains such as function spaces, since its standardness would have to depend on only finitely many of its values, which is not the case. Therefore, it could not be part of the Caliban domain. Instead, our axioms model consequences of the transfer principle. One of these consequences is that the usual operations on the reals (plus, times, reciprocal, max, min, etc.) commute with the standard-part map. For example,

AXIOM
  (x)(y)'rstp (rplus x y)'='rplus (rstp x) (rstp y)'

(The back quotes set off Caliban expressions in metalanguage formulae; (x) and (y) indicate that x and y are universally quantified.) The are similar axioms for the other arithmetic operations. Other axioms give the relationships between the standard-part map and the predicates mentioned above. For example,

AXIOM
  (x)(y)'rclose x y' = '(rstp x) = (rstp y)'

As our main purpose here is to describe our program verification system, we will omit further details of our formalization of non-standard analysis.

We have specified that our hyperfinite ideal interpreter computes results of real arithmetic operations that differ from the ideal ones by an infinitesimal. This state of affairs is modelled by a function, **machine**, of type [NAT]->REAL->REAL. The first argument is to be interpreted as a time stamp, i.e. when the value was computed. Time stamps are represented as lists of natural numbers for the following reason. Each activation record in a context contains a count of the number of steps performed so far in the interpretation of the procedure corresponding to the activation record. The global time of the interpreter can conveniently be taken to be the ensemble of these local times collected together into a list in the order in which they appear in the context. Such a representation of global time turns out to be useful for differentiating the multiple incarnations of local variables that arise when a procedure is activated recursively. The second argument to **machine** is the ideal value approximated by the machine value. The meaning of this function is captured by the following axiom:

AXIOM 'rstp (machine n x)'='(rstp x)'.

5   Symbolic Execution Using Clio

The basic difference between the Ariel approach to symbolic execution and others(cf. [2]) is that Ariel executes symbolically a complete semantic model of the program to be verified. Each state of an Ariel ideal machine represents all pertinent semantic information and each state of a symbolic execution trace is available directly to the theorem prover. The utility of having this much information available is that it allows us to verify *any* program written in the language supported, e.g programs that use aliasing or constructs for which proof rules or predicate transformers are not available. It allows us to verify security or real time properties which require reasoning about of sequences of states. It allows us to prove *total* correctness.

The Ariel system is interactive and integrated with its theorem prover, Clio. Given a specification for a procedure, one verifies it by proving that its symbolic execution trace satisfies its specification. If the program is specified by a precondition and a postcondition, one proves that the final state of such a trace satisfies the postcondition under the assumption that the initial state satisfied the precondition. This is presented to Clio as the assertion that the precondition implies the postcondition. Clio's built-in proof rule for implication causes the precondition to be assumed and a proof of the postcondition to be started; assumption of the precondition causes it to be transformed into a collection of rewrite rules.

The first step Clio takes to prove the postcondition is to send it to the reducer to be put into normal form. This invokes symbolic execution. A typical postcondition might say that the value returned by a function-procedure call (perhaps via a parameter) has some property. Assume, for the sake of discussion, that the postcondition of the procedure we wish to verify is

$$\text{Post 'x' 's' 'v' := 'Return\_state x s' = 'v'}$$

where Return_state is a Caliban function that iterates (in the sense of repeated function application) the next-state function beginning in state s until the current command is a return. Assume x is one of the procedure's parameters and v denotes some machine value. Clio puts Post into normal form by applying to it all the applicable rewrites in its rule base. When the definitions specifying the Ariel abstract machine are compiled by Clio, they are transformed into rewrite rules. So, the process of normalizing Post is essentially a matter of carrying out the repeated application of the next-state function beginning with a rule base that reflects the assumptions of the precondition. If the procedure to be verified consists of no more than a sequence of assignment statements, then when Post is normalized, the store of the resulting state will reflect the effect of the sequence of assignments and the postcondition is proved by "looking up" x in the final store and seeing whether its value is v. Clio should be able to do this automatically.

The usual treatment of conditional statements in symbolic execution is to bifurcate the current path in the symbolic execution tree and attach the conditional statement's guard and its negation to the pathconditions of the respective newly created paths. In the Ariel system, this is handled by a case split. Clio detects the need for a case split automatically, when it encounters a Boolean subexpression whose value it is unable to determine in the process of normalizing an assertion. What happens is that the truth of the subexpression is assumed (and made into rules) and the proof of the assertion continued and similarly for the assumption that the subexpression is *false*. Ariel represents the path condition of other symbolic execution schemes by Clio's rule base, which makes all Clio's rule manipulation facilities available for arguments involving the path condition.

Loops are treated as follows. An assertion is attached to the beginning of the loop statement. This assertion should state

1. that each time control is at the beginning of the loop, the loop guard is *true*,

2. some invariant having to do with what the loop is to compute, and

3. that some function of the variables manipulated by the loop is strictly decreasing.

We then prove that this assertion cannot hold of any unbounded trace of a hyperfinite interpreter. The proof uses a theorem to the effect that in any unbounded trace of a hyperfinite machine, there are two states in which all identifiers are assigned the same values. This contradicts 3. Thus we first prove that the loop terminates provided that the invariance of the loop assertion can be established. Clio makes this into a rule (or rules). The postcondition for the loop should consist of 2 and the negation of 1, i.e.

when the loop ends (because its guard evaluates to *false*), it establishes what we wanted to compute. This is familiar. We prove the postcondition by contradiction. Having assumed its negation, we then prove the loop assertion by induction to get the required contradiction. If parts 2 and 3 of the loop assertion are actually invariant, then the only problem is with a state in which the loop guard is *false*. In this case, symbolic execution is invoked from the state in which control is at the beginning of the loop for the ith time, which allows Clio to prove the postcondition — a contradiction. If, in the state at the beginning of the ith iteration, the loop guard evaluates to *true*, the induction step can be proved directly.

Subroutines may be verified independently. A subroutine is specified by giving a pre- and postcondition for it. Using Clio, one proves that the subroutine terminates in a state satisfying its postcondition when started up in a state satisfying its precondition. Having successfully carried out the verification, one has Clio make the specification into rewrite rules. The latter will essentially be conditional rewrites which add the content of the postcondition to the rule base, whenever it's possible to prove the precondition.

6 Conclusions

We have described Ariel and the approach to program verification upon which it is based. The full paper includes a sample verification of a numerical program. We have built a prototype verification system and are in the process of trying it out on a suite of production numerical routines. These routines implement a continuous simulation of a rocket trajectory. They comprise about 1400 lines of C code (translated from the original FORTRAN) and calculate such things as a solution to the two-body problem, a Runge-Kutta integration, a Newton-Raphson iteration to solve a version of Kepler's equation, etc. Our prototype verifys programs written in a subset of C; we are extending the C system to handle Unix system calls and are in the process of designing a new system for (a subset of) Ada.

7 Acknowledgments

The notion of asymptotic correctness and the use of nonstandard analysis to formalize it is due to Ian Sutherland based on a suggestion of Richard Mansfield. We would like to acknowledge the other members of the Ariel group, past and present, who contributed to the work reported here: Ian Sutherland, Doug Hoover, and Steve Brackin. We thank Garrel Pottinger and an anonymous referee for their comments on an earlier draft of this paper.

References

[1] Mark Bickford, Charlie Mills, and Edward A. Schneider. *Clio: An Applicative Language-Based Verification System.* Odyssey Research Associates Technical Report TR 15-7. April, 1989.

[2] S.L.Hantler and J.C.King. An introduction to proving the correctness of programs. *ACM Computing Surveys*, 8(3):331–353, September, 1976.

[3] Albert E. Hurd and Peter A. Loeb, *An Introduction to Nonstandard Real Analysis.* Academic Press, 1985.

[4] Gordon D. Plotkin. *A Structural Approach to Operational Semantics.* DAIMI FN-19. Computer Science Department. Aarhus University, Denmark. September, 1981.

[5] Ian Sutherland. *A Mathematical Theory of Asymptotic Computation.* Rome Air Development Center Technical Report RADC-TR-87-261, Air Force Systems Command, Griffiss Air Force Base, NY. 1987.

Extension of the Karp and Miller Procedure to Lotos Specifications [1]

Michel Barbeau, Gregor V. Bochmann

Université de Montréal, Département d'IRO

C.P. 128, Succ. "A", Montréal, Canada, H3C 3J7

Abstract

In a companion paper [Barb 90a,b], we proposed a Place/Transition-net (P/T-net) semantics for a subset of Lotos. This subset is such that Lotos specifications can be translated into finite structure Petri nets. It is therefore possible to apply P/T-net verification techniques since they require finite structure. It this paper, we demonstrate that it is possible to apply P/T-net verification methods without an a priori construction of the P/T-net associated to the Lotos specification to be analysed. In particular we consider a well known reachability analysis technique for P/T-nets, namely, the Karp and Miller procedure.

1 Introduction

The goal of this work is to investigate verification techniques for basic Lotos specifications [Bolo 87a]. Basic Lotos is presented in § 4. It can be demonstrated that this language has the computational power of Turing machines. Therefore, generally non-trivial properties are undecidable.

Our verification method is based on Petri nets. Petri net verification techniques are transfered to Lotos. So far, two transfer approaches have been proposed. A first approach consists of translating Lotos specifications into Petri nets and evaluating the properties on the equivalent Petri net models. Presently, there are at least two tools based on this approach ([March 89] and [Gara 90]).

We found that the Lotos to Petri nets translation step is, in many cases, as complex, in terms of time and space, as the verification step itself. In this paper, we propose a second approach which involves no translation from one formalism to another. We adapt Place/Transition-net (P/T-net) verification algorithms to Lotos. We consider a well known P/T-net reachability analysis technique, the Karp and Miller procedure [Karp 69].

The main difficulty in these approaches is to make shure that the P/T-net, that models the Lotos specification, has a finite structure. That is, a bounded number of places and transitions. We model Lotos local process states as Petri net places. We put syntactical constraints so that the number of local process states is bounded. Moreover, the number of alternatives to make a transition, from a given local state, is also bounded. Consequently, the Petri net has a finitre structure. However, these constraints do not bound the number of identical parallel processes. Therefore, finite structure Petri nets does not necessarily mean finite state systems. Consequently, the "Lotos" that can be analyzed with our method is computationaly more powerful than the "Lotos" that can be verified with finite state/transition system based methods (e.g. [Bolo 87b]).

Petri nets can represent, with finite structures, infinite state systems. This means that classical finite state system reachability analysis [Boch 78] is not applicable. Karp and Miller trees and graphs are finite and partial representations of, in general infinite, Petri net reachability trees and graphs. P/T-nets are introduced in § 2. Reachability analysis for P/T-nets is discussed in § 3. Modelling of Lotos by P/T-nets is presented in § 5. The extension of Petri net reachability analysis to Lotos is discussed in § 6.

---

[1]This work has been funded by the Natural Sciences and Engineering Research Council of Canada, the Centre de recherche informatique de Montréal and Bell Northen Research.

2 P/T-nets

We slightly deviate from the usual notation for P/T-nets [Pete 81]. We represent a P/T-net as a tuple (P, T, Act, M_0) where:

- P is a set of places $\{p_1, ..., p_n\}$,

- $T \subseteq \mathcal{N}^P \times Act \times \mathcal{N}^P$, is a transition relation,

- Act is a set of transition labels, and

- $M_0 \in \mathcal{N}^P$, is the initial marking.

A P/T-net has a finite structure if the sets P, T and Act are finite.

$\mathcal{N}$ is the set of non-negative integers. $\mathcal{N}^P$ denotes the set of multi-sets over the set P. A multi-set is a set that can contain multiple instances of the same element. An element $t = (X, a, Y) \in T$ is also denoted as $X - a \to Y$. Its preset $pre(t)$ is X, its postset $post(t)$ is Y and action $act(t)$ is a. The multi-sets X and Y are also called, respectively, the input and output places of t. We denote as $pre(t)(p)$ ($post(t)(p)$) the number of instances of the element p in the preset (postset) of t.

A Petri net marking is also a multi-set. We denote by $M(p_i)$ the number of instances of the element p_i in the multi-set M. A marking M is also denoted as a n-tuple $(M(p_1), ..., M(p_n))$. Instances of the element p_i are also called tokens inside place p_i. The operators $\leq$, $+$ and $-$ denote respectively multi-set inclusion, summation and difference.

$pre(t)(p)$ is the number of tokens that place p must contain to enable transition t. A transition $t \in T$ is enabled in marking M if $pre(t) \leq M$. This is denoted as $M(t >$. An enabled transition can be fired and the successor marking M' is defined as: $M' = M - pre(t) + post(t)$, this is represented as $M(t > M'$.

A P/T-net is illustrated in Fig. 1, places are shown as circles, transitions as bars and tokens as dots inside places. There is a directed edge from place p_i (transition t_j) to transition t_j (place p_i) iff $p_i \in pre(t_j)$ ($p_i \in post(t_j)$). If $pre(t)(p) > 1$ ($post(t)(p) > 1$) we may label the corresponding edge with the value of $pre(t)(p)$ ($post(t)(p)$). The initial marking of this particular net can be denoted as the multi-set $\{p_1, p_2\}$ or as the 6-tuple $(1, 1, 0, 0, 0, 0)$.

3 Reachability Analysis for P/T-nets

With respect to Holzmann's classification [Holz 89], the Karp and Miller tree construction procedure is a *stack search strategy*. It is a depth-first technique that minimizes memory usage at the expense of run time. The whole state space does not have to be maintained in memory, only the path starting from the initial marking to the current marking. In general, Petri nets are not finite state systems. The Karp and Miller tree is also called the coverability tree because for every reachable marking M of a Petri net, there exists a marking M' in the Karp and Miller tree such that $M \leq M'$.

Definition 1 *The coverability tree (CT) associated to a P/T-net $N = (P, T, Act, M_0)$ is a tree, where vertices are labelled with markings of N and edges are labelled with elements in T. The CT can be recursively defined as follows:*

1. The root is labelled with M_0;

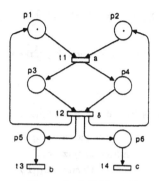

Figure 1: A P/T-net

2. Let x denote a vertex with label M,

    (a) if there exists a vertex y such that $y \prec x$ and $label(y) = M$, then x is a leaf;

    (b) else, the successors of x are in one-to-one correspondence with the elements of the set:

$$S = \{(t, M') : t \in T \wedge M(t > M'\}$$

    Let $(t, M') \in S$, we create the successor vertex z and the edge (x, t, z). The label of z is determined as follows: For $i = 1, ..., n$ (n is the number of places),

        i. if there exists a vertex y such that $y \prec z$, $label(y) \leq M'$ and $label(y)(p_i) < M'(p_i)$ then $label(z)(p_i) = \omega$;

        ii. else, $label(z)(p_i) = M'(p_i)$;

The symbol ω denotes infinity. If the whole CT is maintained in memory, the coverability graph (CG) can be obtained by merging vertices with identical labels. In that case, more memory is required but full connectivity information is stored and can be used to analyse loops.

Example 1: For the net of Fig. 1, the CG is shown in Fig. 2.

With the CT and the CG, the following six problems [Fink 90] become decidable:

1. *Termination* Is the reachability tree finite?

2. *Finiteness* Is the reachability set finite?

3. *Coverability* Given a marking M, is there a reachable marking M' such that $M \leq M'$?

4. *Quasi-liveness* Given an action a, is there a reachable marking M such that a is executed from M ?

5. *Boundedness* Is the number of tokens in a given place bounded?

6. *Regularity* Is the Petri net language recognizable by a finite state automaton?

The language of a Petri net is the set of transition sequences starting from the initial marking. If the language is regular the Petri net can be simulated by a finite state automaton. The Karp and Miller tree and graph constructions do not necessarily detect every deadlock in a nonfinite state system.

$$(1,1,0,0,0,0) - t_1 \rightarrow (0,0,1,1,0,0) - t_2 \rightarrow (1,1,0,0,\omega,\omega) \quad \begin{matrix} \leftarrow t_2 - \\ -t_1 \rightarrow \end{matrix} \quad (0,0,1,1,\omega,\omega)$$

$$\circlearrowleft_{t3,t4} \qquad\qquad\qquad\qquad \circlearrowleft_{t3,t4}$$

Figure 2: Coverability Graph

4 Lotos

A basic Lotos behavior expression is formed out of the following terms:

Inaction	stop
Action prefix	$a; B$
Choice	$B_1[]B_2$
Process instantiation	$p[g_1, ..., g_n]$
Pure interleaving	$B_1|||B_2$
General parallel composition	$B_1|[g_1, ..., g_n]|B_2$
Successful termination	exit
Sequential composition	$B_1 >> B_2$
Disabling	$B_1[> B_2$
Hiding	hide $g_1, ..., g_n$ in B_1

where B, $B1$ and $B2$ are behavior expressions. The semantics of Lotos is given in [Bolo 87a].

This subset of Lotos has the computational power of Turing machines (proved in [Barb 90a]). We conclude that nontrivial properties are generally undecidable. P/T-nets (with finite structures) do not have the computational power of Turing machines as Lotos does. In the rest of this section we define a subset of Lotos, PLotos, that can be modelled by finite structure P/T-nets, and conversely into which P/T-nets can be simulated. The mapping from PLotos to P/T-nets is introduced in the next section whereas P/T-nets simulation in PLotos is discussed in [Barb 90a,b].

We assume that PLotos specifications satisfy the following constraints:

1. *Guarded recursive processes.* A process instantiation term is guarded if it is in the scope of a prefixing operator ";" or in the right sub-expression B_2 of a sequential composition $B_1 >> B_2$ or of a disabling $B_1[> B_2$.

2. *No combination of recursion and general parallel composition.* The general parallel operator "$|[g_1, ..., g_n]|$" is not allowed on recursive paths.

3. *Tail recursion.* The process in which $B_1 >> B_2$ (or $B_1[> B_2$) is defined may not be called from sub-expression B_1. And, in B_1 combination of recursion and parallelism is not allowed.

4. *"Noexit" functionality in pure interleaving.* Operands B_1 and B_2 in parallel composition $B_1|||B_2$ must have the *noexit* functionality.

Paths are defined by choice alternatives and can be illustrated as a tree with behavior expressions labelling nodes, as in the following example:

process $p[a, b, c]$: noexit :=

    $(a; stop|||b; stop)$

    $[]$

    $c; p[a, b, c]$

endproc

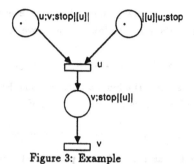

Figure 3: Example

The path number 1 is not recursive whereas the path number 2 is because the process p is recursively called at node labelled "$c; p[a, b, c]$". A process that does not contain parallel operator "$|||$" and "$|[g_1, ..., g_n]|$" on recursive paths can be modelled by a finite state system. If the operator "$|[g_1, ..., g_n]|$" is disallowed on recursive paths, whereas "$|||$" is allowed with functionality *noexit* operands, the system is not finite state but can still be represented by a finite structure P/T-net. The functionality of a behavior B is equal to *exit* iff it terminates with the successful termination action δ, otherwise it is equal to *noexit*, [Bolo 87a].

PLotos specifications are rewritten into simpler forms. Non-recursive paths are expanded, that is, process definitions are substituted for process calls. Then we distinguish every parallel composition $B_1|[g_1, ..., g_n]|B_2$ by labelling the operator with an unique value k. This is represented as $|[g_1, ..., g_n]|_k$.

5 Modelling of Lotos with P/T-nets

The mapping from Lotos to P/T-nets is based on the work of Olderog [Olde 87] for CCSP. In general, a Lotos behavior expression B represents the composition of several concurrent components. The expression B is explicitly decomposed into its parallel components that become tokens when this behavior is activated. Parallel components and states of parallel components are respectively modelled by Petri net tokens and places. The place in which a token is contained denotes the component state. Lotos gates are modelled by Petri net transitions. Tokens, contained in transition input places, represent components synchronized on this gate. Tokens deposited into output places represent the successor components after the transition has occurred. Several tokens, contained in the same place, represent several identical components. This models unbounded process instantiation with finite structure P/T-nets.

For example, the Lotos expression $u; v; stop|[u]|u; stop$ represents two concurrent components. The first component executes actions u and v and then stops. The second component executes action u and becomes inactive. Both components are coupled on gate u and are therefore dependent on each other with respect to the occurrence of u. The decomposition of $u; v; stop|[u]|u; stop$ into components is denoted as the set $\{u; v; stop|[u]|, |[u]|u; stop\}$. In this syntax, we represent explicitly the fact that components are coupled on gate u by concatenating the symbol $|[u]|$ to the right of $u; v; stop$ and to the left of $u; stop$.

Fig. 3 depicts the corresponding P/T-net model. Places modelling states of components are labelled with the corresponding expression components. Transitions are labelled with gate names. The "stop" expression represents inaction and does not appear in the P/T-net. In our construction, edges from place to transition are always one valued and every place has a distinct label. We unambiguously denote a place by its label. The above "u" labelled transition is represented as the triple:

$$\{u; v; stop\|[u]\|, \|[u]\|u; stop\} - u \rightarrow \{v; stop\|[u]\|\}$$

To derive such triples, we define i) a function decomposing PLotos expressions, and ii) a system of inference rules. The head of each rule matches a term of the form:

$$\{p_1, ..., p_m\} - a \rightarrow \{q_1, ..., q_n\}$$

A rule can be used to infer, as a function of place label structures, a transition with preset $\{p_1, ..., p_m\}$, action a and postset $\{q_1, ..., q_n\}$. For instance the rule:

if $M_1 - a \rightarrow M_1'$ and $a \notin \{S, \delta\}$
then $M_1.\|[S]\|_k - a \rightarrow M_1'.\|[S]\|_k$

has been used to infer the transition:

$$\{v; stop\|[u]\|\} - v \rightarrow \{\}$$

We substituted $\{v; stop\}$, u and v to respectively M_1, S and a. M_1' is empty because the decomposition of "stop" is defined as the empty set. We first introduce the decomposition function in §5.1, then we present in §5.2 the inference rules.

5.1 Decomposition Function

The decomposition function is denoted as dec. Its domain is the set of well-formed PLotos behavior-expressions. Its range is the set of all possible multi-sets of place labels.

Let B_1, B_2 denote syntactically correct PLotos behavior expressions, a denote an action name and $S = g_1, ..., g_n$ a list of synchronization gates,

(d1)	$dec(stop)$	$:= \{\}$
(d2)	$dec(a; B_1)$	$:= \{a; B_1\}$
(d3)	$dec(B_1[]B_2)$	$:= \{B_1[]B_2\}$
(d4)	$dec(p[g_1, ..., g_n])$	$:= dec(B_p[g_1/h_1, ..., g_n/h_n])$
(d5)	$dec(B_1|||B_2)$	$:= dec(B_1) + dec(B_2)$
(d6)	$dec(B_1\|[S]\|_k B_2)$	$:= dec(B_1).\|[S]\|_k + \|[S]\|_k.dec(B_2)$
(d7)	$dec(B_1 >> B_2)$	$:= \{B_1 >> B_2\}$
(d8)	$dec(B_1[> B_2)$	$:= \{B_1[> B_2\}$
(d9)	$dec(hide\ S\ in\ B_1)$	$:= hide\ S\ in.dec(B_1)$
(d10)	$dec(exit)$	$:= \{exit\}$

where

- B_p represents the body of process definition p,

- in (d4), $g_1, ..., g_n$ is a list of actual gates,

- $h_1, ..., h_n$ is a list of formal gates,

- $[g_1/h_1, ..., g_n/h_n]$ is the relabelling postfix operator, gate h_i becomes gate g_i $(i = 1, ..., n)$, and

- the expression $dec(B_1).\|[S]\|_k$ denotes $\{x\|[S]\|_k : x \in dec(B_1)\}$, similarly for $\|[S]\|_k.dec(B_2)$ and the expression $hide\ S\ in.dec(B_1)$ denotes $\{hide\ S\ in\ x : x \in dec(B_1)\}$.

The *dec* function is deterministic, taking into account operator precedences. The restriction to guarded recursive processes is required to stop recursion in the *dec* function. The relabelling operator is not user accessible and exists for the semantic description of process instantiation. In Lotos, relabelling is dynamic. Gates are renamed at the execution time. We show in [Barb 90a] that for injective relabelling operators, static and dynamic relabelling are equivalent. For the sake of simplicity, hereafter we consider solely injective relabellings and perform static renaming.

5.2 Inference Rules

This section presents the inference rules of the mapping from PLotos to P/T-nets. The P/T-net $N = (P, T, Act, M_0)$ associated to a PLotos behavior B is such that:

1. $M_0 = dec(B)$, $(\forall p)[M_0(p) > 0 \Rightarrow p \in P]$,

2. if $X \subseteq P$ and $X - a \rightarrow Y$ then $(\forall p)[Y(p) > 0 \Rightarrow p \in P]$, $(X, a, Y) \in T$, $a \in Act$, and

3. only the elements that can be obtained from items 1 or 2 are in P, T and Act.

The transition instances are inferred from the rules bellow.

For all PLotos expressions B_1, B_1', B_2, B_2', action name a, list $S = g_1, ..., g_n$ of synchronization gates and place multi-sets M_1, M_2, M_1', M_2':

(r1) $\{a; B_1\} - a \rightarrow dec(B_1)$

(r2) if $B_1 - a \rightarrow B_1'$
  then $\{B_1[]B_2\} - a \rightarrow dec(B_1')$

(r3) if $B_2 - a \rightarrow B_2'$
  then $\{B_1[]B_2\} - a \rightarrow dec(B_2')$

(r4) if $M_1 - a \rightarrow M_1'$ and $a \notin \{S, \delta\}$
  then $M_1.\|[S]\|_k - a \rightarrow M_1'.\|[S]\|_k$

(r5) if $M_2 - a \rightarrow M_2'$ and $a \notin \{S, \delta\}$
  then $\|[S]\|_k.M_2 - a \rightarrow \|[S]\|_k.M_2'$

(r6) if $M_1 - a \rightarrow M_1'$ and $M_2 - a \rightarrow M_2'$ and $a \in \{S, \delta\}$
  then $M_1.\|[S]\|_k + \|[S]\|_k.M_2 - a \rightarrow M_1'.\|[S]\|_k + \|[S]\|_k.M_2'$

(r7) if $B_1 - a \rightarrow B_1'$ and $a \neq \delta$
  then $\{B_1 >> B_2\} - a \rightarrow \{B_1' >> B_2\}$

(r8) if $B_1 - \delta \rightarrow B_1'$
  then $\{B_1 >> B_2\} - \delta \rightarrow dec(B_2)$

(r9) if $B_1 - a \rightarrow B_1'$ and $a \neq \delta$
  then $\{B_1[> B_2\} - a \rightarrow \{B_1'[> B_2\}$

(r10) if $B_1 - \delta \rightarrow B_1'$
  then $\{B_1[> B_2\} - \delta \rightarrow dec(B_1)$

(r11) if $B_2 - a \rightarrow B_2'$
  then $\{B_1[> B_2\} - a \rightarrow dec(B_2')$

(r12) if $M_1 - a \rightarrow M_1'$ and $a \notin \{S\}$
  then $hide\ S\ in.M_1 - a \rightarrow hide\ S\ in.M_1'$

(r13) if $M_1 - a \rightarrow M_1'$ and $a \in \{S\}$

     then $hide\ S\ in.M_1 - i \rightarrow hide\ S\ in.M_1'$

(r14) $\{exit\} - \delta \rightarrow \{stop\}$

In the "if part" of inference rules (r2), (r3), (r7) (r8), (r9), (r10) and (r11) behavior B_1 (B_2) makes a transition to behavior B_1' (B_2') on action a or δ in accordance with the original Lotos semantics in [Bolo 87a]. Consistency of this P/T-net interpretation of Lotos is formally proved in [Barb 90a].

    Example 2: Consider the following specification[2]:

specification $p1[a, b, c]$: **noexit** :=

        $p2[a, b]|[a]|p2[a, c]$

where

process $p2[x, y]$: **noexit** :=

            $x; exit >> (y; stop|||p2[x, y])$

endproc

endspec

The P/T-net derived from this specification is identical to the net depicted in Fig. 1 with:

$p_1 = a; exit >> (b; stop|||p2[a, b])|[a]|$ $p_2 = |[a]|a; exit >> (c; stop|||p2[a, c])$

$p_3 = exit >> (b; stop|||p2[a, b])|[a]|$ $p_4 = |[a]|exit >> (c; stop|||p2[a, c])$

$p_5 = b; stop|[a]|$ $p_6 = |[a]|c; stop$

6   Coverability Graphs for Lotos

Given a Lotos specification, it is possible to construct an equivalent P/T-net model by successive applications of the above inference rules. This P/T-net then becomes the input of the reachability analysis algorithm to evaluate the properties. In the worst case, the P/T-net can have more vertices and edges than the coverability graph. In our approach we skip the intermediate Lotos to P/T-nets translation step. We derive the coverability graph directly from the Lotos specification then properties are evaluated.

    The syntax of Lotos coverability graphs slightly deviates from the usual syntax for Karp and Miller graphs. Markings are multi-sets of Lotos behavior expression components. We label the root of the graph with the decomposition of the Lotos expression that represents the initial behavior. For example, the decomposition of the initial behavior in Example 2 yields a state represented as the following box:

> $1/a; exit >> (b; stop|||p2[a, b])|[a]|$
> $1/|[a]|a; exit >> (c; stop|||p2[a, c])$

Every line in the box defines the number of instances of one behavior expression component type in the current state. In case there is an infinite number of occurrences, the expression component is paired with the ω symbol.

    We go from one marking to another by application of the inference rules. An inference rule is applicable from one marking if a finite subset of the expression component multi-set matches the preset of the transition in the head of the rule. The successor state is obtained by removing this

---

[2]For the sake of simplicity, labelling of the $|[a]|$ operator is omitted in this example.

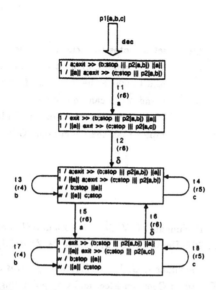

Figure 4: Coverability graph of Example 2

preset from the current state and adding the postset defined by the transition (reformulation of the usual P/T-net firing rule). Every edge is labelled with the number of the inference rule which has been applied to derive the transition and the action name of the transition. The coverability graph of Example 2 is illustrated in Fig. 4.

The six problems stated in § 3 can be solved as follows:

1. *Termination* The reachability tree is infinite if there is at least one circuit in the CG.

2. *Finiteness* The reachability set is infinite if the CG contains one marking and one process p paired with ω.

3. *Coverability* Given a marking M, there exists a reachable marking M' such that $M \leq M'$, if there exists in the CG a marking M'' with $M \leq M''$.

4. *Quasi-liveness* The action a is quasi-live if there exists an edge in the CG labelled with a.

5. *Boundedness* Instantiation of process p is unbounded if there exists a marking M in which $dec(p)$ is paired with ω.

6. *Regularity* The language is regular if every elementary circuit of the CG is labelled by a sequence of transitions $t_1, t_2, ..., t_n$ such that for every place p:

$$post(t_1)(p) - pre(t_1)(p) + post(t_2)(p) - pre(t_2)(p) + \cdots post(t_n)(p) - pre(t_n)(p) \geq 0$$

In general, conclusions can be easily drawn from visual inspection of the CT and the CG. For instance, the language of Example 3 is not regular since:

$$post(t_3)(b; stop||[a]|) - pre(t_3)(b; stop||[a]|) = -1 < 0$$

7 Conclusion

We have presented a reasonable subset of Lotos that can be verified using Petri net reachability analysis techniques. Our method does not require explicit translation from Lotos to Petri nets. Analysis is performed in the Lotos world to which the Karp and Miller procedure is extended. To cope with state space explosion, MCT and MCG can be computed to solve the aforementioned six problems. We experimented the MCT construction procedure and obtained satisfactory results. The MCTs were several times less complex than the nonminimal CTs.

8 References

[Barb 90a] Barbeau, M., Bochmann, G. V. *Deriving Analysable Petri Nets from Lotos Specifications*, Research report no. 707, Dept. d'IRO, Université de Montréal, 1990.

[Barb 90b] Barbeau, M., Bochmann, G. V. *Verification of Lotos Specifications: A Petri Net Based Approach*, Proc. of Canadian Conf. on Elec. and Computer Eng., Ottawa, 1990.

[Barb 89] Barbeau, M., Bochmann, G. V. *Experiences with Automated Verification Tools: Application to Discrete Event Systems*, Proc. of Workshop on Automatic Verification Methods for Finite State Systems, Grenoble, 1989.

[Boch 78] Bochmann, G. V. *Finite State Description of Communication Protocols*, Computer Network 2, (361-372), 1978.

[Bolo 87a] Bolognesi, T., Brinksma, E. *Introduction to the ISO Specification Language Lotos*, Computer Networks and ISDN, Vol. 14, No. 1, (25-59), 1987.

[Bolo 87b] Bolognesi, T., Smolka, S. A. *Fundamental Results for the Verification of Observational Equivalence: A Survey*, Proc. of PSTV VII, Zurich, 1987.

[Fink 90] Finkel, A. *A Minimal Coverability Graph for Petri Nets*, Proc. 11th Int. Conf. on Application and Theory of Petri Nets, Paris, 1990.

[Gara 89] Garavel, H., Najm, E. *Tilt: From Lotos to Labelled Transition Systems*, in P. H. J. van Eijk, C. A. Vissers and M. Diaz (Eds.): The Formal Descrip. Tech. Lotos, North-Holland, 1989.

[Holz 89] Holzmann, G. J. [1989]. *Algorithms for Automated Protocol Validation*, Proc. of Workshop on Automatic Verification Methods for Finite State Systems, Grenoble, 1989.

[Karp 69] Karp, R. M., Miller, R. E. *Parallel Program Schemata*, J. Computer and System Sciences 3, (147-195), 1969.

[Marc 89] Marchena, S., Leon, G. *Transformation from Lotos Specs to Galileo Nets*, in K. J. Turner (Ed.): Formal Description Techniques, North-Holland, 1989.

[Olde 87] Olderog, E.-R. *Operational Petri Net Semantics for CCSP*, LNCS 266, Springer-Verlag, 1987.

[Pete 81] Peterson, J. L. *Petri Net Theory and the Modelling of Systems*, Prentice Hall, 1981.

An Algebra for Delay-Insensitive Circuits
(Abridged Version*)

Mark B. Josephs
Programming Research Group
Oxford University Computing Laboratory
11 Keble Road, Oxford OX1 3QD, U.K.
Phone: +44-865-272574
E-mail: mark%prg.oxford.ac.uk@nss.cs.ucl.ac.uk

Jan Tijmen Udding
Department of Computer Science
Washington University
Campus Box 1045, St. Louis, MO 63130, U.S.A.
Phone: 314-889-6110
E-mail: jtu@cs.wustl.edu

A novel process algebra is presented; algebraic expressions specify delay-insensitive circuits in terms of voltage-level transitions on wires. The approach appears to have several advantages over traditional state-graph and production-rule based methods. The wealth of algebraic laws makes it possible to specify circuits concisely and facilitates the verification of designs. Individual components can be composed into circuits in which signals along internal wires are hidden from the environment.

0   Introduction

A circuit is connected to its environment by a number of wires. If the circuit functions correctly irrespective of the propagation delays in these wires, the circuit is called *delay-insensitive*. Delay-insensitive circuits are attractive because they can be designed in a modular way; indeed no timing constraints have to be satisfied in connecting such circuits together. As a result of the latest Turing Award Lecture [13], the design of delay-insensitive circuits has drawn renewed interest.

The design of delay-insensitive circuits is made difficult by the need to consider situations in which a signal (voltage-level transition) has been transmitted at one end of a wire but has not yet been received at the other end. The algebraic notation presented in this paper may be helpful in the following ways:

---

*The full version of this paper appears in the ACM/AMS DIMACS series and as Technical Report WUCS-89-54, Dept. of C.S., Washington Univ., St. Louis, MO.

1. The functional behaviour of primitive delay-insensitive components can be captured by algebraic expressions.

2. All possible behaviours of the circuit that results when such components are connected together can be determined by symbolic manipulation.

3. The algebra facilitates the precise specification of the circuit that the designer has to build, including obligations to be met by the environment.

4. The algebra supports verification of the design against the specification.

The algebra is based upon Hoare's CSP notation [7]. It adapts the theory of asynchronous processes [8, 9] to the special case of delay-insensitive circuits. The possibilities of *transmission interference* and *computation interference*, characterized by Udding [14, 15], are faithfully modelled in the algebra; the designer is able to reason about these errors, and so avoid them. Underpinning the algebra is a denotational semantics similar to those given in [2, 9]; the semantics is compatible with the failures-divergences model of CSP [1, 7].

Our approach complements that taken by Martin [10, 11] to the design of delay-insensitive circuits. Martin's approach, however, is more general in that he makes use of components that are not delay-insensitive, namely his *isochronic forks*. We have discovered that it is possible to understand many of Martin's circuits by treating the isochronic fork, together with one of the gates to which it connects, as a single primitive delay-insensitive component. (Ebergen [5], on the other hand, has investigated how components that are sensitive to delay can be synchronised to form delay-insensitive circuits.)

The remainder of this paper provides a step-by-step introduction to the algebra. We also prove, as a case study, some of Martin's nontrivial circuit designs to be algebraically correct. Similar verifications have been done by Dill [3, 4]. His verifications, however, are performed at the semantics level rather than at the syntactic/algebraic level. Algebraic calculations are arduous but humanly feasible, as well as mechanisable, and seem to avoid a state explosion.

1 Basic Notions and Operators

A process is a mathematical model at a certain level of abstraction of the way in which a delay-insensitive circuit interacts with its environment. Typical processes are P and Q. A circuit receives signals from its environment on its input wires and sends signals to its environment on its output wires. Thus with each process are associated an alphabet of input wires and an alphabet of output wires. These alphabets are finite and disjoint. Typical names for input wires are a and b; typical names for output wires are c and d. The time taken by a signal to traverse a wire is indeterminate.

In the remainder of this section and section 2, we consider processes with a particular alphabet I of input wires and a particular alphabet O of output wires.

The process P is considered to be "just as good" as the process Q $(P \subseteq Q)$ if no environment, which is simply another process, can when interacting with P determine that it is not interacting with Q. (This is the refinement ordering of CSP [1, 7], also known as the *must* ordering [6].) Two processes are considered to be equal when they are just as good as each other.

The refinement ordering is intimately connected with nondeterministic choice. The process $P \sqcap Q$ is allowed to behave either as P or as Q. It reflects the designer's freedom to implement such a process by either P or Q. (Thus $\sqcap$ is obviously commutative, associative and idempotent.) Now $P \sqcap Q = Q$ exactly when $P \subseteq Q$.

A wire cannot accommodate two signals at the same time; they might interfere with one another in an undesirable way. This and any other error are modelled by the process $\perp$ (Bottom or Chaos). The environment must ensure that a process never gets into such a state. The process $\perp$ is considered to be so undesirable that any other process must be an improvement on it:

Law 0. $P \subseteq \perp$

It follows that $\sqcap$ has $\perp$ as a null element.

We shall mostly be concerned with recursively-defined processes. The meaning of the recursion $\mu X.F(X)$ is the least fixpoint of F. Its successive approximations are $\perp$, $F(\perp)$, $F(F(\perp))$, *etc.* All the operators that we shall use to define processes are continuous (and therefore monotonic); all except for recursion are distributive (with respect to nondeterministic choice).

In earlier approaches to an algebra for delay-insensitive circuits, *cf.* [14, 2, 5], a particular input signal is allowed only when this is explicitly indicated, and otherwise is assumed to lead to interference. This is in contrast with the algebra presented here: an input need not result in interference even though the possibility of such an input has not been made explicit in the algebraic expression. This follows the approach taken in [9] and is more convenient in algebraic manipulation, even though at first it may appear less natural.

Thus we write $a?;P$ to denote a process that must wait for a signal to arrive on $a \in I$ before it can behave like P. It is quite permissible for the environment to send a signal along any other input wire b. Such a signal is ignored at least until a signal is sent along a (or a second signal is sent along b causing interference).

A process that waits for input on a and then for input on b before being able to do anything is actually just waiting for inputs on both a and b, their order being immaterial:

Law 1. $a?;b?;P = b?;a?;P$

Complementary to input-prefixing $a?;P$ is output-prefixing $c!;P$, where $c \in O$. This is a process that outputs on c and then behaves like P. The environment may send a signal on any input wire even before it receives the signal on c; whether or not it can do so safely depends on P.

Two outputs by a process on the same wire, one after the other, is unsafe because of the danger of the two signals interfering with one another before they reach the environment. Also, since any output of a process may arrive at the environment an arbitrary time later, the order in which outputs are sent is immaterial. Therefore, we have the following two laws:

Law 2. $c!;c!;P = \perp$

Law 3. $c!; d!; P = d!; c!; P$

Example 0 Law 2. allows us to prove that $c!; \bot = \bot$. This should not be surprising: the process $c!; \bot$ behaves like $\bot$ after it has output on c; a wholly undesirable state results even before the output has reached the environment.

$c!; \bot$

$=$ { Law 2. }

$c!; c!; c!; \bot$

$=$ { Law 2. }

$\bot$

Finally, as in CSP, prefixing is distributive (with respect to nondeterministic choice). For both input and output-prefixing we have the law

Law 4. $x; (P \sqcap Q) = (x; P) \sqcap (x; Q)$

Example 1 We are now in a position to specify a number of elementary delay-insensitive components, *viz.* the Wire, the Fork, and the C-element.

Consider a circuit with one input wire a and one output wire c. In response to each signal on a, the circuit should produce a signal on c. The precise behaviour of this circuit is given by the following algebraic expression:

$\mu X.\ a?; c!; X$

which we shall refer to as the process W because it can be readily implemented by a wire. Now unfolding the recursion, we have that $W = a?; c!; W$. As in CSP, this equation itself uniquely defines W because its right hand side is guarded.

Next consider the process, with one input wire a and two output wires c and d, defined by the equation $F = a?; c!; d!; F$. This models the behaviour of a fork.

Finally, the Muller C-element repeatedly waits for inputs on wires a and b before outputting on c. It is defined by $C = a?; b?; c!; C$.

When we introduce the *after* operator in the next section, it will become clear that the above expressions do indeed correctly specify the components.

A more general form of input-prefixing is input-guarded choice. Such a choice allows a process to take different actions depending upon the input received. The choice is made between a number of alternatives of the form $a? \rightarrow P$. For S a finite set of alternatives, the guarded choice $[S]$ selects one of them. An alternative $a? \rightarrow P$ can be selected only if a signal has been received on a. The choice cannot be postponed indefinitely once one or more alternatives become selectable.

Choice with only one alternative is no real choice at all:

Law 5. $[a? \rightarrow P] = a?; P$

The environment cannot safely send a second signal along an input wire until the first signal has been acknowledged. Thus the result of sending two signals on a to the process $a?; a?; P$ is $\bot$ rather than P. The process is as useless as a choice with no alternative:

Law 6. $a?; a?; P = a?; [\,] = [\,]$

If two alternatives are guarded on a, then either may be chosen after input has been supplied on a. Indeed, the designer has the freedom to implement only one of the two. This is captured in the following law, where the symbol $\Box$ separates the various alternatives:

Law 7.
$$
\begin{aligned}
& [a? \rightarrow P \ \Box \ a? \rightarrow Q \ \Box \ S] \\
= \ & [a? \rightarrow (P \sqcap Q) \ \Box \ S] \\
= \ & [a? \rightarrow P \ \Box \ S] \sqcap [a? \rightarrow Q \ \Box \ S]
\end{aligned}
$$

Example 2   With the above law we can prove the following absorption theorem. An alternative guarded on a is absorbed by $a? \rightarrow \bot$:

$$[a? \rightarrow \bot \ \Box \ a? \rightarrow P \ \Box \ S]$$

$=$ { combining alternatives using Law 7. }

$$[a? \rightarrow (\bot \sqcap P) \ \Box \ S]$$

$=$ { $\bot$ is the null element of $\sqcap$ }

$$[a? \rightarrow \bot \ \Box \ S]$$

Until a process acknowledges receipt of an input signal, a second signal on the same wire can result in interference. So, for S_0 and S_1 sets of alternatives, we have

Law 8. $[a? \rightarrow [S_0] \ \Box \ S_1] = [a? \rightarrow [a? \rightarrow \bot \ \Box \ S_0] \ \Box \ S_1]$

Example 3   As a matter of fact, we can replace $\bot$ in Law 8. by any process P.

$$[a? \rightarrow [a? \rightarrow \bot \ \Box \ S_0] \ \Box \ S_1]$$

$=$ { absorption law derived in Example 2 }

$$[a? \rightarrow [a? \rightarrow \bot \ \Box \ a? \rightarrow P \ \Box \ S_0] \ \Box \ S_1]$$

$=$ { Law 8. }

$$[a? \rightarrow [a? \rightarrow P \ \Box \ S_0] \ \Box \ S_1]$$

Indeed, if it is unsafe for the environment to send a signal along a particular input wire, it remains unsafe at least until an output has been received. Therefore, we also have the following absorption law.

Law 9. $[a? \to \bot \;\square\; b? \to [a? \to P \;\square\; S_0] \;\square\; S_1] = [a? \to \bot \;\square\; b? \to [S_0] \;\square\; S_1]$

Example 4 With the input-guarded choice we can model somewhat more interesting delay-insensitive components, such as the Merge, the Selector and the Decision-Wait element.

The Merge is a circuit with two input wires a and b and one output wire c. In response to a signal on either a or b, it will output on c:

$$M = [a? \to c!; M \;\square\; b? \to c!; M]$$

We shall discover, in the next section, that this definition implies that it is unsafe for the environment to supply input on both a and b before receiving an output on c.

The Selector is a circuit with one input wire a and two output wires c and d. Upon reception of an input it outputs on one of the two wires:

$$S = [a? \to c!; S \;\square\; a? \to d!; S]$$

Actually, in this case there is no need to use guarded choice. By Laws 5. and 7., an equivalent formulation is $S = a?; ((c!; S) \sqcap (d!; S))$.

Finally, we can define the Decision-Wait element (2×1 in this case). It expects one input change in its row and one input change in its column. It produces as output the single entry which is indicated by the two changing inputs – there are two entries in this case:

$$
\begin{aligned}
DW \;=\; & [r0? \to [r1? \to \bot \;\square\; c? \to e0!; DW] \\
& \square\;\; r1? \to [r0? \to \bot \;\square\; c? \to e1!; DW]\,]
\end{aligned}
$$

(A C-element can be viewed as a 1×1 Decision-Wait element.)

2 More Advanced Constructs *This section has been omitted.*

3 Composition

In this section we define a parallel composition operator. With it we can determine the overall behaviour of a circuit from the individual behaviour of its components. It is understood that if the output wire of one component has the same name as the input wire of another, then these wires are supposed to be joined together; any signals transmitted along such a connection are hidden from the environment. The parallel composition operator is fundamental to a hierarchical approach to circuit design. It permits an initial specification to be decomposed into a number of components operating in parallel, and each of these components can be designed independently of the rest.

The simplicity of the laws enjoyed by parallel composition is one of the main attractions of our algebra. Indeed, in [14] certain restrictions had to be placed on processes before their composition could even be considered; and in [2] the fixed-point definition of parallel composition was rather unwieldy.

Parallel composition is denoted by the infix binary operator $\|$. All the operators we have met so far do not affect the input and output alphabets of their operands; so, for example, in the nondeterministic choice $P \sqcap Q$, we insist that the input alphabet of P is the same as that of Q, and declare that it is the same as that of $P \sqcap Q$. In the parallel composition $P \| Q$, however, the input alphabet of P should be disjoint from that of Q; likewise, the output alphabet of P should be disjoint from that of Q. (These rules prohibit fan-in and fan-out of wires; the explicit use of Merges and Forks is required.) The input alphabet of $P \| Q$ then consists of those input wires of each process P and Q which are not output wires of the other. Similarly, the output alphabet of $P \| Q$ consists of those output wires of each process which are not input wires of the other.

Parallel composition is commutative. It is also associative, provided we ensure that a wire named in the alphabets of any two processes being composed is not in the alphabets of a third process. If one process in a parallel composition is in an undesirable state, then the overall state is undesirable:

Law 28. $P \| \perp = \perp$

When an output-prefixed process $c!; P$ is composed with another process Q, the output is transmitted along c. Depending on whether or not c is in the input alphabet of Q, the signal on c is sent to Q or to the environment:

Law 29. $(c!; P) \| Q = \begin{cases} P \| (Q/c?) & \text{if } c \text{ is in the input alphabet of } Q \\ c!; (P \| Q) & \text{otherwise} \end{cases}$

It remains only to consider parallel composition of guarded choices. The following law specifies the alternatives in the resulting guarded choice.

Law 30. $[S_0] \| [S_1] = [S]$,

> where S is formed from the alternatives in S_0 and S_1 in the following way. For each alternative in S_0 of the form $skip \rightarrow P$, we have $skip \rightarrow (P \| [S_1])$ in S. For each alternative in S_0 of the form $a? \rightarrow P$ with a not in the output alphabet of $[S_1]$, we have $a? \rightarrow (P \| [S_1])$ in S. The alternatives in S_1 contribute to the alternatives in S in a similar way.

Example 5 If one component is able to send a signal that it is unsafe for the other to receive, then their parallel composition is $\perp$.

$(a!; P) \| [a? \rightarrow \perp \square S]$

$= \quad \{ \text{ internal communication on } a, \text{ Law 29. } \}$

$P \| [a? \rightarrow \perp \square S]/a?$

$= \quad \{ \text{ Example ?? and } \perp \text{ null element of parallel composition, Law 28. } \}$

$\perp$

Example 6 We compute a number of simple compositions in this example. Although the resulting behaviours are well-known, it has never previously been possible to give a straightforward algebraic derivation.

Consider first connecting two wires W_0 and W_1 together. Let $W_0 = a?; b!; W_0$ and $W_1 = b?; c!; W_1$. Then, in their parallel composition, signals on b are hidden from the environment.

$W_0 \parallel W_1$

=   { definitions of W_0 and W_1 }

$(a?; b!; W_0) \parallel (b?; c!; W_1)$

=   { one choice is no choice and parallel composition through guarded choice, Law 30., using that b is internal }

$a?; ((b!; W_0) \parallel (b?; c!; W_1))$

=   { internal communication on b, Law 29. }

$a?; (W_0 \parallel (b?; c!; W_1)/b?)$

=   { after through prefixing }

$a?; (W_0 \parallel [b? \to \perp \ \square \ skip \to c!; W_1])$

=   { substituting for W_{0}, and applying Law 30., parallel composition through guarded choice, using that b is internal }

$a?; [\quad a? \to ((b!; W_0) \parallel [b? \to \perp \ \square \ skip \to c!; W_1])$
$\quad\quad \square \ skip \to (W_0 \parallel (c!; W_1))]$

=   { one choice is no choice and absorption as in Example 3 }

$a?; [skip \to (W_0 \parallel (c!; W_1))]$

=   { one choice is no choice and external communication with Law 29. }

$a?; c!; (W_0 \parallel W_1)$

Since this recursion is guarded, we conclude that $W_0 \parallel W_1 = W$.

A more interesting example is the composition of a "one-hot" C-element and a Fork in the following way. The C-element is specified by $C = a?; b?; c!; C$ and the Fork by $F = c?; a!; d!; F$. This is a circuit involving feedback.

$C/a? \parallel F$

=   { Example ?? and definition of F }

$[a? \to \perp \ \square \ b? \to c!; C] \parallel (c?; a!; d!; F)$

=   { one choice is no choice and parallel composition through guarded choice, using that a and c are internal }

$b?; ((c!; C) \parallel (c?; a!; d!; F))$

=   { internal communication on c }

$b?; (C \parallel (c?; a!; d!; F)/c?)$

= { definition of C and after through prefixing }

$b?; ((a?; b?; c!; C) \parallel [c? \rightarrow \perp \; \Box \; skip \rightarrow a!; d!; F])$

= { one choice is no choice, parallel composition through guarded choice, using that a and c are internal, and definition of C }

$b?; [skip \rightarrow (C \parallel (a!; d!; F))]$

= { one choice is no choice, internal communication on a and external communication on d }

$b?; d!; (C/a? \parallel F)$

By uniqueness of guarded recursion, this combination of C-element and Fork behaves just like a wire. Although the Fork signals on a and d "in parallel", this did not lead to a doubling of the number of states which we had to analyse. We could deal with a entirely before d was pulled out of the parallel composition. This technique can be more generally applied and that is why these algebraic manipulations do not lead to a state explosion.

4 A Small Case Study *This section has been omitted.*

5 Conclusion

An algebraic approach has been taken to the specification and verification of delay-insensitive circuits. It has not been necessary to express explicitly all the states that such a circuit can enter; instead the possibility of them arising can be deduced using algebraic laws. This has lead to more concise specifications and shorter proofs than would be possible using other methods. Another simplifying factor has been that, following [14], we do not distiguish between high and low-going transitions; this exposes many symmetries that would not otherwise be apparent. The main advantage of our approach is the ease with which we can compute the parallel composition of components. We have worked through many examples in which we used algebraic laws either to prove further laws or to investigate the behaviour of specified circuits. As a case study, we verified some of Martin's designs, bringing to light interesting facts about his Isochronic Forks, D-elements and Q-elements.

Acknowledgements

We are most grateful to Tony Hoare and Tom Verhoeff for their interest and encouragement. The hospitality of the Department of Computer Science at Washington University helped make it possible for us to collaborate over this research. The work was partially funded by the Science and Engineering Research Council of Great Britain and the ESPRIT Basic Research Action CONCUR.

References

[1] S. D. Brookes and A. W. Roscoe. An Improved Failures Model for Communicating Sequential Processes. *Lect. Notes in Comp. Sci. 197*, 281–305, 1984.

[2] W. Chen, J. T. Udding, and T. Verhoeff. Networks of Communicating Processes and their (De)composition. In J. L. A. van de Snepscheut, editor, *The Mathematics of Program Construction*, number 375 in Lecture Notes in Computer Science, 174–196. Springer-Verlag, 1989.

[3] D. L. Dill and E. M. Clarke. Automatic Verification of Asynchronous Circuits Using Temporal Logic. In H. Fuchs, editor, *1985 Chapel Hill Conference on Very Large Scale Integration*, Computer Science Press, 127–143, 1985.

[4] D. L. Dill. *Trace Theory for Automatic Hierarchical Verification of Speed-Independent Circuits.* PhD thesis, CMU-CS-88-119, Dept. of C.S., Carnegie-Mellon Univ., 1988.

[5] J. C. Ebergen. *Translating Programs into Delay-Insensitive Circuits.* PhD thesis, Dept. of Math. and C.S., Eindhoven Univ. of Technology, 1987.

[6] M. Hennessy. *Algebraic Theory of Processes.* Series in Foundations of Computing. The MIT Press, Cambridge, Mass., 1988.

[7] C. A. R. Hoare. *Communicating Sequential Processes.* Prentice-Hall, 1985.

[8] He Jifeng, M. B. Josephs and C. A. R. Hoare. A Theory of Synchrony and Asynchrony. In *Proceedings IFIP Working Conference on Programming Concepts and Methods*, (to appear), 1990.

[9] M. B. Josephs, C. A. R. Hoare, and He Jifeng. A Theory of Asynchronous Processes. *J. ACM*, (submitted), 1989.

[10] A. J. Martin. Compiling Communicating Processes into Delay-Insensitive VLSI Circuits. *Distributed Computing*, 1:226–234, 1986.

[11] A. J. Martin. Programming in VLSI: From Communicating Processes to Delay-Insensitive Circuits. Caltech-CS-TR-89-1, Department of Computer Science, California Institute of Technology, 1989.

[12] A. W. Roscoe and C. A. R. Hoare. The laws of occam programming. *Theor. Comp. Sci. 60*, 2:177-229, 1988.

[13] I. E. Sutherland. Micropipelines. 1988 Turing Award Lecture. *Communications of the ACM*, 32(6):720–738, 1989.

[14] J. T. Udding. *Classification and Composition of Delay-Insensitive Circuits.* PhD thesis, Dept. of Math. and C.S., Eindhoven Univ. of Technology, 1984.

[15] J. T. Udding. A formal model for defining and classifying delay-insensitive circuits. *Distributed Computing*, 1(4):197–204, 1986.

Finiteness conditions and structural construction of automata for all process algebras

ERIC MADELAINE

INRIA
Route des Lucioles, Sophia Antipolis
06565 Valbonne Cedex (France)
email: madelain@mirsa.inria.fr

DIDIER VERGAMINI

CERICS
Rue Albert Einstein, Sophia Antipolis
06565 Valbonne Cedex (France)
email: dvergami@mirsa.inria.fr

Abstract

Finite automata are the basis of many verification methods and tools for process algebras. It is however undecidable in most process algebras whether the semantics of a given term is finite. We give sufficient finiteness conditions derived from the analysis of the operational rules of the algebra operators. From these rules we also generate the functions that compute automata from terms of the algebra. These constructions allow one to use our verification tools for programs written in many process algebras.

1 Introduction

Verification methods for concurrent systems can be classified in at least three families: theorem proving methods, model-checking, and automata based methods. The first family holds the biggest theoretical power; it may be applied to many sort of undecidable problems and in some sense it can deal with infinite objects. However theorem proving methods have usually a high complexity and there is few hope to make these methods purely automatic. The ECRINS system ([DMdS90]) uses theorem proving methods, together with specialized algorithms, to check for the validity of bisimulation laws in process algebras. The approach is general enough to consider most usual process calculi from the literature; the semantics of the operators is defined in user-defined *calculus description files*, and used by the system to generate specialized behaviour-evaluation algorithms.

We want to apply this parameterized approach for building tools based on automata analysis. The system AUTO ([RS89]) is dedicated to *verification by reduction* of parallel and concurrent programs. AUTO deals only with terms that have finite automata representations. Its main activities are the construction of automata from terms of process

---

[1]This work was partially supported by ESPRIT BRA (n°3006) CONCUR.

[2]The full version of this paper is published in the AMS-DIMACS volume of the Computer Aided Verification Workshop proceedings, R.P.Kurshan ed., 1990

algebras and the reduction and comparison of automata along a large family of equivalences. These activities are mostly intertwined, according to congruence properties of the equivalences that allow for reducing subterms of operators before building any global automaton. This approach cuts off partially the space explosion that causes the well-known limitation of such techniques.

The current AUTO system is using a subset of the MEIJE calculus ([Bo85]) as input language. To ensure that terms have finite representations, we use a two layers structure for input terms. In the lower layer, one can write recursive definitions directly encoding automata: recursive variables correspond to states and transitions are specified through action prefixing and non-deterministic choice (see the example 1 in section 2.2); these are *dynamic* operators, for they build the behaviour of components of a system. In the upper layer, one builds networks of automata using *static* operators (asynchronous parallel composition, renamings of signals, and a restriction operator). Finiteness of automata is guaranteed by forbidding occurrences of the parallel and renaming operators inside the recursive definitions. Observational equivalence appears to be a congruence for the MEIJE parallel and restriction operators, so lower layer automata can be reduced before being composed.

Similar conditions have been given by other authors for other languagaes, e.g. by D. Taubner for CCS, by H. Garavel for Lotos, and by M. Barbeau for Lotos also, but in the case of 1-place Petri-nets models, that may represent strictly more programs than finite automata.

Extending the structural construction of automata to parameterized process algebra, we need new finiteness conditions, computable from the very definitional rules of the operators. Here the splitting between static and dynamic operators makes less sense, as many operators can be used both in dynamic and static positions. Moreover there are operators that are asymmetric; recursion on the left argument of the *enable* and *disable* operators of LOTOS may generate infinite structures, whereas recursion on their right arguments can be used safely. We shall deduce from analysis of their rules which operators may in which position accept a recursive variable as one of its arguments, and which operators are preserving finiteness of automata. The rules analysis also provides us functions associated with each operator for building the automata. Of course the finiteness conditions rely deeply on the format allowed for the operator rules.

In section 2, we give an overview of the concepts from process calculi theory we need in the paper, including a description of the syntax we allow for structural operational semantic rules. In section 3, we discuss finiteness conditions and explain the classification of operators obtained from the analysis of the rules. In section 4 we describe the algorithms for structural construction of the automata, and in the conclusion we describe a prototype system that uses this generic technique and discuss current work.

2 Process Calculi

Process calculi are now a well-accepted generic notion for designing a class of formalisms which share the same definitional principles : CCS [Mi80], SCCS [Mi83], MEIJE [Bo85], ACP [BK86], BasicLOTOS [BB88] to name a few. We shall assume reader's acquaintance with at least one of these languages and its definitional mechanisms.

Process calculi are based on two main types : *actions* and *processes*. Operators take actions and processes arguments into processes, providing a classical algebraic structure. Operational semantics provides interpretation of closed terms into transition systems, with actions as transition labels. Operators and non-closed expressions are then interpreted as transition system transformers; this semantics is defined through behaviour rules in a structural operational style, with a particular format (see [dSi85], [VG88]). We shall describe our format for rules in section 2.4.

Special operators are action renamings and recursive definitions. They are present in all process calculi.

2.1 Actions

In all process algebras actions are themselves structured. This structure is what allows for synchronization and further communication to be handled in relevant operators rules. Just recall the inverse signals in CCS, which meets in synchronization and produce a hidden action τ. In SCCS and MEIJE there is a full commutative monoid of potential simultaneous actions, again containing a group of invertible signals. Actions structure in ACP is more scarce and parametric, while in BasicLOTOS it is only a set of so-called gate names without structure, but for a distinguished termination action δ.

2.2 Recursive definitions

Most process algebras have some sort of recursion operator. In AUTO and ECRINS we use a common recursive definition mechanism for all process algebras. Here is an example of a recursive definition, written with MEIJE operators:

Example 1 _____

```
let rec {x = a:x + b:x +c:stop
    and y = a:y + b:z
    and z = a:z + c:c:y }    in x//y
```

Such recursive definitions are used in AUTO for building finite automata, perhaps composed later on by other operators of the algebra. One can also build infinite structures, not suitable for analysis in AUTO, such as:

Example 2 _____

```
let rec {many-processes = one-process // many-processes}
    in many-processes
```

2.3 Definition of process algebras

A calculus description contains the concrete syntax and abstract syntax definitions of the calculus operators, together with *structural conditional rules* that gives them an operational semantics. The ECRINS *calculus compiler* uses the first part to produce a scanner, a parser and abstract syntax structures for expressions of the calculus. From the semantics part, the compiler will produce the functions for building and combining automata from terms. Here is an example taken from Basic-Lotos:

```
operator disable :: Process Process --> Process
syntax    [> left 4
semantics
        disabling_1        p -- a --> p' & not(a equal d)
                           --------------------------------
                                  p[>q -- a --> p'[>q
        disabling_2        p -- a --> p' & (a equal d)
                           --------------------------------
                                  p[>q -- d --> p'
        disabling_3        q -- a --> q'
                                  ----------------
                                  p[>q -- a --> q'
end
```

2.4   The Conditional Rewrite Rules format

The rules must obey the following format ([DMdS90]):

$$\frac{\{x_j \xrightarrow{u_j} x'_j\}_{J \subset [1..n]} \quad \& \quad P(\{u_j\}_J, a_1, \ldots, a_m)}{Op(x_1, \ldots, x_n, a_1, \ldots, a_m) \xrightarrow{F(\{u_j\}_J, a_1, \ldots, a_m)} T'(\{x_k\}_{[1..n]-J}, \{x'_k\}_J, a_1, \ldots, a_m)}$$

Many definitions in this paper rely on the syntax allowed for the various elements of conditional rules. Let us give precise names to these elements.

Definition 1 *We call:*

- *premises the upper part of the rule and conclusion its bottom part.*

- *subject the term at the left end part of the conclusion. The head operator Op of the subject has process arguments $x_1, \ldots, x_n$ and action parameters $a_1, \ldots, a_m$. Inside P, F and T' some other actions may appear: they are global constants of the calculus and had to be declared as such previously.*

- *formal hypothesis the part of the premises on the left of the & and working formal variables the x_j with $j \in J$.*

- *actions predicate the part of the premises on the right of the &. P belongs to boolean operators closure of the following basic predicates : equality, divisibility, set membership.. over actions terms with synchronization product.*

- *resulting action the label over the arrow of the conclusion. The resulting action F is a function of the formal hypotheses actions (and of the operator action parameters).*

- *resulting process the right end part of the conclusion. The process arguments that appear as formal hypothesis must be transformed as x'_j in the resulting process. This condition does not allow to test a potential future behaviour of a process, without* *making it explicitly perform its action within the considered rule, therefore saving the possibility of choosing another future. This is a restriction to the format in [VG88]. Beside this condition, the resulting process is built from the action and process variables, and the operators of the calculus (excluding recursive definitions).*

3 Finiteness Conditions

The preceding conditional rules defines in a structural manner the semantics function that computes a *transition system* from a (closed) term of a process algebra. This definition is constructive: you can compute each transition of the transition system by building proof trees which nodes are instances of the rules.

Definition 2 *A term of a process algebra has* transition-system finiteness *(TSF) iff the transition system computed from the term using the operators' rules is strongly bisimular to some finite transition system (i.e. with a finite number of states and a finite number of transitions).*

In many process algebras with recursion, this property is undecidable. Sufficient syntactic conditions to ensure FTS will be given in this section, for any process algebra defined using the conditional rewrite rules from the preceding section. As far as possible, these conditions will be expressed in terms of the semantic rules of the operators. We shall use the following notions:

guarded recursion: It is possible to build infinite proof trees for terms containing recursive definitions. Can we found syntactic conditions to guaranty that all proof trees are finite?

non-growing operators: Making the assumption that the arguments of an operator have finite semantics, is it always true that their composition by this operator is a finite automaton? This property holds for most classical operators, but the rule format allows to define exotic operators that create infinitely many states.

sieves: Some unary operators have the nice property that the resulting automaton has exactly the same states than the argument automaton, only some transitions being transformed, erased, or added. We implement their semantics by *sieves*, that is functions that only modify the transitions of the system. Which operators may be implemented in this way?

switches: Inside a recursive definition, the use of recursive variables should be limited in some way, in order to avoid building infinitely many states, or states with infinitely many transitions. Clearly, parallel composition operators and non-alphabetical renamings should be somehow forbidden inside recursive definitions. At which places (defined as occurrences of operator arguments) is a recursive variable allowed to appear?

3.1 Guarded recursion

In order to avoid divergence in the proof tree construction, we introduce as usual a notion of guarded terms. We define here this notion in a rather abstract way, and we shall give a generic algorithm that computes it in a further section. The definition relies on the fact that if a proof tree is infinite, then either it contains a pattern that occurs infinitely, or the subject of its nodes are strictly growing along the infinite branches. The *guarded* property takes care of infinitely repeated patterns, while growing branches will be addressed later.

Definition 3

- *A proof tree is* unguarded *iff the subject of its root is equal to the subject of one of its subtree, or if it has an unguarded subtree.*

- *A term of a process algebra is* unguarded *if it has an unguarded proof tree, or if at least one of its possible reconfigurations is unguarded.*

- *A term of a process algebra is* guarded *if it is not unguarded.*

3.2 Non-growing operators

Growing operators build infinite structures from arguments having finite automata representations. Though no operators of usual process algebras have such a nasty behaviour, we need a syntactical way to ensure that no growing operator is used in a term. In order to obtain infinitely many states in the resulting automaton, one would have to introduce a rule that produces new terms ab infinitum. A natural sufficient condition for ensuring finiteness is to be able to find an order on expressions such that for all rules, the resulting process is not strictly greater than the subject.

It is possible to adapt here many results from the term rewrite system theory, with the difference that we are looking for a non-strict order compatible with our rewrite relation, whereas usual rewrite systems need a strict decreasing order. We give here a simple definition that covers nearly all interesting cases:

We consider families of operators closed under reconfiguration: if an operator belongs to the family, then all operators that occur in the resulting processes of all its rules also belong to the family.

Definition 4 *A family of operators $\{Op_k\}$ is* non-growing *iff there exist a simplification ordering $<$ such that:*

For each rule of each operator, let us denote $Op_k(x_i)$ the subject of the rule and $T_{i_j}[x_i'/x_i]$ its resulting process in which all resulting working variables x_i' have been replaced by their corresponding x_i, then:

either $T_{i_j}[x_i'/x_i] < Op_k(x_i)$

or $T_{i_j}[x_i'/x_i] = Op_k(x_i)$

where $=$ is the syntactic equality on terms.
We say then that all Op_k are $<$-non-growing.

Theorem 1 *Given a family of non-growing operators $F = \{Op_k\}$, an operator $Op_k(x_i)$ of arity n, and n terms T_i having TSF, then $Op_k(T_i)$ has TSF.*

All proofs are in the full paper.

3.3 Sieves

Definition 5 *A sifting operator, or sieve, is an operator with exactly one process argument, which rules obey the following conditions: each rule has exactly one working formal variable (the process argument), may have predicates and any form of resulting action, and the resulting process is obtained from the subject of the rule by substituting the working formal variable by the corresponding resulting process variable.*

This definition includes the *renaming*, *restriction*, and *ticking* operators of MEIJE, the *hiding* operator of TCSP and LOTOS.

The introduction of sieves is two-folded:

- They are an interesting family of automata transformers, acting only on transitions. As such, they can be easily composed and combined with the automata building functions, leading to efficient implementations where no intermediate automata are built for such operators, and only accessble parts are considered.

- They may be used also inside recursive definitions. We need define here a sub-class of sieves such that the language generated by their compositions, modulo some idempotence property, remains finite. Then the states of the generated automaton will be obtained as pairs of a recursive variable and a composition of sieves.

This applies e.g. for alphabetical renamings, hiding, and restriction (for the alphabet of action labels in a term is finite, and the set of all restriction compositions is a finite commutative group).

Of course, it does not apply to ticking or to non-alphabetical renamings, and the MEIJE term: let rec x= a:y and y = b*x in x generates infinitely many states x, b*x, b*b*x, etc.

We need here a *non-growing* definition for resulting action functions:

Definition 6 *An operator rule is action-non-growing iff its resulting action is an action term built only from the following items: the formal action of the rule, the action parameters of the subject, the constant actions of the calculus, and alphabetical renamings (including renaming a label by an invisible action).*

This definition is trivially fulfilled by all operators of BASICLOTOS and CCS, but not by the ticking operator of MEIJE, nor by non-alphabetical renamings.

Definition 7 *A non-growing sieve is a sifting operator which rules are action-non-growing.*

Proposition 1 *Given a finite alphabet of actions, the algebra of all compositions of non-growing sieves has a finite model.*

3.4 Switching operators

We introduce the family of *switching operators* as a generalization of the usual sum operator of CCS/MEIJE.

Definition 8 *Given a family of operators $\mathcal{O}$ and a well founded simplification ordering $\prec$ on expressions generated from $\mathcal{O}$,*
An operator Op in $\mathcal{O}$ is a switching operator *(or simply a* switch*) w.r.t. one of its process arguments "p" iff:*

- *All rules in which "p" is a working formal variable verify the following properties: it has no other premise ("p" works alone) and the resulting process is exactly p'.*

- *All rules where the resulting process contains an occurrence of "p" are non-growing for $\prec$ and are their resulting processes are themselves switches for "p".*

Remarks:

- This includes non-growing operators with no premise at all.

- This definition could be extended by allowing rules in which "p" is working to have as resulting processes T such that $T \prec Op(...,p',...)$, with T being also in some sense a switch for p. However, this would complicate to much both the definition and the related proofs, whereas all classical operators fit the restricted definition we have just given.

The *sum* operator of SCCS, the binary *choice* of BASICLOTOS are switches, but also the *delay* operator of MEIJE and the *disabling* operator of LOTOS for its second argument.

Usual prefixing operators are also switches, including the *action prefix* operators of MEIJE and LOTOS, of course, but also the *enabling* operator of LOTOS for its second process argument. The *internal choice* of TCSP is a switch.

Yet the *external choice* operator of TCSP is not a *switching* operator (see its rules in the annex), because for each of its arguments, it has a rule looking as a *switching* rule, and a rule resembling a *sieve* rule. By the way this operator is one we do not want to be involved in a recursive definition:

Example 3 ──────────────────
```
let rec x = (tau : x) ext-choice a:y in x
```

This term generates the following sequence of resulting processes:

```
let rec {x = tau:x ext-choice a:x}  in x ext-choice a:x
let rec {x = tau:x ext-choice a:x}  in (x ext-choice a:x) ext-choice a:x
...
```

Though this specific case could be reduced (to a finite set of terms) by semantical arguments, the finiteness property may no more be guaranteed at a syntactical level. Semantical arguments for finiteness are out of the scope of this paper.

3.5 Main result

Definition 9 *Given a family of variables V, a term from a process algebra is called a term suitable for recursion on V iff either*

1. *it is a variable from V,*

2. *or it does not contain any variable from V and it has a finite automaton semantics,*

3. *or its head operator is a switching operator for some of its arguments, these arguments are subterms suitable for recursion on V, and all other arguments contains no occurrences of variables in V and have finite automaton semantics,*

4. *or its head operator is a non-growing sieve and its argument is suitable for recursion on V.*

Remark: this definition can be extended to handle nested recursive definitions, by adding an item for any recursive declaration "let rec $\{x_i = e_i\}$ in x_0" such that all e_i are suitable for recursion on $V \cup \{x_i\}$. Such an extension preserves the following theorem, though the proof is still more tedious.

Theorem 2 *Let Proc be a recursive definition "let rec $\{x_i = e_i\}$ in x_0".*
   If Proc is guarded, and if all expressions e_i are terms suitable for recursion on $\{x_i\}$, then the recursive definition has a finite automaton semantics.

Hint : We define a finite set of *states* by induction on the structure of the term, then we prove that the transition system of the term maps in this state space, with a finite number of transitions from each state. The proof is in the full paper.

This property allows us to guarantee that some recursive definition have a finite semantics. In any process algebra, it permits using any combination of nested recursive definitions, and arbitrary closed terms inside recursive definitions. In the case of MEIJE-SCCS, it naturally includes the classical "well-guarded" condition (sums of action-prefix operators). In the case of BASICLOTOS, it allows the occurrence of recursive variables as second arguments of the *enabling* operator and well-guarded occurrences of recursive variables within the second argument of the *disabling* operator.

3.6 Accessibility

All preceding conditions can be restricted to the *accessible* parts of the term. This is not only an optimization issue: considering only accessible parts allows for rejecting less programs, for any violations of a condition inside a non-accessible part of a term will have no consequence on its semantics.

We give in the full paper a sufficient characterization of *potentially accessible* parts of a term, based on the analysis of recursive variable occurrences in the recursive definitions.

4 Algorithms

We have implemented a new version of AUTO using the preceding results. From the rules of the operators, and from their classification in finite, sifting, switching, and non-growing operators, we derive functions that test the syntactical conditions for finiteness, then build the automaton of a term. Due to size constraints, we cannot give here a full description of the generated algorithms. The main ideas are:

- A given term of a process algebra is first checked for guardedness and suitability of its accessible part. Terms that do not satisfy the syntactic conditions are rejected.

- The automaton is built structurally in a bottom-up manner, starting with the recursive definitions at the leaves. There are two main algorithms, one dealing with recursive definitions, and the other one combining several arguments-automata w.r.t. a given operator. Both algorithms take a (composition of) sieve and a reduction congruence as additional parameters.

- The algorithm dealing with a recursive definition maintains a set of states that are pairs of a recursive variable and a sieve, each state associated to a subterm containing recursive variable occurrences. The analysis of such a term generates transitions towards states that have to be compared to already existing states, or added to the set. The finiteness conditions ensure the termination of this computation.

- The composition algorithm is a residual algorithm that traverses both arguments combining their transitions in a manner that depends on the combination operator.

5 Conclusion

The AUTO system we currently distribute is using specific hand-coded algorithms for the operators of the MEIJE0 calculus (stop, prefixing, sum, parallel, restriction, renaming). These algorithms were carefully optimized in order to avoid building parts of product automata that were to be deleted by some restriction operators.

We have built a new prototype of the AUTO system using the generic algorithms of this paper. Tests have been made both for the MEIJE0 calculus and for BASICLOTOS (the prototype has been presented in [MV89]). The MEIJE0 operators are correctly classified by our definitions: the prototype accepts stricty more MEIJE0 programs than the preceeding AUTO system. Some other MEIJE-SCCS operators (see [dSi85]) can be added easily to this syntax, including *ticking*, *interleaving* and the *synchronized product* as non-growing operators. The results are good also for BASICLOTOS operators: the usual finiteness conditions are correctly deduced from the rules. Some limitations of our conditions are listed in the annex. We also obtained efficiency mesurements: this version appears to have the same order of performances than the old version. Moreover, it should be clear that in many cases it allows to build a smaller number of automata (for sieves never require to copy an automaton) and to apply sieves on smaller automata. No optimizations have been done in the first prototype, so the new version is potentially much more efficient than the specialized MEIJE0 version.

This prototype in its BASICLOTOS version is currently integrated in the LOTOS tool environment of the ESPRIT project LOTOSPHERE.

The set of programs accepted by the generic version is of course larger than in the former system, and programs may be written in a much more permissive way: for example parallel compositions may be done in many different ways using various operators and nested recursive definitions are allowed.

The congruence properties of some equivalences versus MEIJE composition operators are also to be generalized. It is very important for space efficiency reasons to apply reductions as deep as possible in the term, in order to create and compose smaller automata. We plan to have the ECRINS system proving congruence laws for various equivalences and various operators, so that the congruence properties can be automatically used in AUTO during the automata construction.

References

[BB88] T. Bolognesi, E. Brinksma, "Introduction to the ISO Specification Language LO-TOS", in *The Formal Description Technique LOTOS*, North-Holland, 1988

[BK86] J.A. Bergstra, J.W. Klop, "Process Algebra: Specification and Verification in Bisimulation Semantics", *CWI Monographs*, North-Holland, 1986

[BS87] T. Bolognesi, S. A. Smolka, "Fundamental Results for the Verification of Observational Equivalence: a Survey", proc. of the IFIP 7th Internaional Symposium on Protocol Specification, Testing, and Verification, North-Holland, 1987

[Bo85] G. Boudol, "Notes on Algebraic Calculi of Processes", *Logics and Models of Concurrent Systems*, NATO ASI series F13, K.Apt ed., 1985

[dSi85] R. De Simone, "Higher-Level Synchronising Devices in Meije-Sccs", *Theoretical Computer Science* 37, p245-267, 1985

[MV89] E. Madelaine, D. Vergamini, "AUTO, a verification tool for distributed systems using reduction of automata", in proceedings of *Forte'89 conference*, Vancouver, North-holland, 1989

[DMdS90] G.Doumenc, E. Madelaine, R. de Simone, *"Proving Process Calculi Translations in ECRINS"*, Technical Repport INRIA RR1192, 1990

[Mi80] R. Milner, *"A Calculus for Communicating Systems"*, Lectures Notes in Comput. Sci. 92, 1980

[Mi83] R. Milner, "Calculi for Synchrony and Asynchrony", *Theoretical Computer Science* 25, p267-310, 1983

[RS89] V. Roy, R. De Simone, *"AUTO - AUTOGRAPH"*, this volume.

[VG88] F.W. Vaandrager, J.F. Groote, "Structured operational semantics and bisimulation as a congruence" CWI report CS-R8845, 1988

[Ve89] D. Vergamini, "Verification of Distributed Systems: an Experiment", in *Formal Properties of Finite Automata and Applications*, LNCS 386, 1990

On Automatically Explaining Bisimulation Inequivalence*

Rance Cleaveland
Department of Computer Science
North Carolina State University
Raleigh, North Carolina 27695-8206
USA

Abstract

This paper describes a technique for generating a logical formula that differentiates between two bisimulation-inequivalent finite-state systems. The method works in conjunction with a partition-refinement algorithm for computing bisimulation equivalence and yields formulas that are often minimal in a precisely defined sense.

1 Introduction

A popular technique for verifying finite-state systems involves the use of a *behavioral equivalence*. In this approach, specifications and implementations are formalized as finite-state machines, and verification consists of establishing that an implementation is equivalent to, in the sense of *behaving the same as*, its specification. A number of equivalences have been proposed in the literature [1, 3, 7, 8, 12, 14], and several automated tools include facilities for computing them [2, 5, 6, 13].

One particularly interesting equivalence is *bisimulation equivalence* [14]. In addition to the fact that a number of other equivalences may be described in terms of it [4], the relation has a *logical* characterization: two systems are equivalent exactly when they satisfy the same formulas in a simple modal logic due to Hennessy and Milner [9]. This fact suggests a useful diagnostic methodology for tools that compute bisimulation equivalence: when two systems are found not to be equivalent, one may explain why by giving a formula satisfied by one and not the other.

The purpose of this paper is to develop a technique for determining a Hennessy-Milner formula that distinguishes two bisimulation-inequivalent finite-state systems. To this end, we show how to use information generated by the partition-refinement algorithm of Kanellakis and Smolka [11] to compute such a formula efficiently. On the basis of our results, the tools mentioned above may be modified to give users diagnostic information in the form of a formula when a system is found not to be equivalent to its specification.

Hillerström [10] also gives a technique for computing a Hennessy-Milner formula. However, his method relies on the use of a backtracking-based algorithm that is less efficient that the more popular partition-based algorithms we are interested in.

The remainder of the paper is organized as follows. The next section defines bisimulation equivalence and examines the connection between it and the Hennessy-Milner Logic. Section 3 then describes a modification of Kanellakis-Smolka bisimulation algorithm that computes distinguishing formulas; a small example is also presented to illustrate the workings of the new algorithm. The final section contains our conclusions and directions for further research.

---

*Research supported by NSF/DARPA research grant CCR-9014775.

2 Transition Graphs, Bisimulations and Hennessy-Milner Logic

Finite-state systems may be represented as *transition graphs*. Vertices in these graphs correspond to the states a system may enter as it executes, with one vertex being distinguished as the start state. The edges, which are directed, are labeled with the actions and represent the state transitions a system may undergo. The formal definition is the following.

Definition 2.1 *A transition graph is a quadruple $\langle Q, q, Act, \rightarrow \rangle$, where:*

- *Q is a set of states (vertices);*

- *$q \in Q$ is the start state;*

- *Act is a set of actions; and*

- *$\rightarrow \subseteq Q \times Act \times Q$ is the derivation relation (set of labeled edges).*

We shall often write $q_1 \xrightarrow{a} q_2$ to indicate that there is an edge labeled a from state q_1 to state q_2; in this case, we shall sometimes say that q_2 is an *a-derivative* of q_1. When a graph does not have a start state indicated, we shall refer to the corresponding triple as a *transition system*. A state in a transition system gives rise to a transition graph in the obvious way: let the given state be the start state, with the other three components of the transition graph coming from the transition system.

Reactive systems [16] compute by interacting with their environment. For such systems, the traditional language equivalence of automata theory is insufficiently discriminating, since the resolution of nondeterministic choices may leave a system in states that react differently to stimuli offered by the environment. Bisimulation equivalence remedies this shortcoming by requiring that equivalent systems have state sets that "match up" appropriately: the start states must be matched, and if two states are matched then they must have matching a-derivatives for any action a either is capable of. These intuitions may be formalized in terms of *bisimulations* on a *single* transition system.

Definition 2.2 *Let $\langle Q, Act, \rightarrow \rangle$ be a transition system. Then a relation $R \subseteq Q \times Q$ is a bisimulation if, whenever $q_1 R q_2$, the following hold.*

1. *If $q_1 \xrightarrow{a} q_1'$ then there is a q_2' such that $q_2 \xrightarrow{a} q_2'$ and $q_1' R q_2'$.*

2. *If $q_2 \xrightarrow{a} q_2'$ then there is a q_1' such that $q_1 \xrightarrow{a} q_1'$ and $q_1' R q_2'$.*

Two states and in a transition system are *bisimulation equivalent* if there is a bisimulation relating them. When q_1 and q_2 are bisimulation equivalent we shall write $q_1 \sim q_2$.

Let $G_1 = \langle Q_1, q_1, Act, \rightarrow_1 \rangle$ and $G_2 = \langle Q_2, q_2, Act, \rightarrow_2 \rangle$ be two transition graphs satisfying $Q_1 \cap Q_2 = \emptyset$. Then G_1 and G_2 are bisimulation equivalent exactly when the two start states, q_1 and q_2, are equivalent in the transition system $\langle Q_1 \cup Q_2, Act, \rightarrow_1 \cup \rightarrow_2 \rangle$. This definition may be generalized to arbitrary transition systems (i.e., ones whoses state sets are not disjoint), at the cost of a slightly more complicated definition for the transition system in which bisimulation equivalence is to be computed. In the remainder of the paper we only consider the problem of determining a formula that distinguishes two inequivalent states in a transition system.

A number of behavioral equivalences may be characterized in terms of bisimulation equivalence on suitably transformed transition systems [4, 5, 6]. For example, if the transition system is *deterministic*, meaning that every state has at most one a-derivative for a given a, then bisimulation equivalence coincides with *language equivalence* from formal language theory. To determine if two states in an arbitrary transition system are language equivalent, it suffices to apply a "determinizing" transformation to the transition system and then determine whether their corresponding states in this new system are bisimulation equivalent. Other equivalences, including testing equivalence [7, 8], failures equivalence [3], and observational equivalence [14], may be computed in an analogous fashion.

$$\begin{aligned}
[\![tt]\!]_T &= Q \\
[\![\neg\Phi]\!]_T &= Q - [\![\Phi]\!]_T \\
[\![\Phi_1 \wedge \Phi_2]\!]_T &= [\![\Phi_1]\!]_T \cap [\![\Phi_2]\!]_T \\
[\![\langle a\rangle\Phi]\!]_T &= \{\, q \in Q \mid \exists q'.\ q \xrightarrow{a} q' \wedge q' \in [\![\Phi]\!]_T \,\}
\end{aligned}$$

Figure 1: The semantics of formulas in Hennessy-Milner Logic.

Bisimulation equivalence also has a *logical* characterization in terms of Hennessy-Milner Logic (HML) [9]: two states are equivalent exactly when they satisfy the same HML formulas. The syntax of HML is defined as follows, where $a \in Act$.

$$\Phi ::= tt \mid \neg\Phi \mid \Phi \wedge \Phi \mid \langle a\rangle\Phi$$

Given a transition system $T = \langle Q, Act, \rightarrow\rangle$, the interpretation of the logic maps each formula to the set of states for which the formula is "true"; Figure 1 gives the formal definition. In the remainder of the paper we shall omit explicit reference to the transition system used to interpret formulas when it is clear from the context. Intuitively, the formula tt holds of any state, and $\neg\Phi$ holds of a state if Φ does not. $\Phi_1 \wedge \Phi_2$ holds of a state if both Φ_1 and Φ_2 do, while the modal proposition $\langle a\rangle\Phi$ holds if the state has an a-derivative for which Φ holds. We shall say that a state q in transition system T *satisfies* formula Φ if $q \in [\![\Phi]\!]_T$.

Let $H(q)$ be the set of HML formulas that a state q satisfies:

$$H(q) = \{\, \Phi \mid q \in [\![\Phi]\!] \,\}.$$

The next theorem is a corollary of a result proved in [9].

Theorem 2.3 *Let* $\langle Q, Act, \rightarrow\rangle$ *be a finite-state transition system, with* $q, q' \in Q$. *Then* $H(q) = H(q')$ *if and only if* $q \sim q'$.

It follows that if two states in a (finite-state) transition system are inequivalent, then there must be a HML formula satisfied by one and not the other. This is the basis of the following definition of distinguishing formula.

Definition 2.4 *Let* $\langle Q, Act, \rightarrow\rangle$ *be a transition system, and let* $S_1 \subseteq Q$ *and* $S_2 \subseteq Q$. *Then HML formula* Φ *distinguishes* S_1 *from* S_2 *if the following hold.*

1. $S_1 \subseteq [\![\Phi]\!]$.

2. $S_2 \cap [\![\Phi]\!] = \emptyset$.

So Φ distinguishes S_1 from S_2 if every state in S_1, and no state in S_2, satisfies Φ. Theorem 2.3 thus guarantees a formula that distinguishes $\{q_1\}$ from $\{q_2\}$ if $q_1 \not\sim q_2$.

Finally, we shall adopt the following criterion in assessing whether a distinguishing formula contains extraneous information.

Definition 2.5 *Let* Φ *be a HML formula distinguishing* S_1 *from* S_2. *Then* Φ *is minimal if no* Φ' *obtained by replacing a non-trivial subformula of* Φ *with the formula* tt *distinguishes* S_1 *from* S_2.

Intuitively, Φ is a minimal distinguishing formula for S_1 with respect to S_2 if each of its subterms plays a role in distinguishing the two.

3 Computing Distinguishing Formulas

In this section, we describe an algorithm for computing bisimulation equivalence [11] and show how to alter it to compute distinguishing formulas. We then consider a small example that illustrates the behavior of the modified algorithm.

```
function split(B, a, B') =
  {{s ∈ B | ∃s' ∈ B'. s →ᵃ s'}, {s ∈ B | ¬∃s' ∈ B'. s →ᵃ s'}} - {∅};

algorithm bisim(Q, Act, →);
  begin
    P₁ := {Q};
    P₂ := ∅;
    while P₁ ≠ P₂ do begin
      P₂ := P₁;
      P₁ := ∅;
      foreach B ∈ P₂ do P₁ := P₁ ∪ split(B, a, B');
    end
  end
```

Figure 2: The partition refinement algorithm for bisimulation equivalence.

3.1 The Kanellakis-Smolka Algorithm

The Kanellakis-Smolka algorithm exploits the fact that an equivalence relation on a set of states may be viewed as a *partition*, or set of pairwise-disjoint subsets (called *blocks*) of the state set whose union is the state set. In this representation blocks correspond to the equivalence classes—so two states are equivalent exactly when they belong to the same block. Beginning with the partition containing one block (representing the trivial equivalence relation consisting of one equivalence class), the algorithm repeatedly *refines* this partition (by splitting blocks) until the associated equivalence relation becomes a bisimulation. In order to determine whether the partition needs further refining, the algorithm looks at each block in turn. If a state in a block B has an a-derivative in a block B' and another state in B does not, then the algorithm splits B into two blocks, one containing the states having an a-derivative in B' and the other containing the states that do not. When no more splitting is possible, the resulting equivalence relation corresponds exactly to bisimulation equivalence on the given transition system. The algorithm is given in Figure 2. Function split is used to split one block with respect to another; notice that $split(B, a, B') = \{B\}$ (i.e. B is not split with respect to a and B') if either all the states in B, or none of them, have an a-derivative in B'. It should also be pointed out that $P_1 = P_2$ exactly when no more splits in P_1 are possible. The worst-case complexity of $bisim$ is $O(|\to| * |Q|)$.

3.2 Generating Distinguishing Formulas

One straightforward way to compute distinguishing formulas is to associate a formula, $\Phi(B)$, with each block B in the partition in such a way that that the following hold.

- $B \subseteq [\![\Phi(B)]\!]$.

- $B' \cap [\![\Phi(B)]\!] = \emptyset$ if $B' \neq B$.

In the initial partition, $\{Q\}$, $\Phi(Q)$ is set to tt. Now suppose a block B is split, i.e. suppose there is an action a and another block B' such that $split(B, a, B') = \{B_1, B_2\}$, with every state in B_1 having a transition into B' and no state in B_2 having one. Then $\Phi(B_1)$ may be set to $\Phi(B) \wedge \langle a \rangle \Phi(B')$, while $\Phi(B_2)$ becomes $\Phi(B) \wedge \neg\langle a \rangle \Phi(B')$. Arguing inductively, it is easy to establish that for any block B, a state satisfies $\Phi(B)$ exactly when it is contained in B. Since two states that are not bisimulation equivalent will eventually wind up in different blocks, it is a simple matter to compute a formula that distinguishes such states: just return the formula associated with one of the containing blocks.

Although intuitively appealing, this approach has a drawback; it generates very large formulas. In general, the size of a formula associated with a block will grow in size as 2^r, where r is defined as the number of iterations of the while loop in $bisim$. In certain cases $r = |Q|$, so the formulas obtained using this method may be exponential in the size of the state space. There is, however, a polynomial-size representation using a set of propositional equations, so this complexity is not as severe as it seems; moreover, it is the case that an exponential-size formula may also be minimal. More importantly, the formulas include a large amount of extraneous information: not only does such a formula distinguish one state from another inequivalent state, it also distinguishes it from every state to which it is inequivalent. In fact, the formulas generated this way are rarely minimal, and because of this, they are not useful from a diagnostic standpoint.

Another Approach

We now describe a better technique for generating distinguishing formulas. The method uses information computed by a slightly altered version of *bisim* that, in addition to computing the partition as described above, retains information about how blocks are split. Then, a postprocessing step constructs a formula that generates a formula distinguishing the states in one block from the states in another.

Bisim is modified as follows. Rather than discarding an old partition after it is refined, the new procedure constructs a "tree" of blocks as follows. The children of a block are the new blocks that result when the block is split; accordingly, the root is labeled with the block Q, and after each iteration of the foreach loop the leaves of this tree represents the current partition. When a block B is split (by $split(B, a, B')$), we place the new block $B_1 = \{ s \in B \mid \exists s' \in B'. \; s \xrightarrow{a} s' \}$ as the left child and the new block $B_2 = \{ s \in B \mid \neg \exists s' \in B'. \; s \xrightarrow{a} s' \}$ as the right child, and we label the arc connecting B to B_1 with a and B'. Recall that every state in B_1 has an a-transition into B' and that no state in B_2 does. If a block is not split during an iteration of the foreach loop, it is assigned a copy of itself as its only child.[1] Figure 5 contains an example of such a tree.

Given a block tree computed by the new version of *bisim*, and two states s_1 and s_2 that are inequivalent and hence in different blocks, the postprocessing step builds a formula $\delta(s_1, s_2)$ that distinguishes $\{s_1\}$ from $\{s_2\}$. Although this formula will not necessarily be minimal either, it will in general be much smaller than the formula computed using the method described above; it is guaranteed to be no larger. The details are as follows.

1. Determine the deepest block P in the block tree such that $s_i \in P$ for $i = 1, 2$. Let L and R be the left and right children of P, with a, B' labeling the arc from P to L. Note that either $s_1 \in L$ and $s_2 \in R$, or vice versa. Let s_L be the state in L, and s_R the state in R.

2. Execute the code in Figure 3. The idea is the following. For each state in B' that is an a-derivative of s_L we will generate a minimal set of formulas satisfied by s_L whose conjunction is satisfied by no derivative of of s_R. We will then take the set yielding the smallest conjunction. *Size* is the variable used to record the size of the current shortest conjunction, while Γ contains the current collection of formulas being built.

3. If $s_L = s_1$ then return $\langle a \rangle \Phi$; otherwise, return $\neg \langle a \rangle \Phi$.

Theorem 3.1 *The formula $\delta(s_1, s_2)$ distinguishes $\{s_1\}$ from $\{s_2\}$.*

Proof. By induction on the depth of the deepest block in the block tree containing both s_1 and s_2.
□

In general, $\delta(s_1, s_2)$ will not be minimal. However, it is possible to characterize situations when $\delta(s_1, s_2)$ will be minimal, as the following result indicates.

Theorem 3.2 *Suppose that in each recursive call to δ generated by $\delta(s_1, s_2)$ ($s_1 \not\sim s_2$) the following holds.*

---

[1] Strictly speaking, this is not necessary; these blocks may be left childless. We include these spurious children to simplify our inductive argument of correctness.

```
Size := ∞;
S_L := { s' | s_L →ᵃ s' } ∩ B';
S_R := { s' | s_R →ᵃ s' };
foreach s'_L ∈ S_L do begin
  Γ := ∅;
  foreach s'_R ∈ S_R do begin
    Φ' := δ(s'_L, s'_R);
    Γ := Γ ∪ {Φ'};
  end;
  foreach Φ_i ∈ Γ do begin
    S_i := { s' ∈ S_R | s' ∉ [[Φ_i]] ∧ ∀Φ_j ∈ Γ. i ≠ j ⇒ s' ∈ [[Φ_j]] };
    if S_i = ∅ then Γ := Γ - {Φ_i};
  end;
  if |∧Γ| < Size then begin
    Size := |∧Γ|;
    Φ := ∧Γ;
  end;
end;
```

Figure 3: Code for generating conjunctions.

1. Let Γ be the set of conjuncts used to build Φ and s'_L the state in s_L that was used to create Γ. Then for each $\Phi' \in \Gamma$ there is an $s_j \in S_R$ such that Φ minimally distinguishes $\{s'_L\}$ from $\{s_j\}$, and $s_j \in [\![\Phi'']\!]$ for all other $\Phi'' \in \Gamma$.

2. $\{ s' \mid s_L \xrightarrow{a} s' \} - B' \subseteq S_R$.

Then $\delta(s_1, s_2)$ is minimal

Proof. By contradiction. Of importance is the fact that Φ is the shortest length formula that δ can build to distinguish an a-derivative of s_L from all a-derivatives of s_R. □

It is also the case that a minimizing procedure can be applied to $\delta(s_1, s_2)$ once it has been computed; the result of this would be a minimal formula. The minimizing procedure is straightforward: repeatedly replace subformulas in the formula by tt and see if the resulting formula still distinguishes s_1 from s_2. If so, the subformula may either be omitted (if it is one of several conjuncts in a larger conjunction) or left at tt. The computational tractability of this procedure remains to be examined, however.

It should be noted that δ may still generate exponential length formulas. However, one may represent such a formula (as a set of propositional equations) in space proportional to $|Q|^3$. This results from the fact that there can be at most $|Q|^2$ total recursive calls generated by the above procedure and the fact that each distinguishing formula is of the form $(\neg)\langle a \rangle \Phi$, where Φ contains at most $|Q|$ conjuncts, each of the form $\delta(s_i, s_j)$ for some s_i and s_j. By saving information appropriately and modifying the procedure for δ so that the semantic information of the formula computed is also returned, we may establish the following bound on the amount of computation needed to compute such a series of equations.

Theorem 3.3 *An equational representation of $\delta(s_1, s_2)$ may be calculated in $O(|Q|^5)$ time, once the tree of blocks has been computed.*

Proof. Follows from the fact that determining the equation for each recursive call of δ requires $O(|Q|^5)$ work. □

We close this subsection with some general remarks about our method. One feature of our approach is that the overhead involved in maintaining the block tree is minimal; nodes need not be labeled with the corresponding sets of states, except at the leaves. Also, the postprocessing step is only invoked after

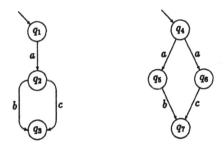

Figure 4: Two inequivalent transition systems.

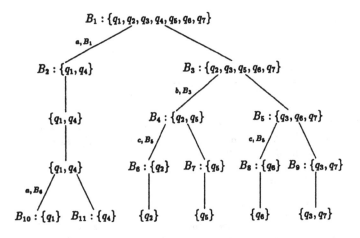

Figure 5: The tree of blocks generated by *bisim*.

equivalence is computed—so if the states of interest are found to be equivalent, then the postprocessing may be avoided altogether.

3.3 An Example

To illustrate our algorithm we consider a well-known example of two transition graphs that are not bisimulation equivalent. Figure 4 shows the transition system that includes the two transition graphs. State q_1 is the start state of one graph, while state q_4 is the start state of the other. Figure 5 contains the tree of blocks generated by the altered *bisim* algorithm. Notice that $q_1 \not\sim q_4$, as they are in different blocks.

In order to build $\delta(q_1, q_4)$, the algorithm first locates the lowest common ancestor of the two blocks (B_2, in this case). The left child is B_{10}, the right child is B_{11}, the action causing the split is a, and the block causing the split is B_6. The formula that will be returned, then, will be

$$\langle a \rangle (\delta(q_2, q_6) \wedge \delta(q_2, q_5));$$

this formula holds of q_1 and not q_4. By repeating this process, it turns out that

$$\delta(q_2, q_6) = \langle c \rangle tt \text{ and}$$
$$\delta(q_2, q_5) = \langle b \rangle tt.$$

So the formula distinguishing q_1 from q_4 is

$$\langle a\rangle(\langle c\rangle tt \wedge \langle b\rangle tt).$$

This formula states that q_1 and q_4 are inequivalent because q_1 may engage in an a-transition and evolve into a state from which both b- and c-transitions are available. Note that this formula is minimal. By way of contrast, the formula generated by the first naive method would be the following.

$$\langle a\rangle tt \wedge \langle a\rangle(\neg\langle a\rangle tt \wedge \langle b\rangle\neg\langle a\rangle tt \wedge \langle c\rangle(\neg\langle a\rangle tt \wedge \neg\langle b\rangle\neg\langle a\rangle tt))$$

This formula is clearly not minimal, since, for example, the formula obtained by substituting tt for subformula $\langle a\rangle tt$ is still distinguishes q_1 from q_4.

4 Conclusions and Future Work

This paper has shown how it is possible to alter partition-refinement based algorithms for computing bisimulation equivalence to compute a formula in the Hennessy-Milner Logic that distinguishes two inequivalent states. The generation of the formula relies on a postprocessing step that is invoked on a tree-based representation of the information computed by the equivalence algorithm. The formulas are often minimal in a certain sense, and the postprocessing step has an $O(|Q|^2)$ effect on the worst-case complexity of the equivalence-checking algorithm.

There are several avenues for future work to be pursued. Clearly, the complexity of the minimization procedure mentioned in passing at the end of Section 3 needs to be analyzed fully; if this procedure is efficient enough, then it may be incorporated into the distinguishing formula generation procedure. Another area of investigation would involve an implementation of our technique; we plan to incorporate this distinguishing formula capability into the Concurrency Workbench [5, 6], a tool for the analysis of finite-state systems. Yet another involves determining appropriate ways of using formulas computed in the course of checking equivalences other than bisimulation equivalence. Of particular interest is *testing (or failures) equivalence* [3, 7, 8]. These equivalences may be characterized in terms of the tests a process may pass and must pass. One method for distinguishing states that are not testing equivalent would be to build a test based on the formula computed by the bisimulation equivalence checker that one state may (or must) pass and that the other must (or may) not. Finally, it may be possible to extend our techniques to the computation of distinguishing formulas in the context of *preorder* checking. Another method of verifying processes involves the use of a behavioral preorder; in this setting, an implementation satisfies a specification if the implementation is "greater than" (intuitively: "behaves at least as well as") the specification. One interesting preorder is the *prebisimulation preorder*, which has a logical characterization in terms of an intuitionistic variant of the Hennessy-Milner logic: one state is "greater than" another if it satisfies all the formulas satisfied by the latter. This property could serve as the theoretical basis for computing diagnostic information in the same way that the logical characterization of bisimulation equivalence served as the theoretical basis for the techniques described in this paper.

Acknowledgement

I would like to thank Henri Korver for spotting errors in, and for his helpful comments on, previous drafts of this paper.

References

[1] Bloom, B., S. Istrail and A. Meyer. "Bisimulation Can't Be Traced." In *Proceedings of the ACM Symposium on Principles of Programming Languages*, 1988.

[2] Boudol, G., V. Roy, R. de Simone and D. Vergamini. "Process Algebras and Systems of Communicating Processes." In *Proceedings of the Workshop on Automatic Verification Methods for Finite-State Systems*, Lecture Notes in Computer Science 407, 1–10. Springer-Verlag, Berlin, 1990.

[3] Brookes, S.D., C.A.R Hoare and A.W. Roscoe. "A Theory of Communicating Sequential Processes." *Journal of the ACM*, v. 31, n. 3, July 1984, pp. 560-599.

[4] Cleaveland, R. and M. Hennessy. "Testing Equivalence as a Bisimulation Equivalence." In *Proceedings of the Workshop on Automatic Verification Methods for Finite-State Systems*, Lecture Notes in Computer Science 407, 11–23. Springer-Verlag, Berlin, 1990.

[5] Cleaveland, R., J. Parrow and B. Steffen. "A Semantics-Based Tool for the Verification of Finite-State Systems." In *Proceedings of the Ninth IFIP Symposium on Protocol Specification, Testing and Verification*, 287–302. North-Holland, Amsterdam, 1990.

[6] Cleaveland, R., J. Parrow and B. Steffen. "The Concurrency Workbench." In *Proceedings of the Workshop on Automatic Verification Methods for Finite-State Systems*, Lecture Notes in Computer Science 407, 24–37. Springer-Verlag, Berlin, 1989.

[7] DeNicola, R. and Hennessy, M. "Testing Equivalences for Processes." *Theoretical Computer Science*, v. 34, 1983, 83-133.

[8] Hennessy, M. *Algebraic Theory of Processes*. MIT Press, Boston, 1988.

[9] Hennessy, M. and R. Milner. "Algebraic Laws for Nondeterminism and Concurrency." *Journal of the Association for Computing Machinery*, v. 32, n. 1, January 1985, 137-161.

[10] Hillerström, M. *Verification of CCS-processes*. M.Sc. Thesis, Computer Science Department, Aalborg University, 1987.

[11] Kanellakis, P. and Smolka, S.A. "CCS Expressions, Finite State Processes, and Three Problems of Equivalence." In *Proceedings of the Second ACM Symposium on the Principles of Distributed Computing*, 1983.

[12] Larsen, K. and A. Skou. "Bisimulation through Probabilistic Testing." Proceedings of the ACM Symposium on Principles of Programming Languages, 1989.

[13] Malhotra, J., Smolka, S.A., Giacalone, A. and Shapiro, R. "Winston: A Tool for Hierarchical Design and Simulation of Concurrent Systems." In *Proceedings of the Workshop on Specification and Verification of Concurrent Systems*, University of Stirling, Scotland, 1988.

[14] Milner, R. *Communication and Concurrency*. Prentice Hall, 1989.

[15] Paige, R. and Tarjan, R.E. "Three Partition Refinement Algorithms." *SIAM Journal of Computing*, v. 16, n. 6, December 1987, 973-989.

[16] Pnueli, A. "Linear and Branching Structures in the Semantics and Logics of Reactive Systems." Lecture Notes in Computer Science 194, 14-32. Springer-Verlag, Berlin, 1985.

Lecture Notes in Computer Science

For information about Vols. 1–454
please contact your bookseller or Springer-Verlag

Lecture Notes in Computer Science

This series reports new developments in computer science research and teaching, quickly, informally, and at a high level. The timeliness of a manuscript is more important than its form, which may be unfinished or tentative. The type of material considered for publication includes

– drafts of original papers or monographs,

– technical reports of high quality and broad interest,

– advanced-level lectures,

– reports of meetings, provided they are of exceptional interest and focused on a single topic.

Publication of Lecture Notes is intended as a service to the computer science community in that the publisher Springer-Verlag offers global distribution of documents which would otherwise have a restricted readership. Once published and copyrighted they can be cited in the scientific literature.

Manuscripts

Lecture Notes are printed by photo-offset from the master copy delivered in camera-ready form. Manuscripts should be no less than 100 and preferably no more than 500 pages of text. Authors of monographs receive 50 and editors of proceedings volumes 75 free copies. Authors of contributions to proceedings volumes are free to use the material in other publications upon notification to the publisher. Manuscripts prepared using text processing systems should be printed with a laser or other high-resolution printer onto white paper of reasonable quality. To ensure that the final photo-reduced pages are easily readable, please use one of the following formats:

Font size (points)	Printing area (cm)	(inches)	Final size (%)
10	13.5 x 20.0	5.3 x 7.9	100
12	16.0 x 23.5	6.3 x 9.2	85
14	18.0 x 26.5	7.0 x 10.5	75

On request the publisher will supply a leaflet with more detailed technical instructions or a TEX macro package for the preparation of manuscripts.

Manuscripts should be sent to one of the series editors or directly to:

Springer-Verlag, Computer Science Editorial I, Tiergartenstr. 17, W-6900 Heidelberg 1, FRG

ISBN 3-540-54477-1
ISBN 0-387-54477-1